Energy, Environment, and Sustainability

AIMS AND SCOPE

This books series publishes cutting edge monographs and professional books focused on all aspects of energy and environmental sustainability, especially as it relates to energy concerns. The Series is published in partnership with the International Society for Energy, Environment, and Sustainability. The books in these series are edited or authored by top researchers and professional across the globe. The series aims at publishing state-of-the-art research and development in areas including, but not limited to:

- Renewable Energy
- Alternative Fuels
- Engines and Locomotives
- Combustion and Propulsion
- Fossil Fuels
- Carbon Capture
- Control and Automation for Energy
- Environmental Pollution
- Waste Management
- Transportation Sustainability

Review Process

The proposal for each volume is reviewed by the main editor and/or the advisory board. The chapters in each volume are individually reviewed single blind by expert reviewers (at least four reviews per chapter) and the main editor.

Ethics Statement for this series can be found in the Springer standard guidelines here https://www.springer.com/us/authors-editors/journal-author/journal-author-helpdesk/before-you-start/before-you-start/1330#c14214

Pravesh Chandra Shukla · Giacomo Belgiorno ·
Gabriele Di Blasio · Avinash Kumar Agarwal
Editors

Renewable Fuels for Sustainable Mobility

Springer

Editors
Pravesh Chandra Shukla
Department of Mechanical Engineering
Indian Institute of Technology Bhilai
Sejbahar, Madhya Pradesh, India

Gabriele Di Blasio
National Research Council of Italy
STEMS
Naples, Italy

Giacomo Belgiorno
PUNCH Torino S.p.A.
Turin, Italy

Avinash Kumar Agarwal
Department of Mechanical Engineering
Indian Institute of Technology Kanpur
Kanpur, Uttar Pradesh, India

ISSN 2522-8366 ISSN 2522-8374 (electronic)
Energy, Environment, and Sustainability
ISBN 978-981-99-1391-6 ISBN 978-981-99-1392-3 (eBook)
https://doi.org/10.1007/978-981-99-1392-3

This Springer imprint is published by the registered company Springer Nature Singapore Pte Ltd.
The registered company address is: 152 Beach Road, #21-01/04 Gateway East, Singapore 189721, Singapore

Preface

Mobility is an essential part of modern life, and conventional fuels represent the main energy sources for transportation. With the increasing pollution levels and the stringent regulation of Greenhouse Gases (GHG), alternative and renewable fuels have also been explored as potential solutions to meet future emissions regulation levels. In this regard, renewable fuels provide a practical alternative supporting the decarbonisation by 2050 of the transportation sector. The most common renewable fuels and their applications in combustion engines are discussed in this book.

The International Society for Energy, Environment and Sustainability (ISEES) was founded at the Indian Institute of Technology Kanpur (IIT Kanpur), India, in January 2014 to spread knowledge/awareness and catalyse research activities in the fields of Energy, Environment, Sustainability and Combustion. Society's goal is to contribute to the development of clean, affordable and secure energy resources and a sustainable environment for society and spread knowledge in the areas mentioned above, and create awareness about the environmental challenges the world is facing today. The unique way adopted by ISEES was to break the conventional silos of specialisations (engineering, science, environment, agriculture, biotechnology, materials, fuels, etc.) to tackle the problems related to energy, environment and sustainability in a holistic manner. This is quite evident in the participation of experts from all fields to resolve these issues. The ISEES is involved in various activities, such as conducting workshops, seminars and conferences, in the domains of its interests. The society also recognises the outstanding works of young scientists, professionals and engineers for their contributions in these fields by conferring them awards under various categories.

Sixth International Conference on 'Sustainable Energy and Environmental Challenges' (VI-SEEC) was organised under the auspices of ISEES from 27 to 29 December 2021, in hybrid mode due to restrictions on travel because of the ongoing COVID-19 pandemic situation. This conference provided a platform for discussions between eminent scientists and engineers from various countries, including India, Spain, Austria, Australia, South Korea, Brazil, Mexico, USA, Malaysia, Japan, Hong Kong, China, the UK, Netherlands, Poland, Finland, Italy, Israel, Kenya, Turkey and Saudi Arabia. At this conference, eminent international speakers presented their views on energy, combustion, emissions and alternative energy resources for sustainable development and a cleaner environment. The conference presented two high-voltage plenary talks by Prof. Ashutosh Sharma, Secretary, DST and Dr VK Saraswat, Honourable Member, NITI Ayog.

The conference included 12 technical panel discussions on energy and environmental sustainability topics. Each session had 6–7 eminent scientists who shared their opinion and discussed the trends for the future. The technical sessions at the conference included Fuels for Sustainable Transport, Challenges for Desalination and Wastewater Treatment and Possible Solutions, Engine Combustion Modelling, Simulation and Sprays, Bioenergy/biofuels, Coal Biomass Combustion for Power Generation, Microbial Processes and Products, Future of IC Engine Technology and Roadmap, Air Pollution and Climate Change: Sustainable Approaches, Sustainable Energy from Carbon Neutral Sources, Biological Waste Treatment and Combustion: Emerging Paradigm and Thermochemical Processes for Biomass. 500+ participants and speakers from around the world attended this 3-day conference.

This conference laid out the roadmap for technology development, opportunities and challenges in Energy, Environment and Sustainability domains. All these topics are very relevant for the country and the world in the present context. We acknowledge the support from various agencies and organisations for conducting the Sixth ISEES conference (VI-SEEC), where these books germinated. We want to acknowledge our publishing partner Springer (special thanks to Ms. Swati Mehershi).

The editors would like to express their sincere gratitude to many authors worldwide for submitting their high-quality work on time and revising it appropriately at short notice. We want to express our special gratitude to our prolific set of reviewers, who reviewed various chapters of this monograph and provided their valuable suggestions to improve the manuscripts.

This book covers the applications of renewable fuels for sustainable transportation. Various fuels and additives are studied and reported in this book from the perspective of sustainable mobility. Methanol, ethanol, butanol, DME, HVO, etc. are discussed in detail for their possible sustainable utilisations in combustion engines. Since spray optimisation for improved combustion and lower emissions is important, optical diagnostic techniques such as Particle Image Velocimetry (PIV), Phase Doppler Interferometry (PDI) technique for spray characterisation and Laser-Induced Incandescence (LII) have been discussed in the chapters. Chapters include recent results and focus on current trends in the transportation sector. The chapters are written by well-known researchers in the field of internal combustion engines and we expect great interest from professionals and post-graduate students for the contents covered herein.

Bhilai, India — Pravesh Chandra Shukla
Turin, Italy — Giacomo Belgiorno
Naples, Italy — Gabriele Di Blasio
Kanpur, India — Avinash Kumar Agarwal

Contents

About the Editors

Dr. Pravesh Chandra Shukla is Assistant Professor in the Department of Mechanical Engineering at Indian Institute of Technology (IIT) Bhilai, India. Dr. Shukla received his Ph.D. from IIT Kanpur and has also worked as Senior Research Associate with the institute. He was a Postdoctoral Researcher in the Division of Combustion Engines, Department of Energy Sciences, Lund University, Sweden. He briefly worked in Ecole Centrale de Nantes, France, in the field of dual fuel combustion. He is a recipient of Young Scientist Award from the International Society for Energy, Environment and Sustainability. Dr. Shukla mainly works in the field of internal combustion engines and alternative fuels for transportation. He worked on the development of additives for high compression ratio heavy duty engines fueled with alcohol. He is involved in investigating the emission characteristics for alternative fuels like biodiesel, HVO and alcohols for conventional and advanced heavy duty compression ignition engines. He has published over 28 technical articles in journals of national and international repute and conference proceedings.

Dr. Giacomo Belgiorno is Technology System Engineer in the Department of Advanced Engineering at PUNCH Torino S.p.A., formerly General Motors Global Propulsion Systems, Torino, Italy. He received his M.S. in Mechanical Engineering from University of Campania Luigi Vanvitelli in 2014, and Ph.D. in Energy Science and Engineering from University of Naples "Parthenope" in 2018. During the Ph.D. program, he worked at Istituto Motori as Research Associate and Guest Researcher in Lund University, Sweden. Subsequently, he worked in CNH Industrial. He has authored three book chapters and more than 20 conference and journal articles.

Dr. Gabriele Di Blasio is currently Research Scientist at the National Research Council of Italy. His main research interest is focused on advanced technologies and fuels for propulsion and energy conversion systems. He has lead various research projects in the field of internal combustion engine technology and fuel development. He has contributed to private and public projects in cooperation with universities, research centres and OEMs. He received his Ph.D. in Mechanical Engineering in 2012.

Formerly Dr. Di Blasio worked as R&D responsible engineer in the industry sector leading projects on the development of dual fuel systems for heavy duty engines. He serves as an editor and reviewer for several indexed journals of national and international repute. He has authored over 70 publications in peer-reviewed journals, conference proceedings, books, book chapters and technical reports. He is a member of SAE International and SAE Engine committee.

Prof. Avinash Kumar Agarwal joined Indian Institute of Technology (IIT) Kanpur in 2001. He worked at the Engine Research Center, UW@Madison, the USA as a Post-Doctoral Fellow (1999–2001). His interests are IC engines, combustion, alternate and conventional fuels, lubricating oil tribology, optical diagnostics, laser ignition, HCCI, emissions, and particulate control, 1D and 3D Simulations of engine processes, and large-bore engines. Prof. Agarwal has published 435+ peer-reviewed international journal and conference papers, 70 edited books, 92 books chapters, and 12200+ Scopus and 19000+ Google Scholar citations. Prof. Agarwal is a Fellow of SAE (2012), Fellow of ASME (2013), Fellow of ISEES (2015), Fellow of INAE (2015), Fellow of NASI (2018), Fellow of Royal Society of Chemistry (2018), and a Fellow of American Association of Advancement in Science (2020). He is the recipient of several prestigious awards such as Clarivate Analytics India Citation Award-2017 in Engineering and Technology, NASI-Reliance Industries Platinum Jubilee Award-2012, INAE Silver Jubilee Young Engineer Award-2012, Dr. C. V. Raman Young Teachers

Award: 2011, SAE Ralph R. Teetor Educational Award-2008; INSA Young Scientist Award-2007, UICT Young Scientist Award-2007, INAE Young Engineer Award-2005. Prof. Agarwal received Prestigious CSIR Shanti Swarup Bhatnagar Award-2016 in Engineering Sciences. Prof. Agarwal is conferred upon Sir J C Bose National Fellowship (2019) by SERB for his outstanding contributions. Prof. Agarwal was a highly cited researcher (2018) and was in the top ten HCR from India among 4000 HCR researchers globally in 22 fields of inquiry.

Part I
General

Chapter 1
Introduction to Renewable Fuels for Sustainable Mobility

Pravesh Chandra Shukla, Giacomo Belgiorno, Gabriele Di Blasio, and Avinash Kumar Agarwal

Abstract Transportation sector is facing new challenges in terms of stringent vehicle tailpipe emissions and CO_2 emission reduction, moving the sector towards decarbonisation. In this view, research in the automotive industry has been developing alternative solutions for the mid- and long-term timeframe. The main pillars are electrification, energy storage, zero-carbon renewable fuels, etc. In the transition phase towards zero tailpipe emission vehicles, improvements in state-of-the-art propulsion systems are still required. Several studies in combustion development through advanced techniques, alternative fuel applications, and after-treatment systems improvements are still facing challenges. Using alternative fuels for internal combustion engines is a sustainable way to use these devices for transportation and to reduce emissions simultaneously. This book includes a detailed review of renewable fuel solutions for sustainable mobility. It covers various renewable fuels such as hydrotreated vegetable oils, methanol, ethanol, butanol, dimethyl ether, and biodiesels. This book emphasises the role of renewable fuels as one of the solutions for sustainable transportation.

Keywords Alternative fuels · Biofuels · Fuel production · Sustainable mobility · Renewable fuels · Internal combustion engines

Mobility is one of the prime requirements of human activities in the modern world. The increasing energy demand of developed countries and the exponential growth of emergent ones have arisen concerns about pollution and CO_2 emissions and their impact on climate change. Policies around the world are defining rules to mitigate

P. C. Shukla (✉)
Department of Mechanical Engineering, Indian Institute of Technology Bhilai, Raipur, India
e-mail: pravesh@iitbhilai.ac.in

G. Belgiorno
PUNCH Torino, Turin, Italy

G. Di Blasio
Istituto di Scienze e Tecnologie per l'Energia e la Mobilità Sostenibili (STEMS), Consiglio Nazionale Delle Ricerche, Naples, Italy

A. K. Agarwal
Department of Mechanical Engineering, Indian Institute of Technology Kanpur, Kanpur, India

P. C. Shukla et al. (eds.), *Renewable Fuels for Sustainable Mobility*, Energy, Environment, and Sustainability, https://doi.org/10.1007/978-981-99-1392-3_1

their impact on the environment and human health (IEA 2022; European Green Deal n.d.). In this regard, pollutant emission regulations are becoming stricter, and tank-to-wheel emissions will be reduced to almost zero in the mid-term. Several solutions can support the transition of the transportation sector towards decarbonisation. Recent regulations move mobility towards carbon-neutral by 2050. The green electrification of vehicle propulsion systems represents a mainstream strategy, and electric vehicles are getting popularity in recent times. Although electric vehicles produce no emissions in the urban environment, these machines use electricity mostly produced using conventional fuels like coal and CNG, which ultimately release tonnes of CO_2 and other pollutants into the environment.

Moreover, mining of lithium, production of lithium batteries and recycling also possess limitations. In a more convenient and cheaper approach, researchers and manufacturers have been working on conventional propulsion system development and novel alternative fuel designs to minimise the transportation sector's environmental impact. Thus, renewable fuels, such as bio-alcohols, biogases, zero-carbon fuels and synthetic fuels, play a crucial role in facing the challenging goals to decarbonise the transport sector (Verhelst and Turner 2019; Sahu et al. 2022; Ianniello et al. 2021). The proposed book extends the scientific and technical discussion and overviews on renewable fuels for clean transportation presented in previous monographs (Blasio et al. 2022a, b; Shukla et al. 2021). This book attempts to study the application of renewable fuels for transport applications, methods for production and basic research techniques for their characterisation in internal combustion (IC) engines. The first section describes the application of renewable fuels, mostly alcohol fuels, in IC engines. The second section deals with fuel production, and finally the last section describes some optical techniques for fuel spray and combustion characterisation. The three sections are, respectively, named (I) renewable fuel applications, (II) renewable fuels production and (III) miscellaneous. A total of 12 chapters are included comprehensive of this introduction.

The first chapter entitled 'Hydrotreated Vegetable Oils for Compression Ignition Engines—the way toward a sustainable transport' deals with the application of Hydrotreated Vegetable Oils (HVO) in modern compression ignition (CI) engines. It highlights the advantages of life-cycle assessment (LCA) and well-to-wheel (WTW). Effects on the combustion process and pollutant emissions are also discussed, demonstrating as HVO is highly compatible with existing engines and fuelling systems. Proceeding with the discussion on the application of alternative fuels, Chap. 2 discusses 'DME as a Green Fuel for Transport Sector'. This chapter primarily deals with the application of DME in IC engines as an excellent option to reduce soot formations. Emission reduction potential is also discussed, demonstrating the superior fuel properties, in terms of cetane number (CN), oxygen content, etc. to improve the fuel atomisation and the combustion process.

Chapter 3 investigates the '*Combustion and Emission Characteristics of Oxygenated Alternative Fuels in Compression Ignition Engines*'. In particular, it assesses the potential of using alcohol fuels in CI engines. An overview of CI engines fuelled with oxygenated fuels (biodiesel and alcohols blends) on combustion characteristics, engine performance and exhaust emissions is presented, demonstrating their effectiveness in reducing CO_2 and particulate emissions. A study on the short-term application of alternative fuels is conducted in Chap. 4, which deals with the '*Functional Use-based Positioning of Conventional Vehicles in conjunction with Alternate Low-Emission Fuels*' in India. The study discusses alternative fuels (natural gas, auto-gas and hydrogen) applicable to IC engines. The study demonstrates that biofuel blends are deployable for all vehicles running on conventional petrol or diesel.

In contrast, the expansion of natural gas usage is constrained by the lack of availability and accessibility beyond a few nodes. A specific application of alcohol fuel is reported in Chap. 5, which discusses the '*Strategies for Efficient Utilisation of methanol in Compression Ignition Engines*' through a review study. It discusses the potential of using methanol in CI engines to improve engine performance and simultaneously reduce emissions, along with dual-fuel technology. Another review study is presented in Chap. 6 which deals with '*The Impact of Renewable Fuels and Fuel Additives (Dodecanol) on Particulate Mass Emission for Sustainable Mobility*'. Experimental and comparative analyses are reported in this study, where dodecanol as a fuel additive is blended in diesel fuel. Results on in-cylinder pressure, hydrocarbon (HC), carbon monoxide (CO) and particulate mass (PM) emissions are presented. Chapter 7 deals with '*A bibliometric review of alcohol-diesel blend in CI engines*', assessing the role of alcohol additives on combustion and performance in CI engines, which were comprehensively reviewed.

Renewability and ease of production of various alternative fuels are promising points where overall CO_2 emission reduction in comparison to the production of conventional fuels. Therefore, Part II is included in this book which mainly focusses on the explanation of the production processes of renewable fuels. Chapter 9 belongs to the fuel production section and deals with the '*Biomass and CO_2-derived fuels through carbon-based catalysis. Recent advances and future challenges*'. The chapter describes the production of liquid transportation fuels from biomass and CO_2. The most promising carbon-based catalysts and processes are presented, which discuss the main challenges to improve the performance of various catalysts and reduce the overall production cost of the fuels. Chapter 10 discusses the '*Waste-to-Energy: Applications and perspectives on sustainable aviation fuel production*'. In particular, it focusses on sustainable waste management methods for waste-to-energy conversion technologies for bio-jet fuel production. Current conversion pathways are further analysed and discussed towards a 'greener' and more sustainable future for the aviation industry.

Most suitable alternative fuels for IC engines defer in their physicochemical fuel properties. Therefore, it is important to study the spray behaviours and other related characteristics before adapting to the commercial level. The last section presents various optical diagnostic techniques for fuel spray and combustion assessments. Chapter 11 explains the '*Feasibility Study of Laser Plasma-Assisted Stratified Combustion and Spray Investigations in a Constant Volume Chamber*'. This chapter explains techniques of spray analysis for gasoline direct injection (GDI) applications such as the Schlieren imaging technique, laser-induced fluorescence, Mie scattering, and phase Doppler interferometry. Fundamentals of laser ignition and associated challenges for the stratified mode of GDI engine operation were also covered. Other optical techniques are also discussed in Chap. 12, titled 'Understanding Combustion in CI Engines for Adoption of Renewable Fuels'. It deals with a detailed explanation of diesel combustion through flame visualisation using various optical diagnostic techniques in constant-volume combustion chambers.

To summarise, the various sections and chapters of the content are organised in three sections: (I) Renewable Fuel Applications, (II) Renewable Fuels Production and (III) Miscellaneous which are listed below:

1. Introduction to Renewables Fuels for Sustainable Mobility.
2. Hydrotreated Vegetable Oils for Compression Ignition Engines—the way toward a sustainable transport.
3. DME as a Green Fuel for Transport Sector.
4. Combustion and Emission Characteristics of Oxygenated Alternative Fuels in Compression Ignition Engines.
5. Functional Use-based Positioning of Conventional Vehicles in conjunction with Alternate Low-Emission Fuels.
6. Strategies for Efficient Utilisation of methanol in Compression Ignition Engines.
7. The Impact of Renewable Fuels and Fuel Additives (Dodecanol) on Particulate Mass Emission for Sustainable Mobility.
8. A bibliometric review of alcohol–diesel blend in CI engines.
9. Biomass and CO_2-derived fuels through carbon-based catalysis. Recent advances and future challenges.
10. Waste-to-Energy: Applications and perspectives on sustainable aviation fuel production.
11. Feasibility Study of Laser Plasma-Assisted Stratified Combustion and Spray Investigations in a Constant Volume Chamber.
12. Understanding Combustion in CI Engines for Adoption of Renewable Fuels.

References

Di Blasio G, Agarwal AK, Belgiorno, G., Shukla, P.C. (2022a). Introduction to Application of Clean Fuels in Combustion Engines. In: Di Blasio G, Agarwal AK, Belgiorno G, Shukla PC (eds) Application of clean fuels in combustion engines. Energy, environment, and sustainability. Springer, Singapore. https://doi.org/10.1007/978-981-16-8751-8_1

Di Blasio G, Agarwal AK, Belgiorno G, Shukla PC (2022b) Introduction to clean fuels for mobility. In: Di Blasio G, Agarwal AK, Belgiorno G, Shukla PC (eds) Clean fuels for mobility. energy, environment, and sustainability. Springer, Singapore. https://doi.org/10.1007/978-981-16-8747-1_1

European Green Deal (n.d.). https://ec.europa.eu/clima/eu-action/european-green-deal_en. Accessed 11 Nov 2022

Ianniello R, Belgiorno G, Di Luca G, Beatrice C, Di Blasio G (2021) Ethanol in dual-fuel and blend fueling modes for advanced combustion in compression ignition engines. In: Shukla PC, Belgiorno G, Di Blasio G, Agarwal AK (eds) Alcohol as an alternative fuel for internal combustion engines. Energy, environment, and sustainability. Springer, Singapore. https://doi.org/10.1007/978-981-16-0931-2_2

IEA. Transport sector CO_2 emissions by mode in the Sustainable Development Scenario, 2000–2030, IEA, Paris, https://www.iea.org/data-and-statistics/charts/transport-sector-co2-emissions-by-mode-in-the-sustainable-development-scenario-2000-2030. Accessed 11 Nov 2022

Sahu TK, Shukla PC, Belgiorno G, Maurya RK (2022) Alcohols as alternative fuels in compression ignition engines for sustainable transportation: a review. Energy Sour Part A: Recov Utiliz Environ Effects 44(4):8736–8759. https://doi.org/10.1080/15567036.2022.2124326

Shukla PC, Belgiorno G, Di Blasio G, Agarwal AK (2021) Introduction to alcohol as an alternative fuel for internal combustion engines. In: Shukla PC, Belgiorno G, Di Blasio G, Agarwal AK (eds) Alcohol as an alternative fuel for internal combustion engines. Energy, environment, and sustainability. Springer, Singapore. https://doi.org/10.1007/978-981-16-0931-2_1

Verhelst S, Turner JW, Sileghem L, Vancoillie J (2019) Methanol as a fuel for internal combustion engines. Progress Energy Combust Sci 70:43–88. ISSN: 0360-1285. https://doi.org/10.1016/j.pecs.2018.10.001

Part II
Renewable Fuel Applications

Chapter 2
Hydrotreated Vegetable Oils for Compression Ignition Engines—The Way Toward a Sustainable Transport

Michele Pipicelli, Giuseppe Di Luca, and Roberto Ianniello

Abstract The COP26 goals rapidly accelerate the shift of road transport to electric vehicles (EVs). However, the global transition to EVs should be assessed carefully. A forced transition to electric mobility without tailored solutions for each case can increase greenhouse gas (GHG) emissions. In this context, low-carbon fuels can be considered a promising short-term solution to efficiently reach the carbon neutrality target. This manuscript aims to highlight the competitive advantages of hydrotreated vegetable oil (HVO) over commercial diesel fuel. Recent works on HVO are considered, ranging from exploring the production processes and spray evolution characteristics to the various engine strategies to highlighting the potential. Greater emphasis was placed on environmental impact assessment, considering the results available for Life Cycle Assessment (LCA) and Well-To-Wheel. The main characteristics and influences of HVO in CI engines are assessed on the combustion process, GHGs, and pollutants emissions. The results show the high potential of the HVO to reduce the impact of the road transport sector actively. It is highly compatible with existing engines and fueling systems while ensuring lower CO_2, CO, THC, PM emissions, and combustion noise levels with similar efficiency and fuel consumption. Additionally, the residual feedstock can assure up to 75% GHG over the whole life cycle. Therefore, sustainable fuels, such as HVO, combined with advanced technologies could not only support the reduction of tailpipe emissions but also benefit the overall CO_2 assessment.

Keywords Hydrotreated vegetable oils (HVO) · Renewable fuels · Compression ignition engine · Life cycle assessment · Greenhouse gas

M. Pipicelli (✉) · G. Di Luca · R. Ianniello
STEMS, Consiglio Nazionale Delle Ricerche, Via G. Marconi, 4, 80125 Naples, Italy
e-mail: michele.pipicelli@stems.cnr.it

M. Pipicelli
Department of Industrial Engineering, University of Naples Federico II, Via Claudio, 21, 80125 Naples, Italy

P. C. Shukla et al. (eds.), *Renewable Fuels for Sustainable Mobility*, Energy, Environment, and Sustainability, https://doi.org/10.1007/978-981-99-1392-3_2

Abbreviations

APAC	Asia-Pacific
BEV	Battery Electric Vehicle
CI	Compression Ignition
CN	Cetane Number
CO	Carbon Monoxide
CO_2	Carbon Dioxide
EGR	Exhaust Gas Recirculation
EMEA	Europe, Middle East, And Africa
EVO	Exhaust Valve Open
EVs	Electric Vehicles
FAME	Fatty Acid Methyl Ester
GHG	Greenhouse Gas
GWP	Global Warming Potential
HD	Heavy-Duty
HVO	Hydrotreated Vegetable Oils
ICEV	Internal Combustion Engine Vehicles
IMEP	Indicated Mean Effective Pressure
LATAM	Latin America
LCA	Life Cycle Assessment
LD	Light-Duty
LHV	Lower Heating Value
NAFTA	North America Free Trade Agreement
NOx	Nitrogen Oxides
PFAD	Palm Fatty Acid Distillate
PM	Particulate Matter
PN	Particulate Number
RED II	Renewable Energy Directive II
THC	Total unburned Hydrocarbon
WLTP	Worldwide Harmonized Light Duty Vehicles Test Procedure

2.1 Introduction

The COP26 global climate summit has defined ambitious goals to mitigate climate change. In particular, to secure global net zero carbon by mid-century, keeping global warming below 2 °C, the global emissions must be reduced by 45% by 2030 compared to 2010. Furthermore, to reach the COP26 targets, another objective is the progressive stop to coal use and coal-firing power plants founding (Arora and Mishra 2021). The prevention of further deforestation and the rise of funds for green and sustainable technologies are also non-negligible objectives. In this context, the transport sector is facing a radical transformation due to the almost worldwide carbon neutrality goal dictated by national and international regulations. In the period 1990–2019, the overall GHG emissions produced by this sector increased considerably, despite improvements in efficiency due to the higher demands. The time target to achieve the goal is for almost all the countries in the year 2050 (Arora and Mishra 2021; Hjelkrem et al. 2020). The main exception is China which has 2060 as the target. The transport sector is responsible for about 23% of the global CO_2 emission (Murdock et al. 2020), and its reduction toward global carbon neutrality is of great concern. The accepted solution for road transport, particularly passenger cars and light commercial vehicles, is the use of fully electric vehicles. However, although they are characterized by higher efficiency, their environmental impact is strictly related to two main factors: the battery pack and the electric energy production method. The tank-to-wheel efficiency of electric vehicles is roughly double that of gasoline and diesel engine-powered ones (Hjelkrem et al. 2020), making it a very attractive solution. However, their carbon intensity is strictly linked to the electric generation method. The GHG equivalent emissions factor related to electric energy production for different energy feedstocks is reported in Table 2.1. Very high variation between the most and the less carbon-intensive feedstock can be noticed with almost two orders of magnitude. The "Renewables 2020—Global status report" (Murdock et al. 2020) states that the transport sector is responsible for 32% of the total energy consumption, of which only 3.3% is from renewable sources. Globally only 11% of the total energy consumption is renewable.

Moreover, in the 2013–2018 period, an increase of 7% in total energy consumption was recorded, of which 15% was from renewable feedstock. On the other side, the full penetration of electric vehicles in the market will increase electric energy consumption by about 20–25%, according to different studies (Teixeira and Sodré 2018; Bellocchi et al. 2018). So in the following years, an increase in electric energy will probably be recorded with renewable energy that may not increase fast enough to follow the requirements.

Table 2.1 GHG emissions factor for various electric energy feedstock

Energy feedstock	Carbon intensity [gCO_2eq/kWh]
Coal	960–1060 (Varga 2013), 975 (Hondo 2005)
Natural gas	500–530 (Varga 2013), 518–608 (Hondo 2005)
Crude oil	620–750 (Varga 2013), 742 (Hondo 2005)
Geothermal	15 (Hondo 2005), 11–47 (Eberle et al. 2017)
Hydro	11 (Hondo 2005)
Wind	29 (Hondo 2005)
Photovoltaic	53 (Hondo 2005), 60 (Constantino et al. 2018)

In 2021 according to the International Energy Agency, a +6% of coal consumption was recorded with an increasing forecast to 2024. So the transition to electric-powered vehicles, if not regulated, can further worsen GHG emissions. Life cycle assessment regarding battery electric vehicles, considering various mixes, shows that the variation in terms of gCO_2/km for a battery electric vehicle (BEV) is so significant that it also comprises conventional vehicles (Faria et al. 2013). This applies to the Polish energy mix, which is based on coal at about 90%. So attention should be paid to the transition period to fully electric vehicles, which can last for decades. In this transition period is essential to search for solutions rapidly disposable to reduce the carbon intensity of the road transport sector. Different analyses have reported that a forced transition to EV in a short time can be counterproductive to fulfilling carbon neutrality and environmental goals. Andersson and Börjesson from the LCA analysis concluded that renewable fuels, including HVO, ethanol, methanol, and other renewable fuels, are a key element to meeting the climate goals in the automotive sector (Andersson and Börjesson 2021), considering the slow rate at which the EU carbon intensity is reduced and the slow substitution rate of the older vehicles. Similar conclusions are given by Ternel et al. (2021), for which biofuels are quick and efficient solutions to reduce GHG emissions of the global transport fleet. However, they should be viewed as complementary to electrification to quickly reach the carbon neutrality target, together with other technologies such as energy management strategies (Beatrice et al. 2022; Giardiello et al. 2022), eco-efficient advanced driving assistance systems (Musa et al. 2021), and vehicular connectivity (Michel et al. 2016). So, renewable fuels can play a crucial role in the decarbonization of the transport sector. Different potential renewable fuels are analysed in the literature for both SI and CI engines. For the latter different fuel families such as alcohols (Pipicelli et al. 2022; Luca et al. 2022; Ianniello et al. 2021; Sahu et al. 2022), ethers (Styring et al. 2021), and esters (Rajasekar and Selvi 2014) were tested in many works. An overview of the fuel classification is reported in Fig. 2.1. In particular, on the left panel is shown a one based on chemical composition and on the right panel, a one based on feedstock. The HVO is a fuel composed of alkanes and belongs to the second generation of renewable fuels. In the figure, a light grey shadowed zone has been used to highlight these classes.

The Fatty Acid Methyl Ester (FAME), a first-generation biofuel, has been commercially available since the 1990s but suffers from many problems such as fuel injection system clogging, cold start problem, crankcase oil dilution, and material compatibility, which limits its use pure (Szeto and Leung 2022). For these reasons, it is usually blended in fossil diesel at 5 or 7% v/v and it is not used in its pure form. The search for new renewable fuel formulations has found in HVO good properties with a potential full substitution ratio to fossil fuel in CI engines. This manuscript analyses the hydrotreated vegetable oils (HVO) capability to decarbonize since its similar properties to diesel fuel assures good retrofitting capabilities. As shown in Fig. 2.1, the HVO is an actual state-of-art sustainable fuel and is already commercially available. Research on HVO fuelled engines is still ongoing. Indeed, being characterized by greater reactivity and a narrower molecular structure, it guarantees a different response to the combustion control parameters than diesel. Therefore,

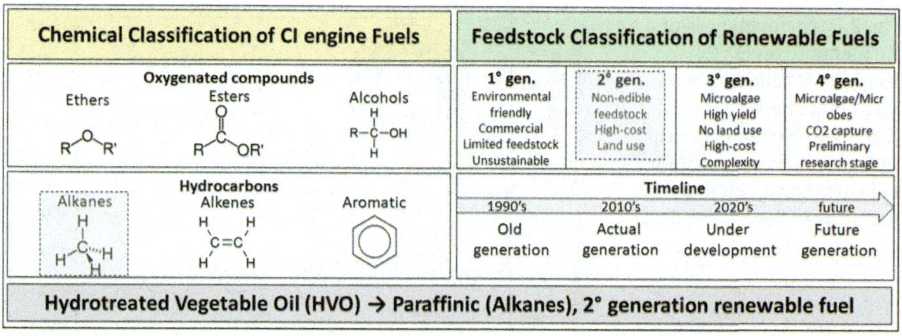

Fig. 2.1 An overview of the classification of fuels. On the left a chemical classification, and on the right one based on feedstocks

although considered a suitable fuel for efficient, low-emission PPCI combustion mode (Hunicz et al. 2022), the following study will focus more on fuel analysis in drop-in mode, using standard engine-control maps, or adopting different recalibration strategies to take further advantage of fuel properties. Synergistic application of alternative and sustainable fuels coupled with advanced combustion architectures (Belgiorno et al. 2020) and injection systems (Beatrice et al. 2019; Blasio et al. 2019), in internal combustion engines has the potential to bring significant benefits also to efficiency, pollutant, and CO_2 emissions (Zhang et al. 2020).

A meta-analysis of the available literature is done using some post-elaboration based on the exported database from Scopus. The search was based on the combination of logic operators (AND, OR, and NOT) and some keywords inside titles and abstracts. In particular, the following research was used: "engine" OR "vehicle" OR "lca" OR "life cycle analysis" AND "hvo" OR "hydrotreated vegetable oil" AND "fuel" AND NOT "hvof". The main synthetic data from the analysis are reported in Fig. 2.2. The first study on HVO is from 2007, with increasing scientific interest starting from 2011, reaching about 30 yearly manuscripts in the last three years. For a fair comparison, 2022 is not considered as it is actually ongoing. A consistent number of citations ranging from 200 to 350 shows the interest of HVO in the renewable fuel field.

Moreover, the manuscripts were grouped into commercial areas: EMEA (Europe, Middle East, and Africa), APAC (Asia-Pacific), LATAM (Latin America), and NAFTA (North America Free Trade Agreement). It is worth to highlight that EMEA is the main commercial area whose research on HVO (about 54% of the total manuscripts) is followed by NAFTA (about 30%). On the other hand, the LATAM shows almost no interest in HVO since its national and international policies focus on bioethanol and Fatty Acid Methyl Ester (FAME) as renewable fuels (Geraldes Castanheira et al. 2014; Renewable Fuels Association Focus Forward 2020).

The following manuscript is structured as follows. First, a brief introduction to HVO chemical-physical properties and production methods are presented. Then, an analysis of the life cycle analysis available in the literature is done. Finally, HVO effects on spray behavior and engine operation, including combustion and emission

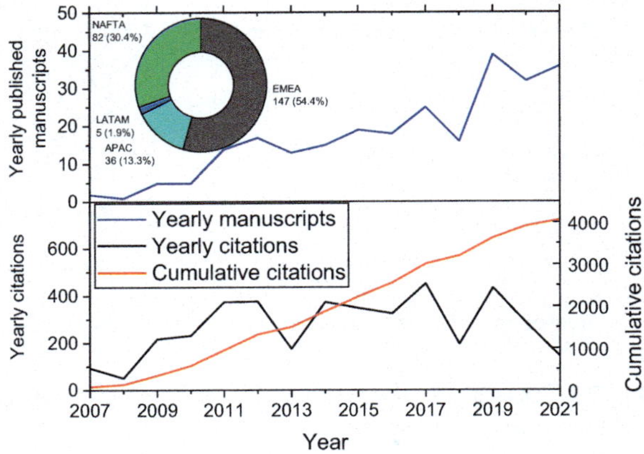

Fig. 2.2 Meta-analysis of the literature on HVO: division of manuscripts based on commercial area, yearly published manuscript on the topic, yearly citation and cumulative citation. *Source data* Scopus database

formation, are discussed. The structural presentation and the data plotting method adopted in the manuscript are intended to conceptualize the HVO effect on the performance and emissions in CI engines and its environmental impact assessment. Most previous reviews and research articles focused on the HVO effects on CI engine performance and emissions characteristics. The manuscript purpose is to provide a snapshot of all aspects related to the environmental impact of HVO regarding both the production process and aspects related to on-road and off-road propulsion.

2.2 Production and Properties of HVO

The term HVO emerged in the last decade when only vegetable oils (e.g. rapeseed, soybean, and corn oil) were used as feedstocks. Nowadays, non-edible feedstocks are preferred for HVO production, such as industrial waste (tall oil and fats) and used cooking oils (Engman et al. 2016). Furthermore, alternative non-food oils such as jatropha, algae, and palm oil are gaining attention to avoid vegetable feedstocks competition with food production due to the scarcity of edible oils (Brahma et al. 2022). For these reasons, the name HVO is no longer accurate in properly describing the fuel origins. Currently, it is common to call HVO with the terms "renewable diesel" and/or "green diesel", with both the terms often used interchangeably (Knothe 2010). However, these terms are not scientifically accurate as they can also refer to other fuels, such as the FAME. Moreover, the original name cannot be easily changed since it is common in the European regulation, fuel standards and biofuel quality recommendations.

The HVO production process can be summarised as reported in Fig. 2.3.

The first stage, the hydrotreatment, takes place in a reactor in a pressure range of 300–500 bar at 300–400 °C. In this first step, hydrogen is added to the liquid feedstock to break down the feedstock triglycerides into various intermediates species, mainly

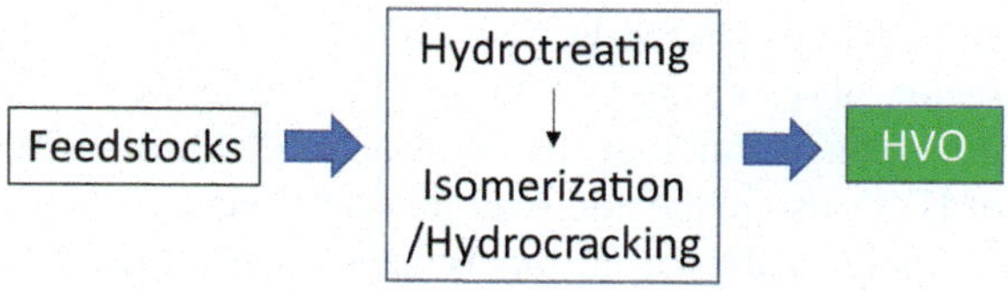

Fig. 2.3 Synthetic description of the production process of the HVO fuel

carboxylic acids, monoglycerides, and diglycerides. The conversion of these intermediate species into alkanes is obtained via three intermediate processes (Šimáček et al. 2010):

- Hydrogenation.
- Hydrodeoxygenation, which results in the removal of oxygen atoms from carboxylic groups in the form of water.
- Hydrodecarbonylation, which leads to the elimination of the carboxylic group in the form of carbon dioxide.

Initially, hydrogen saturates any double bonds of the triglycerides, followed by cleavage to fatty acids and the hydrogenation (hydrogen removes the acid group oxygen as H_2O) of glycerol to propane and water. Finally, the fatty acids are processed through hydrogenation or decarboxylation (the acid group oxygen leaves as CO_2). The split reaction depends on the catalyst used and operating conditions (https://www.etipbioenergy.eu/value-chains/conversion-technologies/conventional-technologies/hydrotreatment-to-hvo). During the process, carbon monoxide, carbon dioxide, water, and propane are obtained as side products (No 2014). The isomerization and hydrocracking process of the alkanes depend on the desired fuel products. For example, the isomerization process is used to improve the low-temperature properties of the final product. Then, the distillation process separates the various desired hydrocarbons products fractions, including HVO, diesel, gasoline, and others (https://www.etipbioenergy.eu/value-chains/conversion-technologies/conventional-technologies/hydrotreatment-to-hvo).

Several industries are involved in HVO production around the world, such as Universal Oil Products (UOP)-Eni (UK, Italy), Neste Oil (Finland), Syntroleum (United States), SK energy (Korea), ConocoPhillips (United States, Ireland), and Nippon Oil (Japan). Each of them uses a different commercial name to indicate their product, e.g. the NExBTL, acronym for "next-generation bio-to-liquid", is the trade name of the product obtained by Neste Oil Corporation (Engman et al. 2016), while for UOP/Eni Ecofining process the trade name is "Green Diesel" (https://www.eni.com/it-IT/attivita/biocarburanti-sostenibili-ecofining-tm.html). The HVO produced by SK energy was referred to as hydrogen-treating biodiesel (No 2014).

As a final product, HVO can be used pure or blended in the existing CI engines due to its similar chemical composition to the reference diesel fossil fuel. As paraffinic fuel, HVO meets the EN 15,940:2016 standard, which regulates HVO and Fischer-Tropsh GTL up to 7.0% in volume of FAME. For higher HVO-diesel blend ratios, the final composition meets the diesel fuel regulations, such as EN590 and ASTM D975. Furthermore, for HVO, the FAME regulations cannot be applicable. Since 2011, with an update of ASTM D7566-14, the HVO has also been approved for the aviation sector, allowing up to 50% of HVO in conventional jet fuel (ETIP Bioenergy—European Technology and Innovation Platform 2020). It demonstrates the potential of HVO also in the non-road transport sector.

Table 2.2 depicts the main characteristic of HVO compared to conventional diesel fuel. From a chemical point of view, the HVO is a paraffinic bio-based liquid composed mainly of paraffin and ISO-paraffin in the range C15–C18 and is free of sulfur and aromatic compounds. These are one of the main polycyclic aromatic hydrocarbons (PAH) precursors (Zubel et al. 2016). Thus HVO can potentially help reduce the particulate matter from its combustion due to lower soot precursors.

The increased amount of paraffinic components in HVO fuel and the absence of aromatics lead to a high cetane number (CN), which indicates a superior fuel ignition quality (Bhardwaj et al. 2013). In addition, the HVO, since it does not contain oxygen, has high oxidation stability, resulting in excellent storage behavior similar to fossil fuel and better than diesel with FAME (Dimitriadis et al. 2018).

The lower boiling point promotes better evaporation of HVO, while the different distillation curves lead to a reduced spray liquid length (Fajri et al. 2022). So an improvement of the air-fuel mixture can be expected to promote lower PM, total hydrocarbons (THC), and CO while slightly higher NO_x.

The HVO density does not meet the European standard EN 590 adopted for diesel fuel, so it can only be used in its pure form on specifically designed vehicles (Murtonen et al. 2009). The regulations usually define fuel density limits since this

Table 2.2 Physical and chemical properties of the selected fuels (Pipicelli et al. 2022; Ping et al. 1996; Hunicz et al. 2020)

Properties	Diesel	HVO
Cetane number [–]	≥51	75
Molar mass [g/mol]	~170	224
Carbon number [–]	12–20	15–18
H/C [–]	1.8	2.17
O/C [–]	0	0
LHV [MJ/kg]	43.0	44.35
(A/F)s [–]	14.5	15.2
Viscosity (@40 °C) [cSt]	2.72	2–4.5
Lubricity (μm corrected wear scar)	315	460
Density (@15 °C) [kg/m^3]	820–845	770–790
Boiling point [°C]	120–400	313
Cloud point [°C]	−5	−5.4
Sulfur content [mg/kg]	6.5	<5
Oxygen content [%wt]	0.77	0
Carbon content [%wt]	87.2	84.6
Hydrogen content [%wt]	12.2	15.4
Aromatics content [%v/v]	23	0

value is coded in the engine control units, and a significant deviation from the nominal value can lead to engine non-optimal operation with higher emissions (Lapuerta et al. 2010). However, a blend of up to 30% of HVO meets the EN 590 requirements, and different fuel with 15% v/v or more is commercially available in Europe (Bohl et al. 2018). Due to the lower density, HVO is characterized by a lower volumetric heating value than market diesel. HVO, thanks to its higher H/C ratio, has a higher heating value per mass unit compared to market diesel.

The lubricity of neat HVO is borderline with common acceptable values of <460 μm, which should assure adequate wear protection for the fuel injection system (Barbour et al. 2000). Some additives should be used to improve its lubricity; however, the same additive adopted for standard winter-grade diesel can be used, which is currently available for oil companies (Hartikka et al. 2012). Moreover, material compatibility is similar to fossil diesel and requires no particular attention for fuel system design and retrofitting (Pellegrini et al. 2015).

As mentioned above, there are different patented production methods of HVO which can be produced from different feedstocks. In the next section, several LCA studies have been reported to appreciate the HVO impact with a cradle-to-grave approach.

2.3 Environmental Impact of HVO

This section discusses the main environmental impact of HVO through an LCA and a well-to-wheel analysis based on the available literature data. In particular, the section focuses on GHG and the related Global Warming Potential (GWP), which represent one of the most relevant problems for which the regulations pose ever-increasing limitations.

The LCA is one of the more adopted techniques for assessing the environmental impact of a product or a process. The European Union, with its Renewable Energy Directive (RED II), had legislated that a life-cycle assessment should address the potential GHG saving of biofuels compared to fossil fuels.

It usually uses different metrics to assess different types of environmental impacts, such as human toxicity, eco-toxicity, loss of biodiversity, and climate change (Keoleian et al. 2006).

The LCA requires defining the process flow, its inputs and outputs, and system boundaries. The main input for the considered process is the fuel feedstock, which can be various for the HVO. The first distinction of the feedstock for the HVO production is between on-purpose and residual ones, which the latter usually shows lower GHG emissions due to neglecting the cultivation phase (Soam and Hillman 2019). Moreover, a further division can be addressed based on edible and not-edible feedstock. The use of edible feedstock such as rapeseed and palm oil is of great concern since the ever-increasing food and biofuels requests, possibly leading to socio-economical implications (Balat 2011). Thus, non-edible sources, such as jatropha oils, are greatly interested (Arvidsson et al. 2011). A general flow process of HVO production is

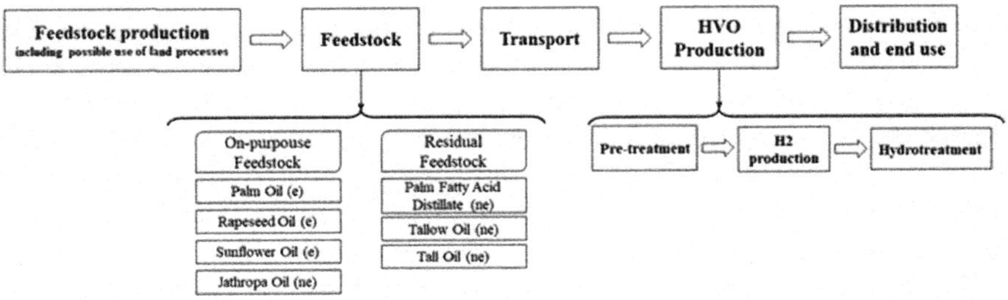

Fig. 2.4 Typical process flow diagram assumed in the literature for HVO LCA. Typical feedstocks are reported, indicating with (e) the edible ones while with (ne) the non-edible ones

depicted in Fig. 2.4. Using residual feedstock can shift the system boundaries of the LCA, avoiding the feedstock production part. The system boundaries should be carefully assessed since they can significantly influence the LCA results, mainly when a feedstock produces more than one product. The inclusion or exclusion of some subprocess is sometimes subjective and can lead to system boundary problems and mismatch (Suh et al. 2004).

In this context, the discussion will be focused only on climate change which is often measured as equivalent carbon dioxide emissions for the functional unit of the product. For fuel, the functional units often used in the literature are the mass of fuel burned, the energy released by combustion, brake energy, and vehicular distance travelled. All the available data reported have been converted in gCO_2eq/MJ of chemical energy of the fuel to avoid dependency on assumed vehicle efficiency, test driving cycle, combustion efficiency, or lower heating value (LHV). These data are shown in Fig. 2.5, where the diesel fuel was added for comparison. The standard diesel fuel GHG emissions vary between 75 and 100 gCO_2eq/MJ, while the HVO shows a most significant variation ranging between 5 and 80 gCO_2eq/MJ. Residual feedstocks such as Palm Fatty Acid Distillate (PFAD), waste and tall oil assure the lowest GHG. Reasoning in terms of mean values, the HVO from residual feedstock assures around −75% of GHG emissions compared to fossil diesel fuel, potentially leading to low carbon transport.

Based on the literature, a general consideration can be drawn regarding the well-to-wheel analysis. Firstly, there is a break-even point in vehicle mileage from which a fully electric vehicle is more sustainable than an internal combustion engine vehicle (ICEV) (including hybrid solutions). The break-even point varies as a function of many factors but among all: electricity production method, fuel and battery technology. Nordelöf et al. (2019) compared through LCA in a heavy-duty application (city buses) various powertrain configurations spacing from full electric to ICEV fueled by diesel and HVO considering 12 years. The results show that the electric bus has the best potential in terms of GHG reduction, up to −90%, in a hypothetical scenario where electricity production has zero carbon intensity. It is also confirmed by Lyng and Brekke (2019). Hoverer in this work it also showed that HVO-fuelled ICEV has less impact than electric if the grid mix GHG intensity is greater than 280 gCO_2eq/kWh. This threshold is further reduced to 200 gCO_2eq/kWh

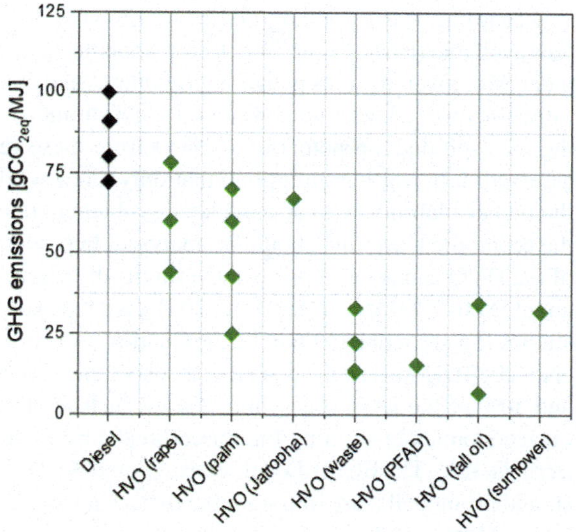

Fig. 2.5 GHG emissions for HVO and diesel fuels by literature LCA data. *Source data* Soam and Hillman (2019), Arvidsson et al. (2011), Källmén et al. (2019), Lyng and Brekke (2019), Liu et al. (2020), Prussi et al. (2020), Benavides et al. (2017), Marek et al. (2016)

if a hybrid powertrain fueled with HVO is assumed. Similarly, assuming a service life of 390,000 km and 10 years, Lyng and Brekke (2019) found that HVO from waste cooking oil can reduce the Global Warming Potential (GWP) by about 70% with respect to fossil fuel and electric vehicles with coal produced electricity.

In general, the HVO assures a reduction in the range of 4–8% (Murtonen et al. 2009; Blasio et al. 2022) of tank-to-wheel GHG emissions compared to diesel, with a reduction of THC, CO, and PM emissions, as detailed in the following section.

The results show that the GHG emissions reduction of HVO can reach 70%, which is still lower than electric vehicles powered by a carbon-intensive electricity mix. The HVO potential reduction shows high variability among feedstock, production process, and different studies. Therefore, there is a need for regulations and governance control to regulate the production process and guarantee the lowest carbon-intensive processes to maximize the potential benefits of HVO. In particular, these highly environmentally friendly results are linked to residual feedstocks, which have limited availability of their nature. Turkey produces 25e3 tons/year of tall oil and a diesel consumption of 15e6 tons/year (Altıparmak et al. 2007). The biofuel yield from tall oil is high, as it can be above 95%, and for a rough calculation, it can be assumed as an ideal 100% (https://www.mdpi.com/2673-4079/2/1/12). Also, considering that all the tall oil produced will be used for HVO production, it can fulfil only about 0.2% of the diesel request. Considering PFAD, according to Neste, there are about 2.5e6 tons/year available globally (PFAD residue from palm oil refining 2020), while only the EU has a demand of about 240e6 tons/year of diesel fuel. The biodiesel yield from PFAD is usually lower (about 80–85%), which leads that the global available PFAD can be used to fulfil less than 1% of the EU demands.

Regarding waste cooking oil, in the EU in 2020, there were potentially 1.15e6 tons available (Ershov et al. 2022), corresponding to about 0.5% of the diesel requirement. Moreover, obvious gathering difficulties and constraints, which require ad-hoc infrastructures, should be considered a constraint to a fully circular economy. So, the mass-production potential of HVO from these feedstocks should be carefully assessed. It is worth highlighting that only a few works treat this problem. Besides the fuel availability, with an associated problem of demand and supply balance, also the economic feasibility should be analysed. In particular, the economic profitability of the HVO can act as a prime mover to its diffusion. The production cost of HVO, about 1800–2200 $/year (Squadrin 2021), is about 3 or 4 times the one of diesel. Some studies foresee a reduction in the production costs of HVO in the next years (Syauqi et al. 2022). However, the production cost of diesel fuel ranges in Europe between 30 and 50% of the retail price (https://www.fuelseurope.eu/knowledge/refining-in-europe/economics-of-refining/fuel-price-breakdown/). In this context, regulations play a crucial role. The EU RED II directive provides a differentiation in fuel taxes, which should promote the use of renewable fuels at the expense of fossil ones (Ershov et al. 2022). Moreover, the HVO should be considered not as a complete substitute for fossil fuels in the actual transport sector but as a more sustainable fuel for the transition to electric mobility in developed countries, long-term solutions for developing countries, and applications less prone to electrification (heavy-duty, off-road, aviation, etc.).

2.4 Application of HVO to CI Engines

HVO can be used in compression ignition engines as a pure fuel, a blended one with traditional fuel, or a blending agent with additives. The research activities for alternative fuels are usually related to a study on the spray characteristics, combustion process, and emissions. This section will first discuss the spray evolution characteristics and then evaluate the fuel properties' effect on the combustion and emissions process for pure HVO.

2.4.1 Spray Characteristics

The evolution of the spray in the combustion chamber strongly influences the combustion process and, consequently, the formation of pollutants (Cardone et al. 2021). Therefore, the characteristics of the spray, in liquid and vapor phases, are fundamental aspects of evaluating the fuel quality. A limited number of manuscripts are available in the literature analyzing the behavior of the HVO spray at engine-like conditions.

The comparison between macroscopic quantities of the spray, such as spray tip penetration and spray cone angle between HVO and diesel, was carried out by Cheng et al. (Zubel et al. 2016). They evaluated the air density effect on spray evolution through the diffusive back-illumination technique in a combustion vessel under non-evaporative conditions. There were no evident differences in spray tip penetration, cone angle, and air-fuel mixing between the two fuels. The results agreed with Sugiyama (Dimitriadis et al. 2018) and Hulkkonen et al. (Bhardwaj et al. 2013), who performed a spray characterization using two different nozzle hole diameters, 0.08 and 0.12 mm and three rail pressures, 450, 1000, and 1980 bar at non-evaporative conditions. In addition, by numerical investigations, Fajri et al. (2022) assessed the effect of HVO properties on spray behavior in non-evaporating and evaporating conditions. HVO and diesel showed very similar behavior even under evaporating conditions. In the evaporating condition, only the droplet sizes for HVO are slightly smaller, justified by the higher vapor concentration due to the lower evaporation temperature, viscosity, and surface tension than diesel. These results are also in agreement with Chen (Murtonen et al. 2009).

Regarding the macroscopic spray characterization of the HVO, however, Bohl et al. (2018) found that the HVO led to a slight increase in the cone angle and lower spray tip penetration due to the lower density, in disagreement with the previous analyses. Also, Fajri et al. showed that HVO has lower maximum liquid penetration compared to diesel fuel (Fajri et al. 2022). It is due to faster evaporation and mixture formation, which can improve engine thermodynamic efficiency, useful for CO_2 reduction. These apparent contrasts are probably the results of different test conditions and nozzle geometry, which can highlight or confuse the different fuel properties (Pastor et al. 2021).

Bjørgen et al. (2020) have characterized the combustion process and in-flame soot phase with HVO and diesel fuel in a compression ignition engine with optical access. The results show a reduction of ignition delay of HVO than diesel according to the fuel cetane number. However, regarding the lift-of-length, no trend was observed in agreement with the ignition delay. A possible reason for this contradiction is the evaporative cooling differences between diesel and HVO since greater evaporation cools down the air-fuel mixture, which leads to a lengthening lift-of-length. Regarding the in-flame soot analysis, the HVO guarantees a soot reduction amount of about 3–6 times compared to diesel fuel, thanks to the absence of aromatics, despite the number of carbon bonds being similar between the two fuels.

In general, all the manuscripts considered conclude that the differences between diesel and HVO, in terms of spray evolution, are insignificant concerning the net effect on the combustion process.

2.5 Combustion and Emissions Characteristics

Although in the literature there are numerous studies regarding HVO, used as a blending agent, mixture, or pure fuel, the following section will report the results in emissions and combustion, referring only to the case of HVO employed as neat fuel.

Based on spray results illustrated in the previous section, the HVO can have the potential to directly replace diesel without any modification to CI engines (Hartikka et al. 2012). This allows its use as a drop-in fuel. However, further improvement in CI engine performances in terms of both emissions, efficiency, and combustion can be achieved with proper recalibration (Blasio et al. 2022; Dimitriadis et al. 2020).

Numerous studies have shown that HVO, characterized by a high cetane number, strongly influences the combustion process, especially at low loads. The high cetane number guarantees a reduction of ignition delay with consequent reduction of soot and benefits in terms of combustion noise. At high and full load, the high in-cylinder temperature has a dominant effect on the combustion process, and the CN effect is negligible, as reported by Bohl et al. (2018) for an HD engine in which virtually identical ignition delay and HRR are obtained for diesel and HVO.

Bhardwaj et al. (2013), through an experimental campaign on a light-duty single-cylinder engine, showed a reduction of particulate matter in the range of 50–65%. Similar results were also observed by Di Blasio et al. (2022) compared the effect of HVO with the same SOI of diesel fuel calibration, the same CA50 and similar HR to the one of the diesel in a single-cylinder research CI engine. They observed that the most effective strategy is dependent on the engine operating point at partial load. Overall, the HVO shows better global performances for all test points. Furthermore, a smoke gain of up to 50% was achieved, proportional to the PN reductions, as confirmed by Dimitriadis et al. (2020), Kuronen et al. 2007).

The PM reduction is also consistent at medium and high loads thanks to the aromatic free fuel composition (Bhardwaj et al. 2013), while the effect of the cetane number is less evident due to the high bulk gas temperatures (Hartikka et al. 2012). Omari et al. (Soam and Hillman 2019) also confirmed the same aspect, who observed an increased tendency of PM at lower loads and a decrease at higher loads. The PM reduction ranges from 25 to 30% using an HVO drop-in strategy at higher loads during the WLTP test cycle. Indeed, not only it generate less PM, but its particles can be oxidized at lower temperatures in the DPF than diesel, as it is free of aromatics and building blocks for the more thermally stable graphite-like material (Szeto and Leung 2022). These advantages lead to a lengthening of the DPF regeneration interval, with further benefits on fuel consumption (Ianniello et al. 2020).

Usually, lower combustion noise can be expected using HVO compared to diesel fuel because of the shorter ignition delay and lower adiabatic flame temperature thanks to the free aromatic and higher CN of HVO. Indeed a shorter ignition delay allows more fuel to be injected during the mixing controlled combustion phase. This implies a more gradual heat release rate and a premixed combination phase peak reduction. It ensures a combustion noise reduction and a lower combustion temperature peak (Liang et al. 2019). Bhardwaj et al. (2013) found a reduction in combustion noise of about 1.5–2.0 dBA reduction in engine noise intensity compared to diesel fuel. Also, Di Blasio et al. (2022) also found at partial load a reduction of combustion noise in the range 1–3 dBA adopting a drop-in calibration strategy. This is also confirmed by Omari et al. in a four-cylinder LD engine (Omari et al. 2017). The indicated thermal efficiency for HVO has a similar value to base diesel fuel. It is supported for both LD and HD engines by Preuß et al. (2021). In addition, Shepel

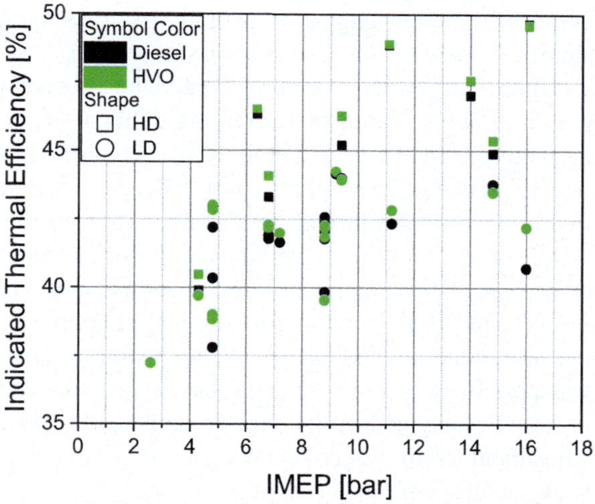

Fig. 2.6 Indicated thermal efficiency for different loads and engines operating conditions. *Source data* Zubel et al. (2016), Hunicz et al. (2020), Di Blasio et al. (2022); Preuß et al. (2021), Preuss et al. (2021)

et al. reported a slight increase in brake thermal efficiency of +0.5% (Shepel et al. 2021), which is confirmed at various loads by Rimkus et al. (Rimkus et al. 2019). The HVO does not give great concern regarding combustion stability. Many works reported values of coefficient of variation of IMEP perfectly in line with the diesel fuel (Blasio et al. 2022; Costa et al. 2022).

The indicated thermal efficiency for HVO has a similar value to base diesel fuel, as reported in Fig. 2.6. The HVO effect on thermal efficiency is strictly linked to the adopted test methodology, such as drop-in, recalibration and optimization, and combustion systems. Negligible differences are found for both LD and HD engines by Preuß (Preuß et al. 2021). In addition, Shepel et al. reported a slight increase in brake thermal efficiency of +0.5% (Shepel et al. 2021), which is confirmed at various loads by Rimkus et al. (2019). The HVO does not give great concern regarding combustion stability. Many works reported values of coefficient of variation of IMEP perfectly in line with the diesel fuel (Di Blasio et al. 2022; Costa et al. 2022).

Regarding NO_x emissions, thanks to the wide availability in the literature, it can be observed that light-duty engines show a slight or negligible reduction (Kousoulidou et al. 2014; Lapuerta et al. 2012) compared to heavy-duty ones (Happonen 2012). The NO_x reduction, as well as for the PN, is mainly due to the CN effect, which strongly influences the ignition delay. It involves a heat release reduction during the premixed phase of the combustion process, with a consequent decrease in combustion temperature and peak firing pressure, conditions unfavorable to the formation of NO_x (Huang et al. 2008). According to Sugiyama et al. (Sugiyama et al. 2011), the benefits of NO_x are more significant in adopting a single injection pulse. It is mainly due to the cetane number, which does not influence the ignition delay reduction in the case of the addition of pilot injection since the gas temperature in the combustion chamber is already very high with the only main one.

However, despite the slight influence on NO_x emissions, the great potential of HVO can be highlighted in overcoming the traditional PM-NO_x trade-off of CI engines. A proper engine recalibration setting for light-duty engines can lead to significant benefits in terms of emissions.

Dimitriadis et al. (2020) investigated the pilot and main injection timings and rail pressure effects, finding that an optimal calibration is to retard the main injection pulse of about 2° CA. Di Blasio et al. (2022) confirmed the substantial reduction in PM with higher EGR levels at ultra-low NO_x, demonstrating that the use of oxygenating fuel would probably provide more margin to the EGR use to reduce emissions further. Aatola et al. (2008) evaluated the combustion timing effect on a heavy-duty engine using pure HVO. Compared to the drop-in condition, further benefits were obtained on NO_x and PM by optimizing the injection times. Omari et al. (2017), thanks to the soot reduction and advanced combustion phasing at medium and high loads, it was possible to optimize the engine calibration setting through an increase in rail pressure, reducing the pilot quantity and the dwell time between the two pulses. This adjustment led to further benefits in emissions by not exceeding the diesel threshold values, emissions, and combustion noise.

In Fig. 2.7, some available data from the literature concerning PM-NO_x trade-offs are reported. LD and HD engine data are shown, and the same line style is used for the same engines and operating points. It is possible to notice that, especially for LD engines, shifting the whole trade-off curve in the ultra-low emissions zone. Figure 2.7 shows that the PM reduction can be exploited to optimize the ERG strategy further. Supposed, on the one side, the reduction of PM guarantees an extension of the DPF regeneration intervals, with consequent reduction of fuel consumption and CO_2 (Omari et al. 2017); on the other one, it can be exploited to reduce NO_x further. The figure shows that it is possible to reduce the NO_x emissions not exceeding the diesel threshold values. This effect is most evident in heavy-duty engines, for which there is a more significant difference between the two fuels in the ultra-low NO_x zone.

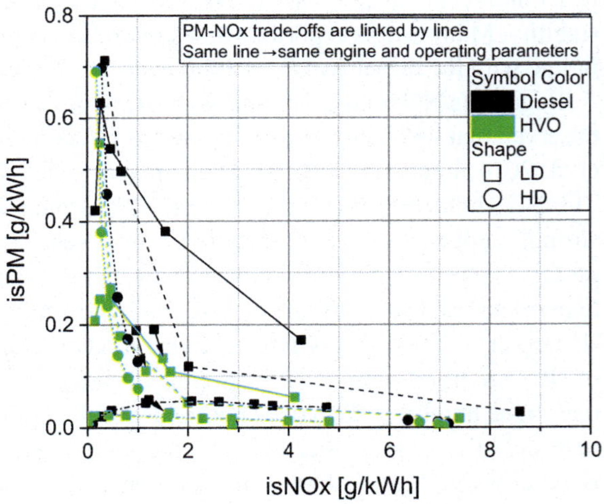

Fig. 2.7 NO_x-PM trade-off for different engines and operating conditions. *Source data* Di Blasio et al. (2022), Dimitriadis et al. (2020), Preuß et al. (2021), Shepel et al. (2021)

Total Hydrocarbons (THC) and Carbon monoxide (CO) do not represent problematic emission components as well as PM and NO_x for CI engines. Thanks to the high cetane number, lower boiling point, and absence of aromatics influence, the HVO favors the reduction of CO and THC, both regarding light and heavy-duty engines. A drop in both indicates an improvement in combustion efficiency (Aatola et al. 2008). Pflaum et al. (2010), through an experimental activity using a four-cylinder diesel engine fuelled with neat HVO, showed a reduction of up to 50% in both CO and THC emissions to diesel. Based on steady-state experimental data, Dimitriadis et al. (Hartikka et al. 2012) found an aligned result: lower THC emissions by up to 45%. A reduction of THC and CO emissions, in the range of approximately 65–70%, was noted over the entire speed load range for HVO due to its shorter ignition delay, as observed with the drop-in condition. Omari et al. (2017), adopting a drop-in HVO strategy, showed a similar THC and CO reduction interval. This unburned emissions reduction was used to optimize the EGR strategy to reduce the NO_x level further.

Below are the NO_x-THC (left panel of Fig. 2.8) and NO_x–CO (right panel of Fig. 2.8) trade-offs to further demonstrate the HVO potential for both engine sizes. As seen previously in Fig. 2.7, further optimizing EGR can reduce NO_x engine out by not exceeding the diesel target. The following graphs show how this optimization does not significantly impact unburnt emissions. For both CO and THC, regardless of the engine, a reasonably constant gap is observed for the entire range considered, particularly for low unburned emissions values.

However, this section discusses combustion-related CO_2, which should not be confused with well-to-wheel or LCA CO_2 emissions. CO_2 reduction is confirmed in the literature in a broad range of conditions. The CO_2 emissions, in general, are reduced thanks to its lower carbon to hydrogen atom ratio (C/H) and increased LHV compared to conventional diesel. The lower C/H ratio shifts the production ratio towards water instead of CO_2, while the increased LHV helps to reduce the quantity

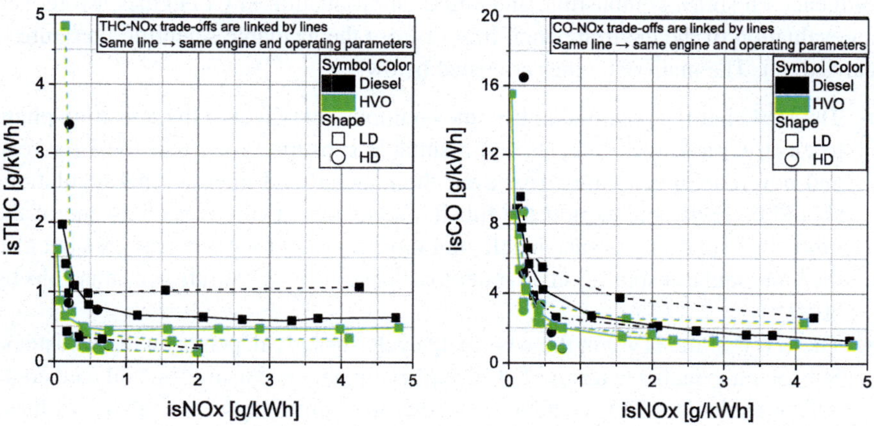

Fig. 2.8 NO_x-THC (on the left) and NO_x–CO (on the right) trade-offs for different engines and operating conditions. *Source data* Zubel et al. (2016), Bhardwaj et al. (2013), Hunicz et al. (2020), Di Blasio et al. (2022)

of fuel used, even considering a lower HVO density. Valeika et al. (2021) found a reduction of CO_2 emission in the range of 3.5–6.7% at a constant engine speed of 2000 rpm and different engine loads. Similar values are reported in Costa et al. (2022) of about 6% for an LD engine, in the range of 4–6% for passenger cars on WLTP driving cycle by Suarez-Bertoa et al. (2019) and up to 4% with 2 HD vehicles by Karavalakis et al. (2016).

Unregulated emissions are rarely investigated in the literature. However, some of the main findings of HVO on these pollutants are briefly listed below. Westphal et al. (2013) found that HVO reduces the PAH while slightly increasing carbonyl emissions are observed, especially for aldehydes, compared to diesel fuel in an HD engine. Hunicz et al. (2020) analyzed the unregulated emissions varying EGR rate in an LD engine. Formaldehydes and acetaldehydes are slightly lower, from without EGR to moderate one (about 17% v/v of O_2 at intake manifold). The differences became greatest at higher EGR levels. For aromatic compounds, until moderate EGR level, the emissions are under 0.02 g/kWh, with a rapid increase proportional to the EGR rate. It is due to the lack of oxygen, which can prevent the oxidation of intermediate combustion species before the Exhaust Valve Open (EVO). McCaffery et al. (2022), testing an HD engine with two different transient test cycles found that carbonyl species strongly depend on test cycles with HVO emission sometimes higher and other times lower than diesel. However, they confirmed the lower PAH emissions for HVO, thanks to the lack of aromatic compounds in fuel formulation.

2.6 Conclusions

In this chapter, an assessment of HVO fuel applied to CI engines is carried out. The work assesses the most important chemical and physical fuel properties with their influence on spray, combustion, and emission production in CI engines. Moreover, an evaluation of the environmental impact from the available scientific literature is carried out. The main outcomes are listed below:

- The HVO has the potential to become an important player in the transition phase to electric passenger vehicles and a suitable substitute for fossil fuels for road and non-road applications. However, the national and international regulations should be revised to exploit the full fuel potential. It can be seen as one of the promising solutions to sustainable mobility, not only for passenger cars but also the long-haul transportation, in order to achieve carbon neutrality in emissions by 2050.
- The GWP of the HVO in the whole life cycle strongly depends on the feedstock. Substantial benefits in terms of gCO_2eq/MJ are obtained using residual feedstock, which results in −75% GHG to fossil diesel. Therefore, the availability of these feedstocks for the mass production of HVO should be addressed.
- In general, no differences in spray evolution are observed between the two fuels; only the droplet sizes for HVO are slightly smaller, thanks to the higher vapor

concentration due to the lower evaporation temperature, viscosity, and surface tension compared to diesel. It demonstrates that HVO can potentially replace diesel directly without any modification to CI engines.

- HVO is a fuel superior to fossil diesel in terms of engine-out emissions (particulate matter, HC, CO, and NO_x), noise, and thermal efficiency for CI engines. The reported benefits mainly derive from its typical basic characteristics, including the absence of cycloalkanes (only n-and ISO-alkanes) and aromatics contents, free of sulfur and other ashes, and less dispersed molecular weights.
- In general, lower combustion noise is obtained thanks to a higher cetane number and the absence of aromatics, which strongly influence the combustion process. Thermal efficiency benefits are not always guaranteed because of strictly depend on engine calibrations.
- At the same operating conditions, slight differences in NO_x emissions can be expected but with lower PM (up to 65%) due to the higher cetane number and lack of aromatic compounds, allowing to overcome the traditional PM-NO_x trade-off.
- Lower THC and CO levels are usually achieved, in the range from 40 to 70%, thanks to the high cetane number, lower boiling point, and absence of aromatics. It allows for further optimize EGR and NO_x emissions.
- Few studies on unregulated emissions have been published so far. However, based on the literature results, the HVO has the potential to reduce the PAH significantly.

References

Aatola H, Larmi M, Sarjovaara T, Mikkonen S (2008) Hydrotreated vegetable oil (HVO) as a renewable diesel fuel: trade-off between NOx, particulate emission, and fuel consumption of a heavy duty engine. SAE Int J Engines 1:1251–1262. https://doi.org/10.4271/2008-01-2500

Altıparmak D, Keskin A, Koca A, Gürü M (2007) Alternative fuel properties of tall oil fatty acid methyl ester–diesel fuel blends. Biores Technol 98:241–246. https://doi.org/10.1016/j.biortech.2006.01.020

Andersson Ö, Börjesson P (2021) The greenhouse gas emissions of an electrified vehicle combined with renewable fuels: life cycle assessment and policy implications. Appl Energy 289:116621. https://doi.org/10.1016/j.apenergy.2021.116621

Arora NK, Mishra I (2021) COP26: more challenges than achievements. Environ Sustain 4:585–588. https://doi.org/10.1007/s42398-021-00212-7

Arvidsson R, Persson S, Fröling M, Svanström M (2011) Life cycle assessment of hydrotreated vegetable oil from rape, oil palm and Jatropha. J Clean Prod 19:129–137. https://doi.org/10.1016/j.jclepro.2010.02.008

Balat M (2011) Potential alternatives to edible oils for biodiesel production—A review of current work. Energy Conv Manag 52:1479–1492. https://doi.org/10.1016/j.enconman.2010.10.011

Barbour RH, Rickeard DJ, Elliott NG (2000) Understanding diesel lubricity, pp 2000-01–1918

Beatrice C, Capasso C, Costa M, Di Blasio G, Di Luca G, Iantorno F, Martoriello G (2022) Model based optimal management of a hybrid propulsion system for leisure boats. J Energy Storage 46:103896. https://doi.org/10.1016/j.est.2021.103896

Beatrice C, Di Blasio G, Pesce FC, Vassallo A, Avolio G, Ianniello R (2019) Key fuel injection system features for efficiency improvement in future diesel passenger cars, pp 2019-01–0547. https://doi.org/10.4271/2019-01-0547

Belgiorno G, Boscolo A, Dileo G, Numidi F, Pesce FC, Vassallo A, Ianniello R, Beatrice C, Di Blasio G (2020) Experimental study of additive-manufacturing-enabled innovative diesel combustion bowl features for achieving ultra-low emissions and high efficiency. SAE Int J Adv Curr Prac Mob 3:672–684. https://doi.org/10.4271/2020-37-0003

Bellocchi S, Gambini M, Manno M, Stilo T, Vellini M (2018) Positive interactions between electric vehicles and renewable energy sources in CO_2-reduced energy scenarios: the Italian case. Energy 161:172–182. https://doi.org/10.1016/j.energy.2018.07.068

Benavides PT, Sun P, Han J, Dunn JB, Wang M (2017) Life-cycle analysis of fuels from post-use non-recycled plastics. Fuel 203:11–22. https://doi.org/10.1016/j.fuel.2017.04.070

Bhardwaj OP, Kolbeck AF, Kkoerfer T, Honkanen M (2013) Potential of hydrogenated vegetable oil (HVO) in future high efficiency combustion system. SAE Int J Fuels Lubr 6:157–169. https://doi.org/10.4271/2013-01-1677

Bjørgen KOP, Emberson DR, Løvås T (2020) Combustion and soot characteristics of hydrotreated vegetable oil compression-ignited spray flames. Fuel 266:116942. https://doi.org/10.1016/j.fuel.2019.116942

Bohl T, Smallbone A, Tian G, Roskilly AP (2018) Particulate number and NOx trade-off comparisons between HVO and mineral diesel in HD applications. Fuel 215:90–101. https://doi.org/10.1016/j.fuel.2017.11.023

Brahma S, Nath B, Basumatary B, Das B, Saikia P, Patir K, Basumatary S (2022) Biodiesel production from mixed oils: a sustainable approach towards industrial biofuel production. Chem Eng J Adv 10:100284. https://doi.org/10.1016/j.ceja.2022.100284

Cardone M, Marialto R, Ianniello R, Lazzaro M, Di Blasio G (2021) Spray analysis and combustion assessment of diesel-LPG fuel blends in compression ignition engine. Fuels 2:1–15. https://doi.org/10.3390/fuels2010001

Constantino G, Freitas M, Fidelis N, Pereira MG (2018) Adoption of photovoltaic systems along a sure path: a life-cycle assessment (LCA) study applied to the analysis of GHG emission impacts. Energies 11:2806. https://doi.org/10.3390/en11102806

da Costa RBR, Roque LFA, de Souza TAZ, Coronado CJR, Pinto GM, Cintra AJA, Raats OO, Oliveira BM, Frez GV, da Silva MH (2022) Experimental assessment of renewable diesel fuels (HVO/Farnesane) and bioethanol on dual-fuel mode. Energy Convers Manag 258:115554. https://doi.org/10.1016/j.enconman.2022.115554

Di Blasio G, Beatrice C, Ianniello R, Pesce FC, Vassallo A, Belgiorno G, Avolio G (2019) Balancing hydraulic flow and fuel injection parameters for low-emission and high-efficiency automotive diesel engines. pp 2019-24–0111. https://doi.org/10.4271/2019-24-0111

Di Blasio G, Ianniello R, Beatrice C (2022) Hydrotreated vegetable oil as enabler for high-efficient and ultra-low emission vehicles in the view of 2030 targets. Fuel 310:122206. https://doi.org/10.1016/j.fuel.2021.122206

Dimitriadis A, Natsios I, Dimaratos A, Katsaounis D, Samaras Z, Bezergianni S, Lehto K (2018) Evaluation of a hydrotreated vegetable oil (HVO) and effects on emissions of a passenger car diesel engine. Front Mechan Eng 4

Dimitriadis A, Seljak T, Vihar R, Žvar Baškovič U, Dimaratos A, Bezergianni S, Samaras Z, Katrašnik T (2020) Improving PM-NOx trade-off with paraffinic fuels: a study towards diesel engine optimization with HVO. Fuel 265:116921. https://doi.org/10.1016/j.fuel.2019.116921

Di Luca G, Pipicelli M, Ianniello R, Belgiorno G, Di Blasio G (2022) Alcohol fuels in spark ignition engines. In: Di Blasio G, Agarwal AK, Belgiorno G, Shukla PC (eds) Application of clean fuels in combustion engines. Springer Nature, Singapore, pp 33–54. https://doi.org/10.1007/978-981-16-8751-8_3

Eberle A, Heath GA, Carpenter Petri AC, Nicholson SR (2017) Systematic review of life cycle greenhouse gas emissions from geothermal electricity. In: National renewable energy lab (NREL), Golden, CO (United States)

Ecofining™: dai rifiuti organici ai biocarburanti. https://www.eni.com/it-IT/attivita/biocarburanti-sostenibili-ecofining-tm.html. Accessed 2 Jul 2022

Engman A, Hartikka T, Honkanen M, Kiiski U, Kuronen L, Lehto K, Mikkonen S, Nortio J, Nuottimäki J, Saikkonen P (2016) Neste renewable diesel handbook. Neste Proprietary Publication, Espoo

Ershov MA, Savelenko VD, Makhova UA, Makhmudova AE, Zuikov AV, Kapustin VM, Abdellatief TMM, Burov NO, Geng T, Abdelkareem MA, Olabi AG (2022) Current challenge and innovative progress for producing HVO and FAME biodiesel fuels and their applications. Waste Biomass Valor. https://doi.org/10.1007/s12649-022-01880-0

ETIP Bioenergy—European Technology and Innovation Platform (2020) Hydrogenated vegetable oil (HVO)

EVC2: Hydrotreatment to HVO. https://www.etipbioenergy.eu/value-chains/conversion-technologies/conventional-technologies/hydrotreatment-to-hvo. Accessed 2 Jul 2022

Fajri H, Clemente Mallada R, Riess S, Strauß L, Wensing M (2022) Mixture formation analysis for diesel, n-dodecane, RME, and HVO in large-scale injector nozzles. In: CO_2 reduction for transportation systems conference. SAE International

Faria R, Marques P, Moura P, Freire F, Delgado J, de Almeida AT (2013) Impact of the electricity mix and use profile in the life-cycle assessment of electric vehicles. Renew Sustain Energy Rev 24:271–287. https://doi.org/10.1016/j.rser.2013.03.063

Fuel price breakdown. In: FuelsEurope. https://www.fuelseurope.eu/knowledge/refining-in-europe/economics-of-refining/fuel-price-breakdown/. Accessed 25 Aug 2022

Geraldes Castanheira É, Grisoli R, Freire F, Pecora V, Coelho ST (2014) Environmental sustainability of biodiesel in Brazil. Energy Policy 65:680–691. https://doi.org/10.1016/j.enpol.2013.09.062

Giardiello G, de Nola F, Pipicelli M, Troiano A, Parascandolo F, Gimelli A, Santoro D, Sessa B (2022) Comparative analysis on fuel consumption between two online strategies for P2 hybrid electric vehicles: adaptive-rulebased (A-RB) versus adaptive-equivalent consumption minimization strategy (A-ECMS). In: WCX SAE world congress experience. SAE International. https://doi.org/10.4271/2022-01-0740

Happonen M (2012) Particle and NOx emissions from a HVO-fueled diesel engine. Tampere University of Technology

Hartikka T, Kuronen M, Kiiski U (2012) Technical performance of HVO (Hydrotreated Vegetable Oil) in diesel engines. SAE International, Warrendale, PA

Hjelkrem OA, Arnesen P, Aarseth Bø T, Sondell RS (2020) Estimation of tank-to-wheel efficiency functions based on type approval data. Appl Energy 276:115463. https://doi.org/10.1016/j.apenergy.2020.115463

Hondo H (2005) Life cycle GHG emission analysis of power generation systems: Japanese case. Energy 30:2042–2056. https://doi.org/10.1016/j.energy.2004.07.020

Huang Y, Wang S, Zhou L (2008) Effects of Fischer-Tropsch diesel fuel on combustion and emissions of direct injection diesel engine. Front Energy Power Eng China 2:261–267. https://doi.org/10.1007/s11708-008-0062-x

Hunicz J, Matijošius J, Rimkus A, Kilikevičius A, Kordos P, Mikulski M (2020) Efficient hydrotreated vegetable oil combustion under partially premixed conditions with heavy exhaust gas recirculation. Fuel 268:117350. https://doi.org/10.1016/j.fuel.2020.117350

Hunicz J, Mikulski M, Shukla PC, Gęca MS (2022) Partially premixed combustion of hydrotreated vegetable oil in a diesel engine: sensitivity to boost and exhaust gas recirculation. Fuel 307:121910. https://doi.org/10.1016/j.fuel.2021.121910

Ianniello R, Di Blasio G, Marialto R, Beatrice C, Cardone M (2020) Assessment of direct injected liquefied petroleum gas-diesel blends for ultra-low soot combustion engine application. Appl Sci 10:4949. https://doi.org/10.3390/app10144949

Ianniello R, Belgiorno G, Di Luca G, Beatrice C, Di Blasio G (2021) Ethanol in dual-fuel and blend fueling modes for advanced combustion in compression ignition engines. In: Shukla PC, Belgiorno G, Di Blasio G, Agarwal AK (eds) Alcohol as an alternative fuel for internal combustion engines. Springer, Singapore, pp 5–27. https://doi.org/10.1007/978-981-16-0931-2_2

Källmén A, Andersson S, Rydberg T, others (2019) Well-to-wheel LCI data for HVO fuels on the Swedish market. Tillgänglig 20. https://f3centrese/app/uploads/f3-23-17_2019-04_Källmén-et-al_Rev_190508_FINAL pdf

Karavalakis G, Jiang Y, Yang J, Durbin T, Nuottimäki J, Lehto K (2016) Emissions and fuel economy evaluation from two current technology heavy-duty trucks operated on HVO and FAME blends. SAE Int J Fuels Lubr 9:177–190. https://doi.org/10.4271/2016-01-0876

Keoleian GA, Spitzley DV (2006) Chapter 7 Life cycle based sustainability metrics. In: Abraham MA (ed) Sustainability science and engineering. Elsevier, pp 127–159

Knothe G (2010) Biodiesel and renewable diesel: a comparison. Prog Energy Combust Sci 36:364–373. https://doi.org/10.1016/j.pecs.2009.11.004

Kousoulidou M, Dimaratos A, Karvountzis-Kontakiotis A, Samaras Z (2014) Combustion and emissions of a common-rail diesel engine fueled with HWCO. J Energy Eng 140:A4013001. https://doi.org/10.1061/(ASCE)EY.1943-7897.0000154

Kuronen M, Mikkonen S, Aakko P, Murtonen T (2007) Hydrotreated vegetable oil as fuel for heavy duty diesel engines, pp 2007-01–4031

Lapuerta M, Rodríguez-Fernández J, Armas O (2010) Correlation for the estimation of the density of fatty acid esters fuels and its implications. A proposed biodiesel cetane index. Chem Phys Lipid 163:720–727. https://doi.org/10.1016/j.chemphyslip.2010.06.004

Lapuerta M, Agudelo JR, Prorok M, Boehman AL (2012) Bulk modulus of compressibility of diesel/biodiesel/HVO blends. Energy Fuels 26:1336–1343. https://doi.org/10.1021/ef201608g

Liang X, Zheng Z, Zhang H, Wang Y, Yu H (2019) A review of early injection strategy in premixed combustion engines. Appl Sci 9:3737. https://doi.org/10.3390/app9183737

Liu CM, Sandhu NK, McCoy ST, Bergerson JA (2020) A life cycle assessment of greenhouse gas emissions from direct air capture and Fischer–Tropsch fuel production. Sustain Energy Fuels 4:3129–3142. https://doi.org/10.1039/C9SE00479C

Lyng K-A, Brekke A (2019) Environmental life cycle assessment of biogas as a fuel for transport compared with alternative fuels. Energies 12:532. https://doi.org/10.3390/en12030532

Marek A, Kardasz P, Karpinski M, Pohrebennyk V (2016) Assessment of the logistic system of fuel life cycle using the LCA method. Agric Eng 20:125–134. https://doi.org/10.1515/agriceng-2016-0050

McCaffery C, Zhu H, Sabbir Ahmed CM, Canchola A, Chen JY, Li C, Johnson KC, Durbin TD, Lin Y-H, Karavalakis G (2022) Effects of hydrogenated vegetable oil (HVO) and HVO/biodiesel blends on the physicochemical and toxicological properties of emissions from an off-road heavy-duty diesel engine. Fuel 323:124283. https://doi.org/10.1016/j.fuel.2022.124283

Michel P, Karbowski D, Rousseau A (2016) Impact of connectivity and automation on vehicle energy use, pp 2016-01–0152

Murdock HE, Gibb D, Andre T, Sawin JL, Brown A, Andre T, Appavou F, Brown A, Ellis G, Epp B, Gibb D, Guerra F, Joubert F, Kamara R, Kondev B, Levin R, Murdock HE, Sawin JL, Seyboth K, Skeen J, Sverrisson F, Wright G, Corcoran F, Yaqoob H, Assoum D, Gicquel S, Hamirwasia V, Ranalder L, Satzinger K, Williamson LE, Findlay K, Swenson A, Urbani F, Froning S, Reise N, Mastny L (2020) Renewables 2020—Global status report. France

Murtonen T, Aakko-Saksa P, Kuronen M, Mikkonen S, Lehtoranta K (2009) Emissions with heavy-duty diesel engines and vehicles using FAME, HVO and GTL fuels with and without DOC+POC aftertreatment. SAE Int J Fuels Lubr 2:147–166. https://doi.org/10.4271/2009-01-2693

Musa A, Pipicelli M, Spano M, Tufano F, De Nola F, Di Blasio G, Gimelli A, Misul DA, Toscano G (2021) A review of model predictive controls applied to advanced driver-assistance systems. Energies 14. https://doi.org/10.3390/en14237974

No S-Y (2014) Application of hydrotreated vegetable oil from triglyceride based biomass to CI engines—A review. Fuel 115:88–96. https://doi.org/10.1016/j.fuel.2013.07.001

Nordelöf A, Romare M, Tivander J (2019) Life cycle assessment of city buses powered by electricity, hydrogenated vegetable oil or diesel. Transp Res Part d: Transp Environ 75:211–222. https://doi.org/10.1016/j.trd.2019.08.019

Omari A, Pischinger S, Bhardwaj OP, Holderbaum B, Nuottimäki J, Honkanen M (2017) Improving engine efficiency and emission reduction potential of HVO by fuel-specific engine calibration in modern passenger car diesel applications. SAE Int J Fuels Lubr 10:2017-01–2295. https://doi.org/10.4271/2017-01-2295

Pastor JV, García-Oliver JM, Micó C, García-Carrero AA (2021) An experimental study with renewable fuels using ECN Spray A and D nozzles. Int J Engine Res 14680874211031200. https://doi.org/10.1177/14680874211031200

Pellegrini L, Beatrice C, Blasio GD (2015) Investigation of the effect of compression ratio on the combustion behavior and emission performance of HVO blended diesel fuels in a single-cylinder light-duty diesel engine. SAE International, Warrendale, PA. https://doi.org/10.4271/2015-01-0898

(2020) PFAD residue from palm oil refining. In: Neste worldwide. https://www.neste.com/products/all-products/raw-materials/pfad-residue-palm-oil-refining. Accessed 5 July 2022

Pflaum H, Hofmann P, Geringer B, Weissel W (2010) Potential of hydrogenated vegetable oil (HVO) in a modern diesel engine, pp 2010-32–0081

Ping WD, Korcek S, Spikes H (1996) Comparison of the lubricity of gasoline and diesel fuels. SAE International, Warrendale, PA

Pipicelli M, Di Luca G, Ianniello R, Gimelli A, Beatrice C (2022) Alcohol fuels in compression ignition engines. In: Di Blasio G, Agarwal AK, Belgiorno G, Shukla PC (eds) Application of clean fuels in combustion engines. Springer Nature, Singapore, pp 9–31. https://doi.org/10.1007/978-981-16-8751-8_2

Preuß J, Munch K, Denbratt I (2021) Performance and emissions of renewable blends with OME3-5 and HVO in heavy duty and light duty compression ignition engines. Fuel 303:121275. https://doi.org/10.1016/j.fuel.2021.121275

Preuss J, Munch K, Denbratt I (2021) Effect of injection strategy and EGR on particle emissions from a CI engine fueled with an oxygenated fuel blend and HVO, pp 2021-01–0560

Prussi M, Yugo M, DE PL, Padella M, Edwards R (2020) JEC well-to-wheels report v5. Publications Office of the European Union, Luxembourg

Rajasekar E, Selvi S (2014) Review of combustion characteristics of CI engines fueled with biodiesel. Renew Sustain Energy Rev 35:390–399. https://doi.org/10.1016/j.rser.2014.04.006

Renewable Fuels Association Focus Forward (2020) Ethanol industry outlook

Rimkus A, Žaglinskis J, Stravinskas S, Rapalis P, Matijošius J, Bereczky Á (2019) Research on the combustion, energy and emission parameters of various concentration blends of hydrotreated vegetable oil biofuel and diesel fuel in a compression-ignition engine. Energies 12:2978. https://doi.org/10.3390/en12152978

Sahu TK, Shukla PC, Belgiorno G, Maurya RK (2022) Alcohols as alternative fuels in compression ignition engines for sustainable transportation: a review. Energy Sour Part a: Recov Utiliz Environ Effects 44:8736–8759. https://doi.org/10.1080/15567036.2022.2124326

Shepel O, Matijošius J, Rimkus A, Duda K, Mikulski M (2021) Research of parameters of a compression ignition engine using various fuel mixtures of hydrotreated vegetable oil (HVO) and fatty acid esters (FAE). Energies 14:3077. https://doi.org/10.3390/en14113077

Šimáček P, Kubička D, Šebor G, Pospíšil M (2010) Fuel properties of hydroprocessed rapeseed oil. Fuel 89:611–615. https://doi.org/10.1016/j.fuel.2009.09.017

Soam S, Hillman K (2019) Factors influencing the environmental sustainability and growth of hydrotreated vegetable oil (HVO) in Sweden. Bioresour Technol Rep 7:100244. https://doi.org/10.1016/j.biteb.2019.100244

Squadrin G (2021) NWE HVO, SAF at records on feedstocks prices, demand | Argus Media. https://www.argusmedia.com/en/news/2220328-nwe-hvo-saf-at-records-on-feedstocks-prices-demand. Accessed 25 Aug 2022

Styring P, Dowson GRM, Tozer IO (2021) Synthetic fuels based on dimethyl ether as a future non-fossil fuel for road transport from sustainable feedstocks. Front Energy Res 9

Suarez-Bertoa R, Kousoulidou M, Clairotte M, Giechaskiel B, Nuottimäki J, Sarjovaara T, Lonza L (2019) Impact of HVO blends on modern diesel passenger cars emissions during real world operation. Fuel 235:1427–1435. https://doi.org/10.1016/j.fuel.2018.08.031

Sugiyama K, Goto I, Kitano K, Mogi K, Honkanen M (2011) Effects of hydrotreated vegetable oil (HVO) as renewable diesel fuel on combustion and exhaust emissions in diesel engine. SAE Int J Fuels Lubr 5:205–217. https://doi.org/10.4271/2011-01-1954

Suh S, Lenzen M, Treloar GJ, Hondo H, Horvath A, Huppes G, Jolliet O, Klann U, Krewitt W, Moriguchi Y, Munksgaard J, Norris G (2004) System boundary selection in life-cycle inventories using hybrid approaches. Environ Sci Technol 38:657–664. https://doi.org/10.1021/es0263745

Sustainable Chemistry | Free Full-Text | Synthesis of Biodiesel from Tall Oil Fatty Acids by Homogeneous and Heterogeneous Catalysis. https://www.mdpi.com/2673-4079/2/1/12. Accessed 4 July 2022

Syauqi A, Halimatussadiah A, Purwanto WW Assessing and choosing the optimum blend of biodiesel, hydrogenated vegetable oil, and petroleum diesel based on sustainability in Indonesia. Biofuels Bioprod Biorefining n/a. https://doi.org/10.1002/bbb.2402

Szeto W, Leung DYC (2022) Is hydrotreated vegetable oil a superior substitute for fossil diesel? A comprehensive review on physicochemical properties, engine performance and emissions. Fuel 327:125065. https://doi.org/10.1016/j.fuel.2022.125065

Teixeira ACR, Sodré JR (2018) Impacts of replacement of engine powered vehicles by electric vehicles on energy consumption and CO_2 emissions. Transp Res Part d: Transp Environ 59:375–384. https://doi.org/10.1016/j.trd.2018.01.004

Ternel C, Bouter A, Melgar J (2021) Life cycle assessment of mid-range passenger cars powered by liquid and gaseous biofuels: comparison with greenhouse gas emissions of electric vehicles and forecast to 2030. Transp Res Part d: Transp Environ 97:102897. https://doi.org/10.1016/j.trd.2021.102897

Valeika G, Matijošius J, Górski K, Rimkus A, Smigins R (2021) A study of energy and environmental parameters of a diesel engine running on hydrogenated vegetable oil (HVO) with addition of biobutanol and castor oil. Energies 14:3939. https://doi.org/10.3390/en14133939

Varga BO (2013) Electric vehicles, primary energy sources and CO_2 emissions: Romanian case study. Energy 49:61–70. https://doi.org/10.1016/j.energy.2012.10.036

Westphal GA, Krahl J, Munack A, Rosenkranz N, Schröder O, Schaak J, Pabst C, Brüning T, Bünger J (2013) Combustion of hydrotreated vegetable oil and jatropha methyl ester in a heavy duty engine: emissions and bacterial mutagenicity. Environ Sci Technol 47:6038–6046. https://doi.org/10.1021/es400518d

Zhang T, Eismark J, Munch K, Denbratt I (2020) Effects of a wave-shaped piston bowl geometry on the performance of heavy duty Diesel engines fueled with alcohols and biodiesel blends. Renew Energy 148:512–522. https://doi.org/10.1016/j.renene.2019.10.057

Zubel M, Bhardwaj OP, Heuser B, Holderbaum B, Doerr S, Nuottimäki J (2016) Advanced fuel formulation approach using blends of paraffinic and oxygenated biofuels: analysis of emission reduction potential in a high efficiency diesel combustion system. SAE Int J Fuels Lubr 9:481–492. https://doi.org/10.4271/2016-01-2179

Chapter 3
DME as a Green Fuel for Transport Sector

Ayush Tripathi and Avinash Kumar Agarwal

Abstract Transport sector is one of the largest contributor to global greenhouse gas (GHG) emissions. GHG emissions are the leading cause of global warming. Various government bodies have imposed strict emission legislation across the worldwide on industries with large carbon footprints. Therefore, the transport sector is greatly strained to meet stringent emission norms. Crude oil refined products such as diesel and gasoline are the primary fuels used in the transport sector. Greener e-fuels with low carbon footprint are suitable alternatives to replace the fossil fuels, which can reduce the life cycle CO_2 emission and meet emission norms. Dimethyl ether (DME) is one such greener alternative to diesel. Diesel-fuelled CI engines are majorly emitting the NO_x and soot emissions. DME can reduce NO_x and soot emissions simultaneously. The molecule of DME does not contain a C–C bond. Therefore, it resists the formation of soot precursors. High-temperature zones are the ideal zones for NO_x formation in an engine combustion chamber. The high latent heat of vaporisation of DME reduces the in-cylinder temperatures; hence, decreases the NO_x formation. DME also shows a high cetane number (CN), molecular oxygen, shorter ignition delay, and superior atomisation characteristics than baseline mineral diesel. However, lower viscosity, high vapour pressure, longer fuel injection delay due to higher compressibility, and lower lubrication properties of DME makes it difficult to be used by diesel fuel injection equipment (FIE). Certain modifications are required in diesel FIE to operate on DME. This chapter briefly discusses the various modifications required in FIE to utilise DME. The effects of superior properties of DME on various engine parameters are also described.

Keywords GHG emissions · E-fuels · DME · Life cycle analysis · Soot · Life cycle · Fuel injection equipment (FIE)

A. Tripathi · A. K. Agarwal (✉)
Engine Research Laboratory, Department of Mechanical Engineering, Indian Institute of Technology Kanpur, Kanpur 208016, India
e-mail: akag@iitk.ac.in

P. C. Shukla et al. (eds.), *Renewable Fuels for Sustainable Mobility*, Energy, Environment, and Sustainability, https://doi.org/10.1007/978-981-99-1392-3_3

Abbreviations

ASTM	American Society for Testing and Materials
aTDC	After Top Dead Centre
BSFC	Brake Specific Fuel Consumption
BTE	Brake Thermal Efficiency
bTDC	Before Top Dead Centre
CAD	Crank Angle Degree
CHR	Cumulative Heat Release
CH_4	Methane
CI	Compression Ignition
CN	Cetane Number
CNG	Compressed Natural Gas
CO	Carbon Monoxide
CO_2	Carbon Dioxide
DEE	Diethyl Ether
DME	Dimethyl Ether
DS	Dry Soot
EGR	Exhaust Gas Recirculation
EGT	Exhaust Gas Temperature
EPA	Environmental Protection Agency
FIE	Fuel Injection Equipment
GDP	Gross Domestic Product
GHG	Greenhouse gases
HP	High Pressure
HRR	Heat Release Rate
HTR	High-Temperature Reactions
H_2	Hydrogen
IC	Internal Combustion
IMEP	Indicated Mean Effective Pressure
LHV	Lower Heating Value
LNT	Lean NO_x Trap
LP	Low Pressure
LPG	Liquified Petroleum Gas
LTR	Low-Temperature Reaction
MMT	Million Metric Ton
NBR	Nitrile Butadiene Rubber
NOAA	National Oceanic and Atmospheric Administration
NOx	Oxides of Nitrogen
NTP	Normal Temperature Pressure
ON	Octane Number
OSHA	Occupational Safety and Health Administration
PAHs	Polycyclic Aromatic Hydrocarbons
PCCI	Premixed Charge Compression Ignition

PM	Particulate Matter
PPCCI	Partially Premixed Charge Compression Ignition
RoPR	Rate of Pressure Rise
SCR	Selective Catalytic Reduction
SoI	Start of injection
SOx	Oxides of Sulphur
SO_2	Sulphur Dioxide
UK	United Kingdom

3.1 Introduction

The transport sector plays a vital part in the GDP growth of any country. This sector contributes ~6.4% to the GDP of India (India Planning Commission 2007). The average annual growth rate of the transport sector is much higher than India's GDP growth rate. The transport sector consumes crude oil-origin fuels such as diesel and gasoline. The transport sector is responsible for ~13.5% of energy-related CO_2 emissions in India (New Climate Institute 2020). India is the third-biggest CO_2 emitter in the world, after China and USA. India committed to achieve the carbon neutrality by 2070 at Glasgow's COP26 climate change conference. Therefore, Indian government aims to reduce and eventually eliminate tailpipe emissions from the transport sector without impacting the economy. In 2019, India imported crude oil worth US$ 105 billion (https://energy.economictimes.indiatimes.com/news/oil-and-gas/indias-crude-oil-import-bill-set-to-rise-20-to-105-billion-in-fy19/63881639). Crude oil imports accounted for ~4% of the GDP. Reducing the crude oil import would help cut down the import bill and increase the forex reserve in India.

Rapid depletion of fossil sources due to rising energy consumption has become a primary global concern in the twenty-first century since the reserves are finite. At the present consumption rates, all known fossil fuel reserves will deplete in the next 45 years (https://www.worldometers.info/oil/#:~:text=World%20Oil%20Reserves&text=The%20world%20has%20proven%20reserves,levels%20and%20excluding%20unproven%20reserves). A higher consumption rate is particularly concerning because fossil fuel production/burning emits greenhouse gases (GHG), which cause global warming. Worldwide CO_2 emissions are rapidly increasing (Fig. 3.1) due to anthropogenic activities.

According to the National Oceanic and Atmospheric Administration (NOAA), this has resulted in an average global temperature rise of 0.8 °C over the last 100 years. Environmental Protection Agency (EPA) claims that the sea level has risen by 20 cm since 1970 (https://www.livescience.com/37057-global-warming-effects.html). At the current rate of GHG emissions, several weather forecast models anticipate a 2 °C rise in average earth temperature, which might cause fast melting of the polar ice. This forecast indicates a 3-feet rise in the sea level by the end of this century (https://www.livescience.com/37057-global-warming-effects.html; https://clintonwhitehouse5.archives.gov/Initiatives/Climate/next100.html). This sea level rise can sweep away many island nations and large coastal towns and submerge large land areas worldwide, creating mayhem.

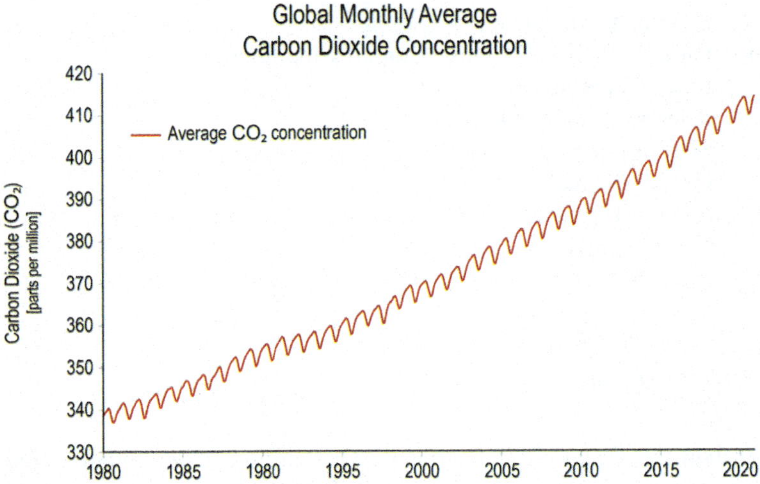

Fig. 3.1 Increase in global CO_2 emissions since 1980 (https://www.globalchange.gov/browse/indicators/atmospheric-carbon-dioxide)

Extreme weather occurrences such as storms, droughts, and flooding would be exacerbated by rising global average temperature (Fig. 3.2) (https://www.livescience.com/37057-global-warming-effects.html). Globalising carbon-neutral fuels and sustainable technologies at the right moment could help to reduce the global warming by 0.50 °C (https://www.europarl.europa.eu/news/en/headlines/society/20190926STO62270/what-is-carbon-neutrality-and-how-can-it-be-achieved-by-2050). For sustainable development, the production of alternative transport fuels should also be carbon neutral.

Fig. 3.2 Average global temperature rise (https://en.wikipedia.org/wiki/Global_surface_temperature)

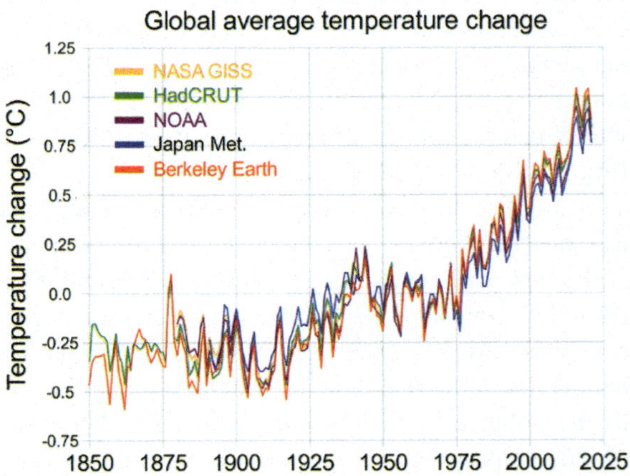

Many nations are implementing plans to cut net GHG emissions and achieve the carbon neutrality by 2050. Like Norway and the UK, many European countries want to phase out fossil-fuel-powered vehicles by 2035 (https://www.roadtraffic-technology.com/features/european-countries-banning-fossil-fuel-cars/). There is a push to reduce the net GHG emissions by at least 55% within 2030 and achieve the carbon neutrality by 2050 (https://eur-lex.europa.eu/legal-content/en/TXT/?uri=CELEX%3A52021PC0559). The use of clean alternate fuels which emits fewer GHGs than conventional fuels is one way to achieve the carbon neutrality. GHG emissions from the transport industry are significant. Throughout worldwide many governments are enforcing the strict emission standards to reduce GHG emissions from the transport sector. Europe has implemented EURO emission norms for internal combustion (IC) engine-powered vehicles to reduce the GHG emissions. In Europe, new on-road vehicles must comply with EURO-VI emission norms (https://eur-lex.europa.eu/legal-content/en/TXT/?uri=CELEX%3A52021PC0559).

3.2 Challenges for Compression Ignition (CI) Engines

CI engines are a popular choice for heavy-duty transport power plants due to their superior thermal efficiency, lower fuel consumption, and high power output than SI engines (https://www.worldometers.info/oil/#:~:text=World%20Oil%20Reserves&text=The%20world%20has%20proven%20reserves,levels%20and%20excluding%20unproven%20reserves). The CI engines are mostly fuelled by diesel. Diesel-fuelled CI engines emit high NO_x and soot. The main cause of increased soot emission is an inadequate fuel-air mixture forming several oxygen-deficient and fuel-rich zones. In the fuel-rich zones, long-chain C–C bonds aid in soot formation. This can be resolved by using an oxygenated fuel such as DME. CI engines require various exhaust gas after-treatment systems to comply with stringent emission norms. A significant reduction in pollutants is necessary to comply with future emission standards and to compete with the BEVs. DME is an excellent replacement for diesel for reducing NO_x emissions. Soot precursors hardly form due to the absence of a C–C bond in the molecular structure of DME. Therefore, DME combustion produces very low soot. These characteristics make DME a superior alternative for CI engines to replace the mineral diesel.

3.3 Alternative Fuel Scenario

India implemented Bharat stage-VI emission norms to regulate emissions from IC engine-powered vehicles in April 2020. The transport industry must also investigate new, alternative, low-carbon fuels to reduce GHG emissions and comply with prevailing and upcoming stringent emission norms. Low carbon fuels are the such kind of fuels which produce less CO2 while undergoing the combustion. These fuels do not require the large modifications in existing infrastructure. These low carbon fuels lie in the category of electro fuels (e-fuels) whenever produced via carbon capture methods because CO2 is being consumed during the production process. Therefore, manufacturing of e-fuels is carbon negative process. While performing the life cycle analysis, carbon footprint of e-fuels is near zero unlike low carbon fuels whose life cycle carbon footprint is lower than crude oil derived conventional fuels but significantly higher than e-fuels. Therefore, e-fuels like e-DME, e-Methanol, e-Methane are the suitable choice for transport industry to meet the goal of becoming carbon-neutral by 2050. European union exempt e-fuels from 2035 ban on new sales of combustion-engine cars.

Using locally available alternative low-carbon fuels to replace the diesel and gasoline effectively can reduce the dependence on crude oil. Such alternative fuels include Dimethyl ether (DME), Methanol, Ethanol, Diethyl ether (DEE), Green Hydrogen (produced from renewable sources), Hydrotreated Vegetable Oil, etc. These fuels are produced in India using different methods and feedstocks, e.g., methanol from coal gasification, ethanol from sugarcane, and DME from dehydration of methanol. India has a vast reserve of high-ash coal and adequate sugarcane production. These fuels can be produced in abundance from these resources, reducing the nation's dependence on crude oil imports. Adding 15% methanol to gasoline can reduce crude oil import by at least 15%. This would result in a 20% reduction in GHG emissions, including the reductions in NO_x, PM, and SO_x emissions, improving urban air quality. By mixing 20% DME in LPG, Rs 6000 crore can be saved annually (https://www.niti.gov.in/methanol-economy). India has a 2 MMT per annum installed capacity for methanol production, and DME production facilities are under commissioning. In collaboration with Israel, the Indian government is setting up five methanol and DME production plants. According to national green hydrogen mission, Indian government is planning to develop the green hydrogen production of 5 MMT (Million Metric Tonne) per annum. However, India doesn't have any production facility for Hydrotreated Vegetable Oil.

3.4 DME: An Alternative Fuel

Regulatory bodies impose emission legislations to cut GHG emissions from the road transport sector. It also helps achieve the mission objective of restricting the global temperature rise within the 1.5 °C limit. With this perspective, automotive manufacturers are working on producing more efficient engines while meeting the prevailing emission legislation. Among the two distinct engine types, diesel-fueled CI engines emit higher soot and NO_x than SI engines (Kim et al. 2008a). Due to their trade-off, soot and NO_x emissions cannot be reduced simultaneously (Kim et al. 2008a). By lowering the in-cylinder temperature, technologies such as exhaust gas recirculation (EGR) minimize NO_x emissions (Yoon et al. 2013). However, EGR deteriorates the combustion and increases soot emissions from CI engines. DME is the most suitable alternative to reduce soot and NO_x emissions simultaneously. Soot

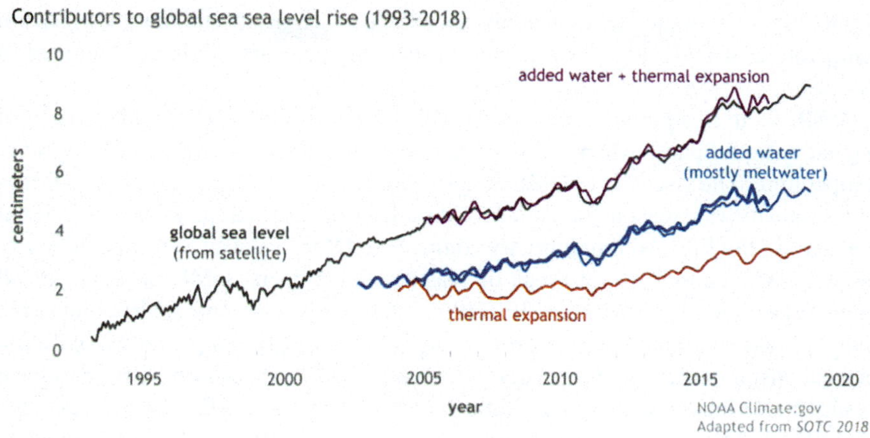

Fig. 3.3 Average global sea level rise (https://www.climate.gov/media/12868)

precursor formation is suppressed in the DME engine due to the absence of C–C bonds in its molecule. Therefore, soot emissions from DME engines are negligible. In addition, EGR significantly reduces NO_x emissions without increasing the soot in DME combustion. This section discusses the properties, advantages, and drawbacks of DME use in CI engines (Fig. 3.3).

DME-fuelled vehicles have shown promising performance and emission characteristics compared to conventional diesel. Essential parameters for ensuring the sustainability of alternative fuels include their production from renewable sources, compatibility with engine components, economic viability, lower overall carbon footprint, and higher H/C ratio (Valera and Agarwal 2019). Other properties such as fuel-bound oxygen and lack of C–C bond make them more attractive. DME is the simplest of all ethers, having fuel-bound oxygen without a C–C bond. DME may meet the requirement of a sustainable CI engine alternate fuel. It can reduce emissions of NO_x and PM. However, substantial adjustments are required in CI engines, particularly in FIE (Fig. 3.4).

DME has physical properties similar to LPG (Fig. 3.5). LPG is mainly propane and butane. Storage, transport and fuel handling of DME are practically equivalent to LPG due to their almost similar physico-chemical properties (Park and Lee 2014). The ignition delay of DME is significantly lower than diesel (Fig. 3.6). Superior spray atomisation characteristics of DME led to better fuel-air mixing and short ignition delay. The low boiling point and viscosity of DME causes the quick spray atomization and fast vaporisation of fine size droplets. Therefore, an efficient mixing

Fig. 3.4 Molecular structure of DME

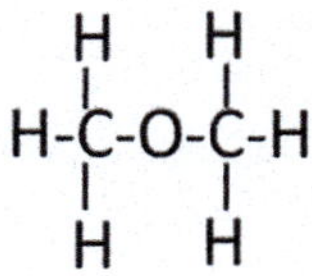

of fuel vapors with surrounding air takes place easily. High CN and low auto-ignition temperature of DME plays a vital role in reducing the ignition delay compared to diesel.

DME is an opaque gaseous ether at 0.1 MPa and 298 K. DME converts from vapour to liquid phase when compressed to pressure above 0.5 MPa at atmospheric temperature. The viscosity of DME is less than a tenth of conventional diesel. DME's low viscosity is a reason for poor lubricity, leading to friction, wear and internal leakages in the FIE. Adding lubricity additives to DME is a way to resolve lubricity issues. DME harms several types of plastics and rubbers. DME dissolves in all known elastomers used in diesel FIE. Teflon and Buna-N rubber are DME-compatible materials (MacFarlane 1965). However, the creep and stiffness properties of Teflon make it unsuitable as a sealing material for use in the FIE (Sorenson 2001). Sivebaek et al. (2001) observed premature failure of diesel FIE as DME lacks to offer the same levels of viscosity or lubricity as mineral diesel. It may be challenging to

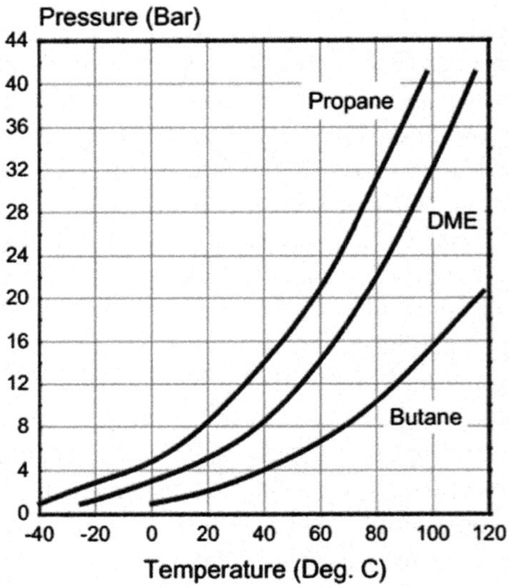

Fig. 3.5 Variations in vapour pressure of propane, DME and butane w.r.t. temperature (Park and Lee 2014)

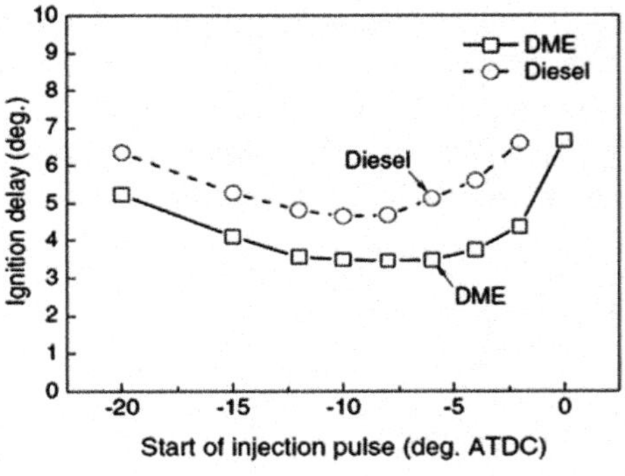

Fig. 3.6 Ignition delay of diesel and DME at varying fuel injection timings (Park and Lee 2014)

increase the lubricity of DME to acceptable levels. However, modifying the pump designs to accept DME could be a feasible technical solution. DME has a bulk modulus of 0.637 MPa, which is ~1/3 of mineral diesel (1.49 MPa) (Kim et al. 2008b). The compressibility of DME is higher than diesel due to its lower elastic modulus (Fig. 3.12). Therefore, a larger compression energy is required by the fuel pump for DME than mineral diesel. In a closed system, the compression energy required for DME is 3.2 times more than dodecane at a compression pressure of 25 MPa at 323 K (Park and Lee 2013). Due to a large cetane number and low injection delay of DME, its ignition occurs earlier than mineral diesel. The combustion process gets completed faster for DME compared to baseline diesel. Flash boiling provides greater fuel and air mixing due to the low boiling point of DME (Kim et al. 2007a).

Spray tip penetration for DME is lesser than mineral diesel due to its low boiling point and kinetic viscosity which leads to improved atomisation and fast droplet mass dispersion compared to diesel. The finer DME droplets are expected to decelerate faster than diesel while approaching downstream. Therefore, DME has short spray tip penetration than diesel fuel under the same fuel injection conditions (Fig. 3.7). The high evaporation rate and small droplet size of DME also leads to superior charge mixing and mixture homogeneity. This leads to improved combustion efficiency and low combustion noise than mineral diesel. Spray tip penetration of DME decreases as the surrounding pressure increases primarily due to the high ambient gas density suppresses the spray growth (Park and Lee 2014; Kim et al. 2009). Spray cone angle of DME is marginally higher than diesel. DME spray has less radiative heat loss compared to diesel due to the non-luminous flame. DME spray evaporation rate is higher than spray breakup rate which avoids the secondary spray breakup in DME, unlike diesel.

DME is a fluorocarbon propellant substitute due to its low vapour pressure, strong solvency properties and pleasant atmospheric behaviour (Sorenson 2001).

Figure 3.8 shows the global share of DME according to its usage. Most DME is used as an LPG-DME blend. A similar LPG and DME vapour pressure makes it suitable for blending both (Fig. 3.5).

DME is widely used in aerosol sprays, and its second-largest application is in the aerosol industry. The transport sector utilises only a tiny fraction of globally

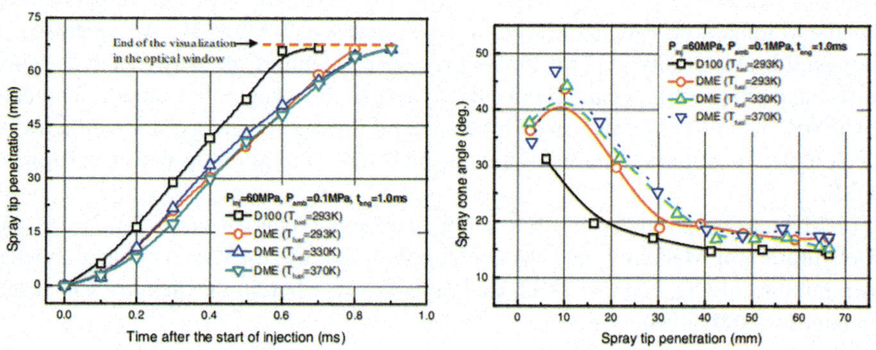

Fig. 3.7 Spray tip penetration and spray cone angle of diesel and DME (Park et al. 2010)

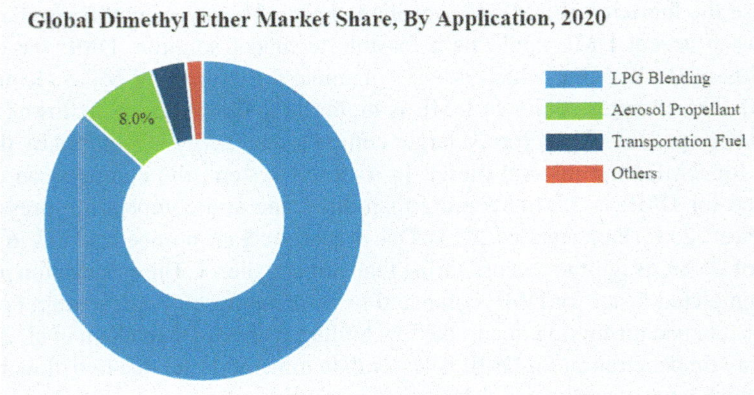

Fig. 3.8 Global market share of DME by its applications (https://www.fortunebusinessinsights.com/dimethyl-ether-market-104309)

produced DME. However, DME utilisation in transport sector is increasing rapidly due to its superior combustion and emission characteristics. The C/H ratio of the fuel affects the water content in the exhaust and hence, DME-fuelled engines show more water in exhaust than in mineral diesel (Oguma et al. 2005).

3.5 DME: Production Routes

China is the biggest producer of DME. Japan also has DME production facilities. Many other countries, namely North America, Uzbekistan, Indonesia, and Trinidad and Tobago, are planning to install DME production facilities. Sweden is home to the world's first bio DME plant. Annually ~5 to 9 million tons of DME is produced across the worldwide (Szybist et al. 2014). Several renewable and non-renewable methods are available for DME production (Kumar and Agarwal 2021). Using suitable catalysts, DME can be produced from fossil feedstocks such as coal and natural gas (Jang et al. 2009). Moreover, DME can be produced from renewable energy sources such as biomass (via methanol route), solar, and wind (Fig. 3.9). An electrolyser powered by electricity generated using renewable sources of energy such as wind farms and photovoltaics can produce the hydrogen in a syngas mixture for synthesis of DME. Therefore, DME can also be called a renewable energy storage pathway (Falco 2017). Approximately 1.4 tons of methanol is required to produce 1 ton of DME.

DME can be produced by two major pathways (Agarwal et al. 2017): (a) the Dehydration of Methanol; and (b) Direct conversion to DME from syngas. Syngas is a mixture of CO, H_2, CO_2, CH_4 and other gases, wherein the share of CO is the maximum (~30–60%).

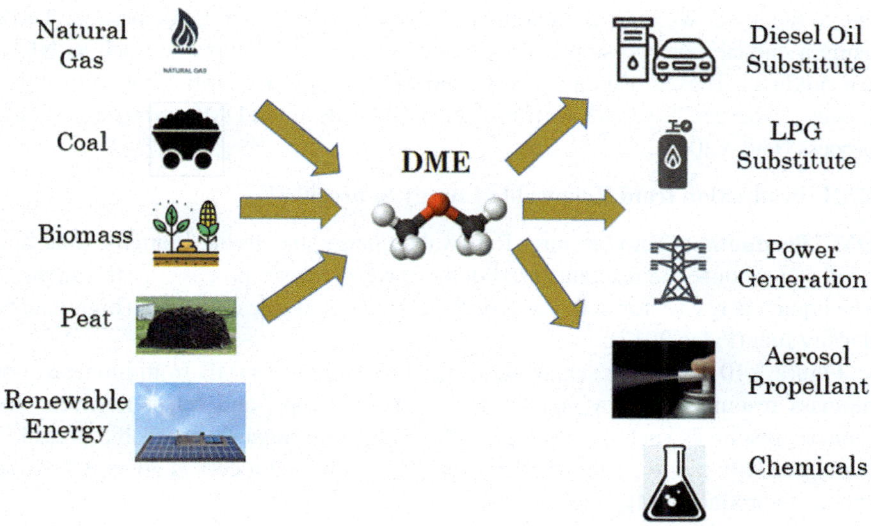

Fig. 3.9 Different production pathways and applications of DME

(a) **Dehydration of Methanol**: Methanol is first produced from syngas and then dehydrated into DME. This is the two-step process, wherein methanol production from syngas and dehydration of methanol to produce DME occurs in different reactors. This process is shown in Eqs. 3.1, 3.2 and 3.3, wherein the first two reactions show methanol production from syngas, and the third reaction shows methanol dehydration to produce DME.

$$CO + 2H_2 \rightarrow CH_3OH \tag{3.1}$$

$$CO_2 + 3H_2 \rightarrow CH_3OH + H_2O \tag{3.2}$$

$$2CH_3OH \rightarrow CH_3OCH + H_2O \tag{3.3}$$

(b) **Direct DME Conversion**: In this process, DME production is done using bi-functional catalysts such as Cu/ZnO. This is a one-step DME production process, given by Eq. 3.4.

$$2CO + 6H_2 \rightarrow CH_3OCH + 3H_2O \tag{3.4}$$

Syngas is fed at 5 MPa pressure and 270 °C for direct production of DME because it requires a high kinetic reaction rate (Falco 2017). This reaction is highly exothermic; therefore, extreme caution is needed to avoid any chances of an accident. A direct synthesis reaction is preferred over an indirect synthesis reaction for DME production due to the lesser wastage of natural gas. Most companies worldwide produce DME using a one-step or two-step DME production process. Companies such as NKK, Air products, Topsoe, Korea Gas Co., and JFE Ho produce DME using a one-step process, whereas Toyo, MGC, Lurgi, and Udhe produce DME using a two-step process (Falco 2017).

DME Production from Renewable Energy Sources

DME Production From biomass: Renewable energy sources like biomass, solar, and wind can produce the reactants used in the DME synthesis process. DME functions as a liquid energy vector in this approach, storing renewable energy in a high-energy density fuel (Falco 2017).

Figure 3.10 shows the schematic of the production of DME from biomass. For biomass resources such as agro-residues, energy crops, and forest residues. For biomass resources such as organic trash, sewage and manure, anaerobic digestion combined with pyrolysis produce syngas. The synthesis process is governed by the syngas to produce DME.

DME Production from Solar Energy: Syngas is a mixture of CO and H_2. Hydrogen for the syngas can be produced using an electrolyser, as shown in Fig. 3.11.

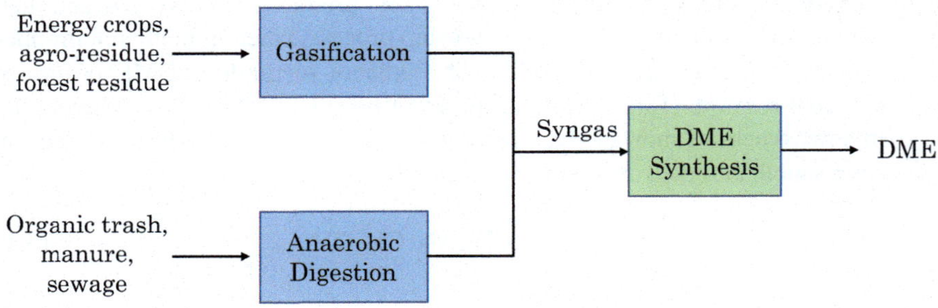

Fig. 3.10 DME production from biomass

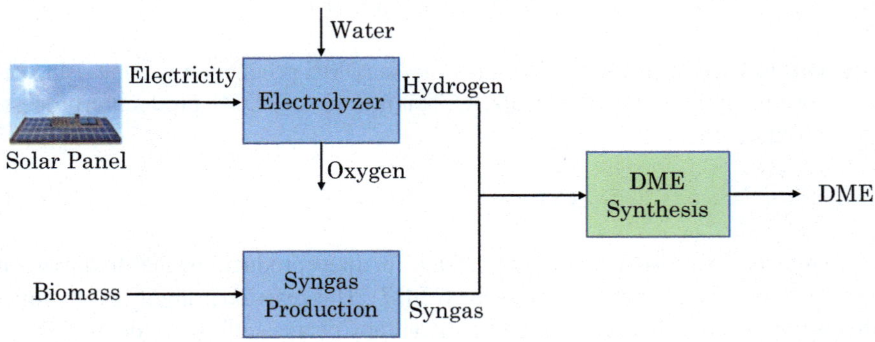

Fig. 3.11 DME production from solar energy

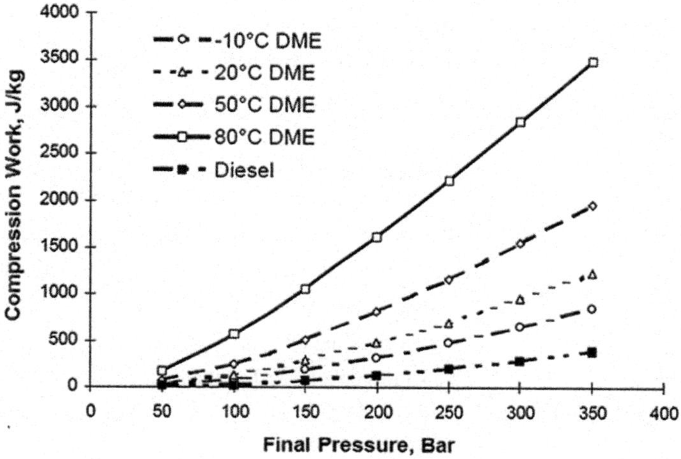

Fig. 3.12 Compression work for diesel and DME at different temperatures (Park and Lee 2013)

The electrolyser operates using solar power-generated renewable electricity. The hydrogen then mixes with the CO/CO_2 produced by biomass gasification to correct the stoichiometry, which is suitable for DME production. Renewable energy, such as solar energy, can be stored as high-energy density liquid fuel DME. DME, like liquid fuels, can be more easily stored and transported than hydrogen, which has several issues related to its storage and distribution. Therefore, on a large scale, DME can easily emerge as a suitable carrier and storage medium of renewable energy and its use in the transport sector (Fig. 3.12). DME produced from renewable energy sources is also termed as e-DME.

3.6 DME: Health, Environmental and Safety Effects

DME is a colourless liquid or gas-phase fuel, depending on the storage conditions. Mild exposure to DME vapours causes eye, nose, and throat irritation. DME's overexposure can cause headaches, dizziness, light-headedness, and even unconsciousness. Direct skin contact with liquid DME may cause serious frostbite (https://nj.gov/health/eoh/rtkweb/documents/fs/0758.pdf). However, many studies have concluded that DME is a non-carcinogen (MacFarlane 1965). Long-duration underground parking of DME fuelled vehicles should be avoided because DME vapours can replace the oxygen in the parking space. This may cause suffocation to humans and animals. DME is flammable, and there is a risk of explosion if an open flame is thrown on the ground, where DME vapours may be accumulated. The health effects of DME are listed in Table 3.1.

Well-to-wheel SO_2 and NO_x emissions from DME-fuelled engines are lower than mineral diesel. Particulate emissions are also lower in the case of DME. SO_2 and

Table 3.1 Health effects of DME (MacFarlane 1965)

Item	Health effects on humans
Principal routes of exposure	Inhalation, eye contact, skin contact
Acute toxicity	None
Inhalation	May cause central nervous system depression with nausea, headache, dizziness, vomiting and incoordination
Eyes	May irritate. Contact with liquid may cause frostbite
Skin	May irritate. Contact with liquid may cause frostbite
Skin absorption hazard	No known hazard in contact with skin
Ingestion	Not an expected route of exposure
Chronic effects	No known effect based on the information available
Aggravate medical conditions	Skin disorders, effect on the central nervous system, respiratory disorders

particulates are responsible for winter fog. Therefore, DME's winter and summer smog potential is lower than mineral diesel. Emissions of benzene, toluene, xylene, and PAHs are lower for DME because a molecule of DME contains only two carbon atoms connected by oxygen. Soil contamination is unlikely to happen with DME because DME is stored in a pressurised vessel and remains in the vapour phase above −25 °C. DME leak will likely vaporise into the atmosphere rather than contaminate soil or water (MacFarlane 1965). DME has lower reactivity in photochemical reactions. Therefore, DME has significantly lower ozone-forming reactivity than all organic compounds explored for fuel usage (Sorenson 2001; Bowman and Seinfeld 1995). DME does not form formaldehyde and other oxygenated compounds in the photochemical reactions in the environment (Japar et al. 1990). Formaldehyde emissions are higher in DME combustion than in mineral diesel. Formaldehyde is a carcinogenic, poisonous, and allergic compound (Tripathi et al. 2022). Diesel oxygen catalysts (DOC) used in DME-fuelled engines can reduce formaldehyde levels to zero.

3.7 Properties of DME

DME is the best alternative to diesel, considering its superior physicochemical properties over diesel (Table 3.2). Low autoignition temperature and high CN of DME make it an excellent replacement to diesel in CI engines. Low vapour pressure of DME facilitates better fuel-air mixing, thus, increasing combustion efficiency and reducing HC and CO emissions than baseline diesel. However, the low kinematic viscosity of DME causes wear and leakage in FIE (Table 3.2).

Table 3.2 Properties of DME and diesel (Tripathi et al. 2022)

Properties	DME	Diesel
Chemical Structure	CH_3OCH_3	C_8–C_{25}
Molecular Weight (g/mol)	46.07	190–220
Density (l) (kg/m^3) @25 °C	668	856
Density (g) (kg/m^3) @25 °C	1.918	–
Flammability range in air (% v/v)	(3.4–18)	(1–6)
Boiling point (°C)	−23.6	180–370
Autoignition temperature (°C)	350	250
Cetane number (CN)	55–62	40–55
Octane number (ON)	–	–
Lower heating value (LHV) (MJ/kg)	27.6–28.8	42.5–43.2
Latent heat of vaporization (kJ/kg)	460–470	270
Stoichiometric ratio	8.9	14.3
Oxygen (% w/w)	34.8	0
Flash point (°C)	−41	>55
Kinematic viscosity (l) cSt @30 °C	<0.1	3.0–7.5
Ignition energy (mJ)	0.29	20
Vapour pressure (kPa) @20 °C	516.76	1–10

3.8 Advantages and Challenges of DME

DME is gaining much interest among researchers as an ultra-clean, renewable CI engine fuel. DME has better atomisation characteristics, high CN, oxygen content (34.8% w/w), latent heat of vaporisation, and low self-ignition temperature (Kanta et al. 2022) (Mehra and Agarwal 2022). The ignition delay difference between diesel and DME is ~1.3–2.3° CA (Park and Lee 2013). During combustion polycyclic aromatic hydrocarbons (PAHs) are formed in the combustion chamber. DME is a non-toxic fuel that degrades quickly in the troposphere. The C–O bond is present in DME, and due to its lower bond energy than the C–H bond, it breaks more easily. This results in the superior ability of the pyrolysis mechanism to initiate chain reactions at low temperatures, leading to auto-ignition at lower temperatures for DME (Park and Lee 2014; Agarwal et al. 2017). DME can utilise infrastructure designed for propane storage and transport (Morsy et al. 2006). Due to the superior evaporative characteristics of DME spray, the wall-wetting effect reduces in DME-fueled engines.

DME also has some challenges, such as ultra-low viscosity, low self-lubrication properties, high vapour pressure, low critical point, and so on, all of which make it difficult for conventional FIS (Agarwal et al. 2017). DME has a lower calorific value and bulk modulus, one-third of mineral diesel. We can't utilise the same sealing materials for DME as we do for traditional diesel FIE. The compatibility of DME is good with most metals and alloys but not with polymeric compounds (Kumar and Agarwal 2021). Numerous parts of the FIE leak due to the ultra-low viscosity, necessitating the employment of different sealing materials. Many components corrode due to a lack of self-lubrication properties. The viscosity of DME-diesel blended fuels containing 25% DME is lower than ASTM specification for diesel. These difficulties can be addressed by adding viscosity improvers and lubricity additives to DME. However, this will not be enough to bring DME's viscosity up to standard fuels. Due to its low viscosity, DME does not form a hydrodynamic fluid layer, which causes metal-to-metal contact and wear. Various studies have shown that when DME content is greater than 50% in any blended or dual-fuel operation, none of the additives or fuels provides appropriate viscosity, resulting in the corrosion of many components, such as the plunger of a high-pressure pump (Chapman et al. 2004). This issue can be partially handled by building a DME-dedicated FIE rather than adding different additives to DME to make it compatible with traditional diesel-dedicated FIEs. DME has a low modulus of elasticity, making it five times more compressible than diesel. This results in greater compression effort for the fuel pump to pressurise the DME (Park and Lee 2014; Arcoumanis et al. 2008).

DME has a low enthalpy of combustion than diesel. Lower spray penetration and faster DME ignition lead to inadequate air entrainment in the DME spray. This might cause deteriorated combustion, particularly at high loads and speeds, leading to significant CO emissions and high fuel consumption in DI DME-fueled CI engines. The lower acoustical velocity caused by DME's higher compressibility also causes a longer injection delay than diesel fuel (Huang et al. 2009). Sulphur content in DME is zero. However, SO_2 emissions from DME fuelled engines are higher than diesel. DME with a high concentration (500 ppm) of fatty acid-based lubricant emits more SO_2 than DME with a low concentration (50 ppm) of fatty acid-based lubricants (Oguma et al. 2005). Although, these fatty acids and ester-based lubricants have no sulphur content. Engine oil seems to be the source of SO_2. The chances of DME leakage into the high-pressure fuel pump are large due to the low viscosity of DME. The mechanism for collecting the gaseous DME from the gallery of plunger supplies it back to the fuel return line after liquifying it. As a result, the leaking DME collecting mechanism might have collected the gaseous DME and lubricating oil to lubricate the HP fuel pump. Therefore, SO_2 emissions generally increase in a DME engine (Oguma et al. 2005).

3.9 Engine Hardware Modifications for DME Adaptation

Hardware modifications are required in the existing diesel-fueled CI engines to make them compatible with DME fueling. Significant changes in the FIE and sealing materials, along with a change in the material, are required.

DME Storage and Handling System: Vapour pressure of DME varies from 4 to 10 bars (abs) during normal weather conditions in India (Table 3.3). DME storage tank and its various components, such as valves, must withstand pressure as high as 10 bar. DME generates more vapours than diesel in the FIE (Sorenson 2001).

DME is a gas at NTP due to its high vapour pressure. DME becomes liquid at ~6 bar absolute pressure at normal temperature. Table 3.3 shows DME's vapour pressures, corresponding temperatures, and other thermodynamic properties.

Unlike diesel, DME must be pressurised using either a low-pressure (LP) pump or compressed nitrogen. It ensures DME supply in the liquid phase and avoids vapour lock in the fuel lines. DME has high compressibility since it is near the critical state (Sorenson 2001). Due to the high compressibility of DME, the plunger of high-pressure (HP) pump would not be able to generate the required pressure, and most of its energy would be wasted in compressing DME. Hence, a high-pressure DME supply is required at the inlet of the HP pump to avoid this problem. Fuel heating may occur when engine runs at high temperatures, increasing compressibility and a drop in fuel density. This creates difficulties in maintaining a full engine load due to the lack of fuel supply. Therefore, DME-powered diesel engines' fuel cooling or temperature management is necessary (Sorenson 2001). This arrangement will ensure the supply of liquid DME. Another alternative is to supply the vapourised DME back into the DME storage tank. Then, DME will be compressed and liquefied again by the LP pump before supplying to the feed pump.

Material Compatibility: Replacement of various sealing parts of FIS with parts made from DME-compatible materials is paramount to its widespread usage. Otherwise, its use in existing engines causes engine parts to wear and fuel leakage. Kass and Daw (2016) performed Hansen solubility analysis for various materials with DME, diesel and its blends. Styrene butadiene rubber (SBR) and nitrile butadiene rubber (NBR) showed similar solubility potential for DME, diesel and blends of both fuels. Therefore, no degradation of these materials is expected. However, fluorocarbon became more incompatible with DME upon increasing the DME concentration in diesel-DME blends. Polyester resins were highly incompatible with both fuels and showed high solubility with diesel and DME. Table 3.4 shows the compatibility of different materials with DME.

The compatibility of DME is good with metals and alloys because DME does not decompose into acids (https://afdc.energy.gov/files/pdfs/3608.pdf). The most common hose and seal material are NBR. Three manufacturers considered NBR a DME-compatible material, but Allorings did not consider it DME-compatible.

Lubricity Dosing Arrangement: The kinematic viscosity of DME is around one-tenth of diesel. Diesel acts as a lubricant for the FIE. However, DME causes the wear of FIE components due to the lack of viscosity and self-lubricating characteristics. Adding lubricity and viscosity-improving additives to the DME is necessary to avoid leakage and wear of engine and FIE components. Hansen et al. (2000) observed that the "Lubrizol LZ539N" lubricity additive has the best wear-reducing characteristics. DME does not have a recognisable odour. Therefore, odour additives must be added to DME to detect any leakage and to minimise the chances of fatal accidents. Hansen et al. (2000) used Ethyl mercaptan as an odour improver in DME due to its almost similar boiling point to DME and better blending characteristics with DME. A 1000 ppm lubricity additive (Lubrizol LZ539N) and 20 ppm of odour additive Ethyl mercaptan were initially added to DME to avoid the adverse effects on exhaust emissions. Unlike DME, lubricity and odour additives are in the liquid phase during

Table 3.3 Thermodynamic properties of DME (Teng et al. 2001)

Pressure (abs) (bar)	Temperature (°C)	Density (kg/m^3)		Enthalpy (kJ/kg)		Latent heat (kJ/kg)	Entropy (kJ/kg K)		Surface tension (N/m)
		Liquid	Vapour	Liquid	Vapour		Liquid	Vapour	
1.0	−25.0	723.42	2.327	98.513	566.03	467.52	0.881	2.766	0.01778
1.013	−24.8	723.16	2.345	98.963	566.25	467.29	0.884	2.767	0.01776
2.0	−8.28	701.33	4.443	137.48	584.97	447.50	1.152	2.842	0.01551
3.0	2.81	686.00	6.522	164.07	597.42	433.36	1.319	2.891	0.01402
4.0	11.3	673.78	8.581	184.52	606.46	421.95	1.443	2.927	0.01288
5.0	18.4	663.39	10.64	201.21	613.36	412.14	1.541	2.956	0.01194
6.0	24.4	654.21	12.69	215.33	618.74	403.40	1.624	2.980	0.01114
7.0	29.7	645.90	14.75	228.80	624.22	395.42	1.695	3.001	0.01044
8.0	34.5	638.24	16.83	241.11	629.13	388.02	1.758	3.020	0.00981
9.0	38.9	631.09	18.92	252.30	633.36	381.06	1.814	3.036	0.00924
10.0	42.9	624.35	21.03	262.61	637.06	374.45	1.865	3.051	0.00871
11.0	46.6	617.95	23.16	272.20	640.33	368.13	1.912	3.064	0.00823
12.0	50.1	611.82	25.30	281.20	643.24	362.05	1.955	3.076	0.00778
13.0	53.4	605.92	27.48	289.69	645.85	356.16	1.996	3.087	0.00736
14.0	56.5	600.23	29.67	297.76	648.20	350.44	2.033	3.097	0.00697
15.0	59.5	594.70	31.90	305.46	650.33	344.86	2.069	3.106	0.00660
16.0	62.3	589.31	34.15	312.85	652.25	339.40	2.102	3.114	0.00625
17.0	65.0	584.04	36.43	319.98	654.01	334.04	2.134	3.122	0.00591
18.0	67.6	578.89	38.74	326.86	655.62	328.75	2.164	3.130	0.00560
19.0	70.1	573.82	41.09	333.55	657.08	323.53	2.193	3.136	0.00530

(continued)

Table 3.3 (continued)

Pressure (abs) (bar)	Temperature (°C)	Density (kg/m^3)		Enthalpy (kJ/kg)		Latent heat (kJ/kg)	Entropy (kJ/kg K)		Surface tension (N/m)
		Liquid	Vapour	Liquid	Vapour		Liquid	Vapour	
20.0	72.5	568.83	43.47	340.06	658.43	318.37	2.221	3.143	0.00501
21.0	74.8	563.90	45.88	346.41	659.66	313.25	2.248	3.148	0.00474
22.0	77.0	559.03	48.33	352.63	660.79	308.16	2.273	3.154	0.00447
23.0	79.2	554.20	50.82	358.73	661.83	303.09	2.298	3.159	0.00422
24.0	81.3	549.40	53.35	364.74	662.78	298.04	2.322	3.163	0.00398
25.0	83.3	544.63	55.92	370.67	663.66	292.99	2.345	3.167	0.00375
26.0	85.3	539.88	58.53	376.53	664.46	287.93	2.368	3.171	0.00353
27.0	87.3	534.75	61.18	382.34	665.20	282.86	2.391	3.176	0.00331
28.0	89.2	528.41	63.88	388.11	665.88	277.77	2.414	3.181	0.00306
29.0	91.0	522.77	66.63	393.85	666.50	272.65	2.436	3.185	0.00285
30.0	92.8	517.48	69.43	399.58	667.06	267.48	2.457	3.188	0.00265
31.0	94.6	512.33	72.28	405.31	667.58	262.27	2.478	3.191	0.00247
32.0	96.3	507.19	75.18	411.05	668.05	257.00	2.498	3.194	0.00229
33.0	98.0	501.97	78.13	416.82	668.47	251.65	2.518	3.196	0.00212
34.0	99.6	496.63	81.14	422.63	668.86	246.23	2.538	3.198	0.00196
35.0	101.2	491.17	84.21	428.49	669.20	240.71	2.557	3.200	0.00180
36.0	102.8	485.59	87.35	434.43	669.51	235.08	2.577	3.202	0.00165
37.0	104.4	479.88	90.54	438.40	667.73	229.33	2.596	3.204	0.00151
38.0	105.9	474.06	93.80	443.60	667.04	223.44	2.616	3.205	0.00138

(continued)

Table 3.3 (continued)

Pressure (abs) (bar)	Temperature (°C)	Density (kg/m³)		Enthalpy (kJ/kg)		Latent heat (kJ/kg)	Entropy (kJ/kg K)		Surface tension (N/m)
		Liquid	Vapour	Liquid	Vapour		Liquid	Vapour	
39.0	107.4	468.11	98.31	449.50	666.88	217.38	2.635	3.206	0.00123
40.0	108.9	462.04	102.80	455.86	667.00	211.14	2.654	3.207	0.00110
41.0	110.3	455.82	107.20	462.46	667.15	204.69	2.674	3.208	0.00097
42.0	111.7	449.40	111.62	469.14	667.13	197.98	2.693	3.207	0.00086
43.0	113.2	442.75	116.19	475.74	666.74	190.99	2.712	3.206	0.00075
44.0	114.5	435.78	121.05	482.14	665.80	183.66	2.731	3.205	0.00065
45.0	115.9	428.41	126.39	488.23	664.15	175.93	2.750	3.202	0.00055
46.0	117.2	420.52	132.41	493.96	661.66	167.70	2.769	3.198	0.00045
47.0	118.6	411.96	139.38	499.32	658.17	158.86	2.788	3.194	0.00036
48.0	119.9	402.55	147.61	504.34	653.59	149.25	2.808	3.188	0.00028
49.0	121.2	392.02	157.56	509.18	647.79	138.61	2.829	3.181	0.00020
50.0	122.4	380.01	169.84	514.11	640.68	126.57	2.852	3.172	0.00013
51.0	123.7	365.90	185.38	519.77	632.16	112.39	2.878	3.162	0.00007
52.0	124.9	348.43	205.60	527.61	622.15	94.54	2.912	3.150	0.00003
53.0	126.1	323.66	232.88	542.49	610.58	68.08	2.965	3.136	0.000004
53.7	127.0	258.51	258.51	601.46	601.46	0.000	3.125	3.125	0.00000

Table 3.4 Material compatibility of different sealing materials with DME (Kass and Daw 2016)

Material class	Manufacturer				
	Mykin	Geotech	Hargraves	Allorings	Parr
Elastomers					
NBR&SBR	Satisfactory	Acceptable		Not recommended	Acceptable
Neoprene	Doubtful	Unsatisfactory			
Fluorocarbon	Fair		Acceptable	Not recommended	Acceptable
Fluorosilicone			Acceptable	Recommended	Acceptable
Silicone				Recommended	Acceptable
Plastics					
PTFE (Teflon)		Resistant			

NTP conditions. Thus, DME must be in the liquid phase for better mixing of DME and additives. DME should be stored at a higher pressure than its vapour pressure. Fuel must be circulated continuously in the fuel tank to avoid the settling down of additives after prolonged dosing. DME is supplied to the consumption tank from the main DME storage tank using a feed pump. Odour and lubrication additives are injected into the DME fuel line using an additive dosing pump and are mixed with the DME flowing in the fuel lines. A circulation pump is installed after the dosing pump to help with continuous fuel circulation, preventing improper mixing.

3.10 DME Fuelled Vehicle Development Projects

3.10.1 DME FIE Development for Heavy-Duty Trucks

DME is easier to liquefy than other fuels, such as CNG and Hydrogen. Therefore, liquid DME can easily be stored in the vehicle. The DME engine for a 20-ton GVW truck was developed under the Ministry of Land, Infrastructure and Transport of Japan (Tsuchiya and Sato 2006). In the DME FIE, the diameter of the nozzle hole of the fuel injector was increased to supply a higher amount of DME. A feed pump was used to compress the DME to moderate the pressure. DME was provided from the feed pump to the modified BOSCH HP pump. HP pump compressed the DME to a pressure higher than the fuel injector's nozzle opening pressure. DME was injected into the combustion chamber by the fuel injector, and some amount of DME was allowed to flow to the DME fuel tank via fuel return lines. Since the DME combustion is smoke-free, heavy EGR was introduced to reduce NO_x emissions. The exhaust gas was taken from the exhaust manifold upstream of the turbine and delivered to the upstream compressor assembly. NO_x emissions were reduced to one-fourth of Japan's long-term emission regulation limit (2005).

3.10.2 DME FIE Development for City Bus

The main hurdles for DME-dedicated FIE development are high vapour pressure, compressibility, and low viscosity of DME. The common rail type system is best suited to handle all these features of DME to develop dedicated FIE for city buses (Hansen et al. 2000). The electrical control unit (ECU) monitors the DME distribution to system. The fuel level is checked in both tanks at the time of the ignition event. DME is allowed to flow from the fuel tank with a large quantity of DME to the tank with a lower quantity to ensure even fuel distribution in both tanks. DME is pumped with a pressure of 10 bar higher than the vapour pressure. The purge system of the purge tank operates on the compressed air from the brake system of the bus. A separate fuel line connects the purge tank to the FIE. A level indicator in the purge tank calculates the amount of liquid DME. If the DME quantity is more than 15 L, this will be indicated on the driver's interface. If the pressure inside the purge tank reaches higher than the permissible limit, the pneumatic control unit activates the purge compressor. The compressor compresses the gaseous DME and supplies it to the fuel tank until the pressure inside the purge tank reaches the permissible limit (Hansen et al. 2000). Arrangement for cooling of circulating DME was necessary. A separate circuit was provided to handle the engine's water-to-air charge-air cooling. The water temperature is adequate for DME cooling. Afterwards, the city bus was operated on the DME. Significantly lower PM and NO_x emissions were observed while operating the city bus on DME. HC and CO emissions were reduced while using an oxidation catalytic converter.

3.10.3 DME FIE Development for Engines

The FIE system customised for DME was tested in the engine. With the changed SoI timing of DME and unaltered SoI timing of diesel, BTE was slightly greater for diesel than DME at lower loads. The BTE of diesel and DME were comparable at medium-to-high loads (Oguma et al. 2004). EGT was higher for diesel than DME at all engine operating conditions. Due to the lower LHV of DME, lesser energy was released during combustion than diesel. Therefore, lower in-cylinder temperature and EGT were observed for DME. Huang et al. (2009) reported the lowest BSFC for both fuels in the speed range, where maximum torque was obtained. As engine load increased, the BSFC of diesel and DME reduced. BSFC for DME was lower than mineral diesel at lower loads but comparable at higher loads. The LHV of DME is ~2/3rd that of mineral diesel. As a result, each cycle requires 1.9 times the diesel volume in the combustion chamber to maintain the same diesel engine power output from DME fuelling. This led to a long fuel injection duration and large DME consumption at high engine speeds. Therefore, the BSFC of DME increased drastically and became equivalent to the 1.9 times the BSFC of diesel at high engine loads.

3.11 Combustion Characteristics of DME

The engine's performance and exhaust emissions depend on combustion characteristics. Therefore, improving the in-cylinder combustion can improve engine performance and reduce the emissions. This section compared combustion parameters such as in-cylinder pressure, rate of pressure rise (RoPR), heat release rate (HRR), cumulative heat release (CHR), indicated mean effective pressure (IMEP), ignition delay and combustion phasing for DME and diesel engines.

3.11.1 In-Cylinder Pressure and RoPR

Lower peak pressure of the DME-fueled engine was observed than mineral diesel for the same fuel injection volume since DME has a lower calorific value (28.43 MJ/kg) than diesel (42.5 MJ/kg) (Fig. 3.13) (Park and Lee 2014; Kim et al. 2007a). More DME must be injected to improve the combustion to match the diesel energy input (Park and Lee 2014). Many studies have reported a lower peak in-cylinder pressure for DME than baseline diesel (Huang et al. 2009; Longbao et al. 1999; Sorenson and Mikkelsen 1995). Huang et al. (2009) reported that the DME's SoI delay was substantially longer than the reduction in ignition delay; thus, the SoC of DME retarded compared to diesel. Hence, the rise of in-cylinder pressure and $RoPR_{max}$ is also retarded for DME than diesel. SoI of DME was delayed due to the requirement of high compression energy for DME HP pump than diesel. Engine noise was also low for DME combustion due to low peak in-cylinder pressure and $RoPR_{max}$ (Sorenson and Mikkelsen 1995; Christensen et al. 1997; Huang et al. 1999). The effects of changes in SoI timings and EGR rates on the in-cylinder pressure were also reported by researchers.

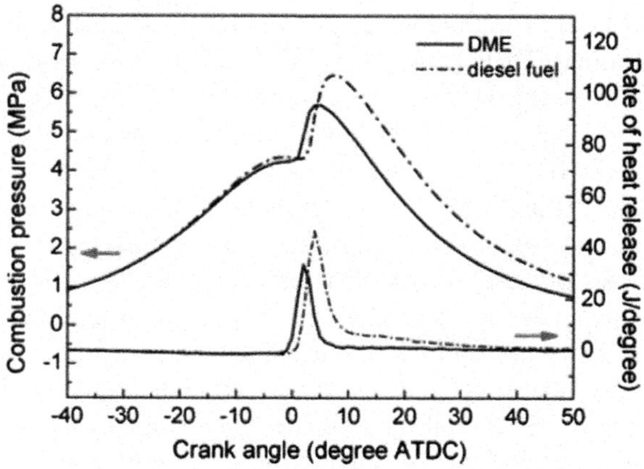

Fig. 3.13 In-cylinder pressure and heat release rate variations of DME and baseline diesel at SoI timing of 6° bTDC (Kim et al. 2007a)

Kim et al. (2008b) compared in-cylinder pressures at 8° bTDC and 2° bTDC SoI timings for DME and diesel. Since DME spray ignited more quickly than other fuels, DME-fueled engines exhibited a higher maximum in-cylinder pressure at 8° bTDC SoI timing. For this, increased chemical and thermal effects, heat capacity and dilution effects were responsible since they lengthened the ignition delay and enabled superior charge mixing. At the retarded injection timing of 2° bTDC, a higher premixed spike was observed than the injection timing of 8° bTDC for DME. Retarded SoI timing of fuel increased the ignition delay, increasing the accumulated energy between SoI and SoC. Therefore, a higher premixed spike was observed at 2° bTDC SoI timing. Yoon et al. (2013) showed that a high EGR rate lowered the in-cylinder temperatures and pressures during the expansion stroke.

3.11.2 HRR and CHR

DME has a lower HRR than mineral diesel in the premixed combustion phase. Due to its low ignition delay, a small DME quantity accumulates in the combustion chamber before the SoC. This results in lesser combustion noise and lower NO_x emissions in DME combustion than diesel. The HRR_{max} in the diffusion combustion phase was also lower for DME than diesel due to its rapid mixing and high diffusivity (Longbao et al. 1999). At the same fuel SoI timing, the fuel injection delay was longer for DME due to the high compression work requirement than diesel. Therefore, the SoC and HRR of DME were significantly retarded than diesel (Huang et al. 1999). Huang et al. (2009) also reported a lower peak of HRR for DME than diesel. The rise in HRR was retarded for DME than diesel.

DME burns faster because of its short ignition delay, resulting in fast and increased heat release. DME produced more heat than mineral diesel at the same injection timing, same fuel energy input, and fixed engine speed. Premixed combustion affects heat release, which is critical for CHR in DME combustion of a homogeneous charge. Kim et al. (2008b) analysed the HRR of mineral diesel and DME at various SoI timings. HRR_{max} was lower and advanced for DME at 8° bTDC SoI timing due to decreased energy supply from DME than diesel. Once SoI was retarded to 2° bTDC, the HRR_{max} for DME combustion was greater than diesel. Retarded SoI timing of DME increased the ignition delay, increasing the accumulated energy between the SoI and SoC. The temperature of the charge reduced rapidly, and the reaction rate slowed when the ignition was delayed beyond the TDC, lowering the peak HRR of diesel combustion. Kim et al. (2008a) compared three injection scenarios: a single injection near TDC, an early single injection for PCCI combustion, and a dual injection for partially premixed charge compression ignition (PPCCI) combustion. The dual injection includes one early injection followed by a late injection after the TDC. The overall mass of fuel injection was the same for all three circumstances to keep the same equivalence ratio. The traditional single injection caused dispersed combustion and a single peak in the HRR curve. There were two peaks in the HRR curve for PCCI combustion. Low-temperature combustion caused the initial peak,

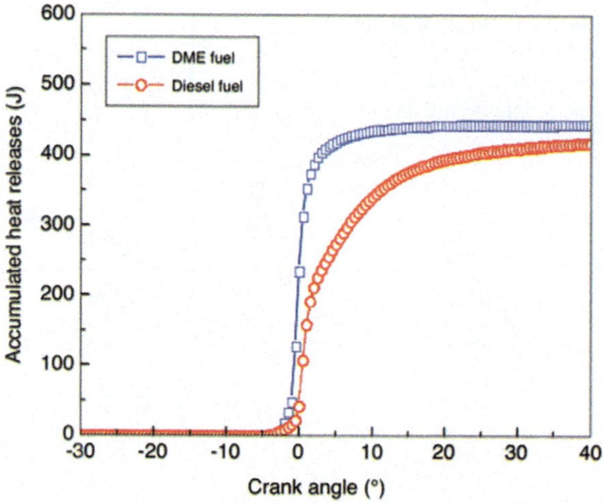

Fig. 3.14 CHR of diesel and DME (Kim et al. 2012)

followed by a short igniting delay due to the fuel's negative temperature coefficient behaviour. A premixed high-temperature reaction was responsible for the second large peak. When the equivalence ratio increased, both LTR and HTR advanced. Two peaks were attributed to initial early injection, like PCCI combustion, while the third peak was due to diffusion combustion of late second fuel injection, similar to PCCI combustion. The initial two peaks of PPCCI combustion were lower than the peaks of PCCI combustion due to a lower amount of fuel injected (fuel injection is split in two parts in PPCCI combustion).

DME had a larger injected fuel mass than diesel, which resulted in a higher CHR for a given calorific value (Fig. 3.14). CHR of DME and diesel became stable after around 5° and 30° aTDC, respectively. DME exhibited faster atomisation and evaporation characteristics than diesel. DME rapidly mixes with the surrounding air and burns faster than diesel. This resulted in DME's earlier end of combustion than mineral diesel (Kim et al. 2012). Hence, the CHR of DME became stable at 25° CA before the diesel CHR.

3.11.3 IMEP

Kim et al. (2008b) detected a reduction in IMEP at advanced SoI timings due to increased negative effort during the compression stroke. A higher IMEP means greater combustion efficiency. Despite the fast ignition of DME, it has a greater IMEP than mineral diesel over practically the entire range of injection duration (Fig. 3.15).

DME combustion creates a non-luminous flame, producing lower radiation and cooling losses than diesel. DME combustion efficiency is also higher than diesel. These are some of the possible reasons for higher IMEP during DME combustion. PCCI combustion significantly increases IMEP than conventional CI combustion

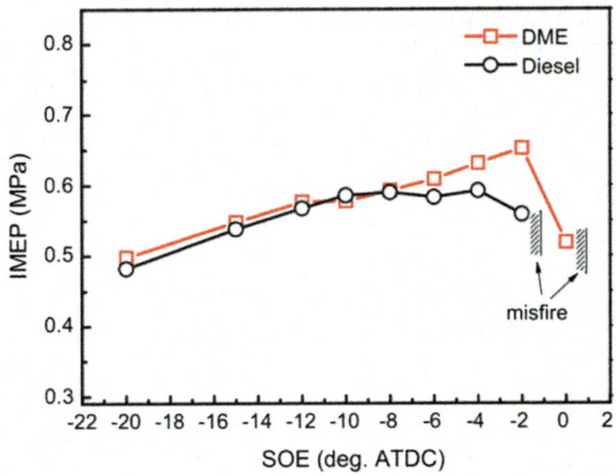

Fig. 3.15 IMEP of diesel and DME at varying start of energizing and constant energy input (Park and Lee 2013)

(Kim et al. 2008a). Yoon et al. (2013) reported that due to long ignition delay and combustion duration, the introduction of EGR increased the IMEP because the heat of combustion was released between the end of the compression stroke and the beginning of expansion stroke.

3.11.4 Ignition Delay and Combustion Phasing

The ignition delay of DME is significantly lesser than mineral diesel (Fig. 3.6) (Park and Lee 2014; Kim et al. 2008b; Huang et al. 1999, 2009). High cetane number, superior spray atomisation characteristics and low autoignition temperature are responsible for DME's short ignition delay (Longbao et al. 1999). Ignition delay strongly depends on temperature. Lower temperature increases the ignition delay (Park and Lee 2014). However, DME's fuel injection delay is higher than diesel due to its higher compressibility. Kass and Daw (2016) reported that the injection delay of DME decreased more compared to diesel with increasing load. The ignition delay of DME was reduced more than diesel with increasing engine load. The high compressibility of DME was responsible for the difference in acoustics velocity and injection delay between both test fuels (Huang et al. 2009).

Ignition delay of DME can be expressed as;

$$\tau_{id} = \mathbf{0.022P^{-0.316}e^{(4206/T)}}$$

Where; τ_{id} = ignition delay (ms), P = in-cylinder pressure (bar), T = in-cylinder gas temperature (K) (Park and Lee 2014).

3.11.5 Fuel Line Pressure, Injection Duration and Injection Delay

Christensen et al. (1997) reported that the peak fuel line pressure (P_{line_max}) of DME was significantly lower than diesel. At the start of DME's fuel injection, the pressure rise was lower than diesel. The plunger of the HP pump displaces the fuel through the nozzle of the fuel injector. A greater amount of compression work of the HP pump was used to compress the DME than diesel. Therefore, DME's pressure built up for the same plunger movement was lower. The study also reported lower fuel injection velocity and longer fuel injection duration of DME than baseline diesel. Fuel injection duration increased upon increasing the plunger stroke. A large amount of DME was injected due to its low calorific value to maintain the same power output as the diesel engine. Therefore, an increase in plunger stroke increased the fuel injection duration of DME compared to mineral diesel (Huang et al. 1999). P_{line_max} and RoPR of DME were substantially lower than diesel under identical fuel supply advance angle and load (Huang et al. 2009). DME's high compressibility and low nozzle opening pressure are perhaps the reason for low fuel line pressure. The speed at which the sound wave moves through a medium is called its acoustic velocity. Acoustics velocity increased in DME with increasing engine load. Because of low acoustic velocity, the pressure wave takes longer to propagate in DME than in diesel. Therefore, the DME injection delay is longer than mineral diesel (Huang et al. 1999). Fuel injection duration of DME is lesser as compared to diesel because lower kinetic viscosity of DME led to the rapid flow of fuel through the injector nozzle, therefore substantially higher flow rate as compared to diesel (Kim et al. 2007a).

3.12 Emission Characteristics of DME

There is a growing concern about tailpipe emissions from automobiles due to their adverse health and environmental effects. Therefore, governments impose strict emission legislation on the road transport industry. This section compares the formation and reduction techniques of various regulated, unregulated, particulates and trace metal emissions from diesel and DME engines.

3.12.1 Regulated Emissions

Tailpipe emissions regulated CO and HC emissions from gasoline engines in 1959 in California for the first time. IC engine emissions are now regulated in many nations worldwide. NO_x, CO, HC and PM are the regulated emissions. This section compares the regulated emissions from diesel and DME-fueled engines.

NOx Emissions: Combustion temperature, presence of oxygen, and residence time at peak combustion temperature are the major parameters influencing NO_x production, according to the Zeldovian diesel-fueled engines due to itch mechanism (Zhao et al. 2014; Zhu et al. 2012). NO_x emissions increase where excess oxygen and high peak temperature are available in the combustion chamber. Many studies on NO_x emissions from DME-fueled CI engines are inconsistent. Some studies claim that DME generates lower NO_x than diesel-fueled engines due to its high CN, low heating value, and high latent heat capacity (Zhu et al. 2012; Mukherjee et al. 2022). In addition, retarded SoI timing of DME can further reduce the NO_x emissions upon DME combustion (Fig. 3.16). Under the same diffusion condition, DME combustion produces lower NO_x than diesel. In contrast, NO_x emissions increase for DME under the same HRR condition (Park and Lee 2014). Since the release of free oxygen from DME is limited, the influence of intermolecular oxygen on NO_x emissions is insignificant.

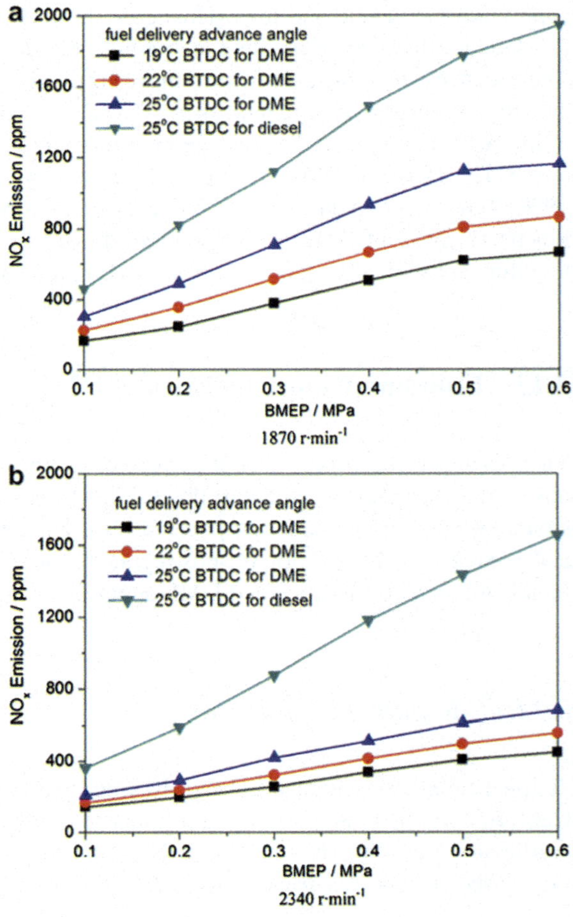

Fig. 3.16 NO_x emissions from diesel and DME at varying engine speeds and SoI timings (Zhu et al. 2012)

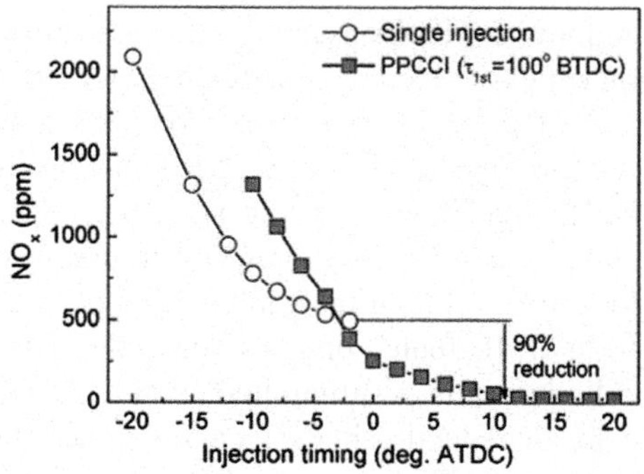

Fig. 3.17 Effect of single fuel injection and PPCCI strategies on NO_x emissions from DME fuelled engine (Kim et al. 2008a)

In contrast, many studies reported that DME-fuelled engines emitted higher NO_x than diesel-fuelled engines (Kim et al. 2007a; Thomas et al. 2014). Faster ignition and combustion of DME enhanced the in-cylinder temperature, raising NO_x emissions more than diesel (Park and Lee 2013). DME has a quick combustion response due to its high CN and excellent spray atomisation characteristics. As a result, DME has a higher HRR and HRR_{max} than diesel. The fuel-bound oxygen of DME also acts as an oxidiser, speeding up the NO_x formation reactions and raising the peak combustion temperature. As a result, the DME engines emit high NO_x emissions. Kim et al. (2007b) also showed high NO_x emissions from DME-fueled engines across the selected SoI timings because the charge temperature increased in the combustion chamber due to faster HRR caused by rapid combustion. NO_x emissions increased at 2° bTDC SoI timing. NO_x emission further increased at advanced SoI timing of 4° bTDC. Two popular technologies, low-temperature combustion (LTC) and EGR, can reduce the NO_x emissions from DME-fueled engines to the same or even lower levels than diesel engines. For LTC approaches like PCCI, DME-fueled engines emitted low NO_x (Fig. 3.17) (Kim et al. 2008a). Exhaust gas after-treatment methods include urea selective catalytic reduction (SCR), lean NO_x traps (LNT), HC-SCR, and the combination of EGR and catalysts are also effective in limiting NO_x emissions (Park and Lee 2014).

When the fuel injection in a CI engine is advanced enough, combustion occurs in a lean, premixed mode, which produces extremely low NO_x while simultaneously increasing HC emissions and fuel consumption. PPCCI was used to obtain the advantages of PCCI combustion while removing its drawbacks. PCCI is combined with conventional CI combustion in the PPCCI combustion, wherein an early pilot injection is employed to premix a portion of fuel, and a second injection occurs after the TDC.

HC and CO Emissions: Unburnt HC and CO emissions are the intermediate compounds formed during the combustion of complex hydrocarbons in favourable conditions. They are further oxidised to CO_2 and H_2O. High combustion temperatures accelerate the oxidation of HC and CO, lowering their concentration in the exhaust. The chain termination of the hydrocarbon oxidation process causes HC emissions (Jang et al. 2009; Oh et al. 2010). Under fuel-rich and imperfect fuel-air mixing conditions, HC forms from partial or unburned fuel. Many studies reported that DME-fueled engines emitt significantly less HC than diesel-fueled engines (Fig. 3.18).

DME has superior spray atomization, evaporation characteristics, and a shorter ignition delay, so it isn't easy to generate over-rich regions due to short mixing time. DME's fuel bound oxygen avoids fuel-rich zones and lowers the HC emissions (Park and Lee 2014). DME spray penetration is lower than mineral diesel. Hence, DME is less susceptible to the wall-wetting effect. This also results in lower HC emissions from DME-fueled engines since diesel spray is more likely to impinge upon the cylinder walls, resulting in incomplete combustion due to the wall heat transfer and a drastic reduction in the local in-cylinder temperature inhibiting HC oxidation. Park and Lee (2013), Kim et al. (2008b) and Youn et al. (2011) reported that DME-fueled engines emitted lesser HC and CO than diesel-fueled engines (Fig. 3.18).

PCCI combustion requires earlier fuel injection than a conventional CI engine. If the piston shape is not modified, using PCCI combustion in a CI engine fuelled by DME increases HC and CO emissions. Low in-cylinder pressure can increase spray penetration and fuel deposition on the cylinder walls (Kim et al. 2008a). The PPCCI combustion is a hybrid of PCCI and conventional combustion that reduces NO_x, CO and HC emissions from DME engines at the same time. Kim et al. (2007b) reported

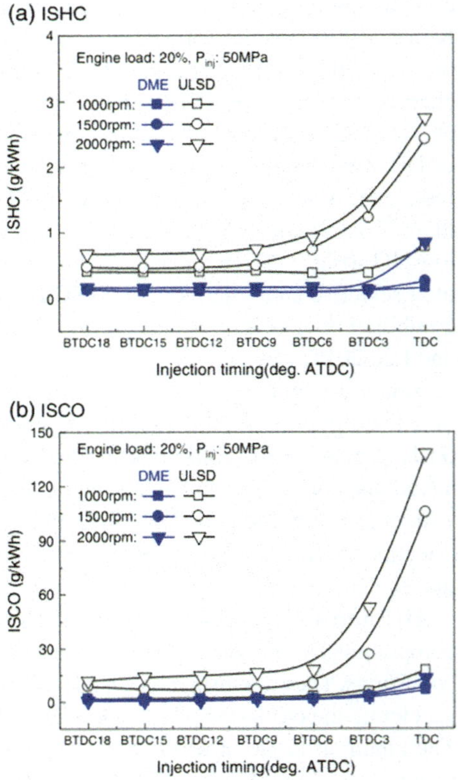

Fig. 3.18 ISHC and ISCO emissions from diesel and DME at varying fuel injection timings and engine speeds (Youn et al. 2011)

that PPCCI combustion generates higher HC and CO emissions than conventional single-injection combustion. If the main injection is delayed after the TDC, HC emissions rise higher than in the single injection case for all pilot injection timings. HC emissions increased for main injection timing before 20º bTDC with advanced pilot injection. Despite the dilution effect of EGR, Yoon et al. (2013) found that the increased oxygen level and reduced C–H ratio caused complete combustion, reducing the unburned CO and HC emissions. However, for all the test cases and EGR rates, CO and HC emissions rose at advanced SoI timing before 25° bTDC. This causes a significant fuel quantity to be trapped in crevices during prolonged ignition delay, as well as an increased risk of unburned HC emissions due to the crevice effect and wall-wetting at high EGR rates with earlier SoI timings. High EGR rates may reduce combustion temperature, increasing HC emissions. Furthermore, by lowering the combustion temperatures, CO–CO_2 conversion slows down (Thomas et al. 2014). Incomplete combustion occurs when the fuel-air mixture is excessively rich or lean, thus increasing CO emissions (Park and Lee 2014). Generally, it has been observed that DME generates lesser CO than diesel engines because of the absence of C–C bonds, high fuel oxygen and low C/H ratio. These characteristics allow far more efficient mixing and oxidation of intermediate species. Using larger injector holes and lower fuel injection pressure (FIP) than diesel results in higher CO emissions from DME-fueled engines since DME combustion requires a long injection duration. At a low local equivalence ratio, high atomization behaviour and a short ignition delay of DME results in too-lean mixture regions, which increases CO emissions. The β-scission of methoxy-methyl radicals during the combustion of DME produces formaldehyde (CH_2O), which also causes high CO emissions from DME engine than an equivalent diesel engine. CO emissions were reduced with advanced main injection and retarded pilot injection timings, according to Kim et al. (2007b), but they remained higher than with conventional single injection conditions.

Oguma et al. (2005) reported lower THC emissions from DME combustion than diesel. Ethylene (C_2H_4), ethane (C_2H_6), Propylene (C_3H_6), 1–3 butadiene (C_4H_6), benzene (C_6H_6) and toluene (C_7H_8) emissions were measured as a part of the measurement of THCs. DME operations exhibited lower ethylene, ethane, and toluene levels than diesel operations. Since these hydrocarbons can act as nuclei for PM, soot gets reduced with DME operations compared to diesel operations. However, Propylene, 1,3-butadiene, and benzene emissions were higher in DME combustion than in diesel.

Soot Emissions: Soot generation is inhibited by the absence of C–C bonds in the molecular structure of DME (Mukherjee et al. 2022). When such fuels are burned, the precursors for soot formation, such as aromatic and acetylenic species, are not produced. These aromatic species cause soot to evolve and expand with the help of acetylenic species (Karpuk et al. 1991). Ethylene (C_2H_4) and propargyl (C_3H_3) are major soot precursors. With an increase in fuel oxygen, the formation of these soot precursors decreases (Kim et al. 2011). High concentrations of free radicals drive carbon oxidation after the addition of DME, decreasing the carbon availability for synthesising soot precursors. Soot generation thrives in fuel-rich, high-temperature

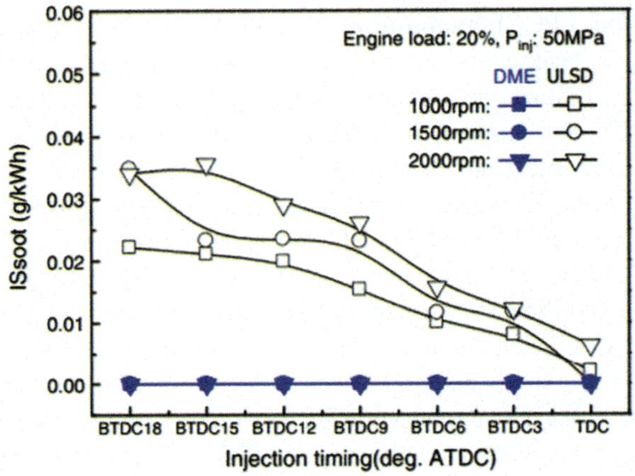

Fig. 3.19 Soot emissions from diesel and DME at varying fuel injection timings (Youn et al. 2011)

zones with temperatures ranging from 1500 to 2500 K (Park and Lee 2014; Zhao et al. 2014).

DME combustion produces virtually no soot (Fig. 3.19) (Park and Lee 2013). Due to the addition of lubricity additives and lubrication oil in fuel and oil pumps, minor amounts of soot are occasionally discovered from the DME combustion. Yoon et al. (2013) reported that soot emissions from DME-fueled engines remained minimal even when EGR was introduced due to the presence of fuel oxygen.

The reaction flow of DME is shown in the diagram under typical fuel-rich conditions (Fig. 3.20). Many researchers analysed the effects of the addition of DME and isomers of butanol on soot formation. Xu et al. (2022) reported that adding DME and butanol significantly reduced the aromatic species formation. However, DME was more effective in reducing aromatic species formation. Zhang et al. (2019) reported that the PAH formation was lower in DME than in ethanol. According to kinetic analyses, DME, with its symmetric C–O–C structure, differed from ethanol in synthesising hydrocarbon intermediates larger than C_1. As a result, DME-doped flames had lower C_2H_2 concentration than ethanol-doped flames. Therefore, DME suppressed the formation of soot precursors more efficiently than ethanol.

3.12.2 Unregulated Emissions

Formaldehyde (HCHO), acetaldehyde (CH_3CHO), and formic acid (HCOOH) are reported to be higher in DME than in diesel-fueled engines (Fig. 3.21) since it contains higher fuel oxygen and lubricity additives than diesel. Introducing DOC can reduce formaldehyde to insignificant levels; however, catalysts do not affect acetaldehyde emissions.

The lubricity additives blended with DME must be carefully monitored since they can impact the formation of these oxygen-based compounds. The reduction in PM could be offset by an increase in other pollutants (MacFarlane 1965; Park and

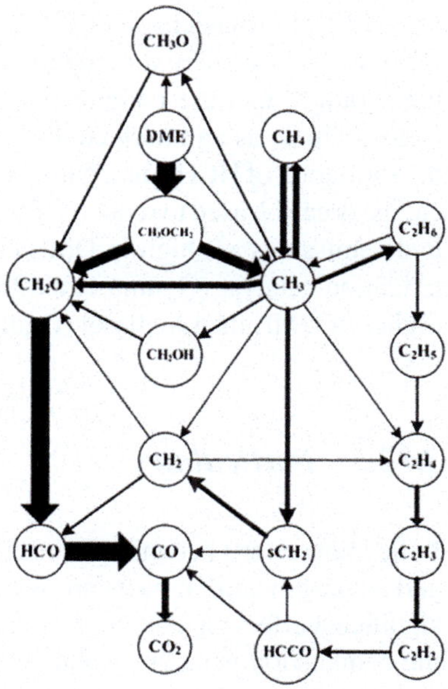

Fig. 3.20 Reaction flow diagram of soot formation during DME combustion (Xu et al. 2022)

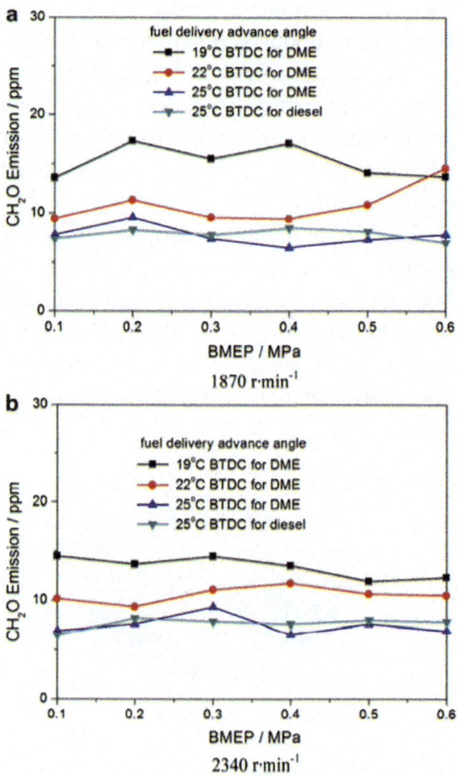

Fig. 3.21 Formaldehyde emissions from diesel and DME at varying engine speed and SoI timings of DME (Zhu et al. 2012)

Lee 2013). Formaldehyde can be produced as an intermediary in the DME engine. Also, it produces partial oxidation products in the exhaust. The average in-cylinder temperature, mixture composition and residence time influence the formation of formaldehyde as an intermediate. Outlet gas temperature and residence time affect the formation of HCHO as a partial oxidation product (Park and Lee 2013). Oxidation catalysts can be used to reduce formaldehyde to a negligible level. However, acetaldehyde emissions are high in DME combustion than in diesel, regardless of using an oxidation catalyst (Oguma et al. 2005; Jie et al. 2010). Also, methane emission is higher in DME than in diesel (Oguma et al. 2005).

3.12.3 Particulates

Policymakers are becoming conscious of regulating PM emissions as their health, and environmental impacts are severe. Especially small sized PM emissions have a significant effect on ambient air quality and human health. In locally fuel-rich zones, the combustion generates solid carbon, which is oxidised. A small fraction of fuel and lubricating oil evaporate and emerge in the exhaust as volatile or soluble organic molecules (Kittelson 1998). These soluble organic molecules contain polycyclic aromatic compounds (Quality of Urban Air Review Group 1996).

The structure of solid particulates is shown in Fig. 3.22. In locally fuel-rich zones, solid carbon is generated after combustion. A large part of the particle mass exists in the accumulation mode in the diameter range of 0.1–0.3 μm. Carbonaceous agglomerated particles and associated adsorbed materials lie in this size range (Kittelson 1998).

The size of particle determines the environmental implications of engine exhaust particles in numerous ways. It affects the particle's atmospheric residence time, optical qualities, surface area and health effects. Particles in the 0.1–10 μm diameter range have the longest residence time of approximately one week in the atmosphere. Smaller particles are removed from the atmosphere via diffusion and coagulation processes, whereas larger particles are quickly removed by settling. The average

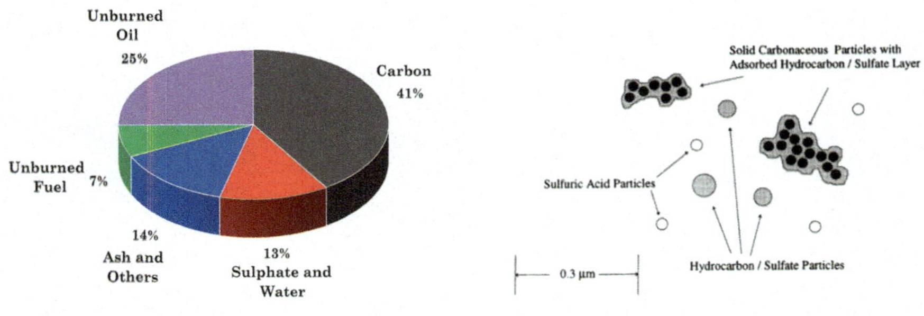

Fig. 3.22 Mineral diesel origin particulate composition (Kittelson 1998)

atmospheric residence time for nanoparticles is around 15 min (Quality of Urban Air Review Group 1996). Coagulating particles in the accumulation mode is the principal mechanism for removing these small particles (Fig. 3.23).

Diesel exhaust particles have an average surface area of around 100 m^2/g (https://www.missoulacounty.us/home/showpublisheddocument/3945/635809567464330000). The individual nuclei that make up the agglomerates have nearly all their surface area for adsorption. Thus, the size of the individual nuclei in the agglomerates, rather than the agglomeration size, determines the surface area of diesel particles.

PM emissions are significantly low for DME combustion than diesel due to 35% (w/w) fuel oxygen and absence of C–C bonds (MacFarlane 1965). Therefore, DME-powered vehicles can meet all current global soot emission regulations without using any DPF or trap. According to studies, 99% of PM emitted by a DME engine are in the nanoparticle size range, which are more harmful to human health than larger particles. However, researchers do not indicate whether tailpipe emissions risk human health regarding absolute volume and count of PM particles (Li et al. 2008).

Figure 3.24 shows that DME's particle number concentration is lower than mineral diesel under all engine operating conditions. PM from a diesel engine mainly originates from fuel combustion and lubricating oil burning in a DME engine. The consumption of lubricating oil is substantially lower than the consumption of diesel. As a result, the PM number concentration from DME combustion is significantly lower than diesel for all particle diameters. As engine speed increases, the in-cylinder flow in the chamber increases, promoting air-fuel mixing. Therefore, a more homogeneous oxygen distribution suppresses the dry soot (DS) formation. As a result, particle number concentration decreases with increasing engine speed for both diesel and DME (Wei et al. 2014).

PN concentration in the accumulation mode increases for diesel combustion with increasing the engine load at constant engine speed. At high loads, an increase in

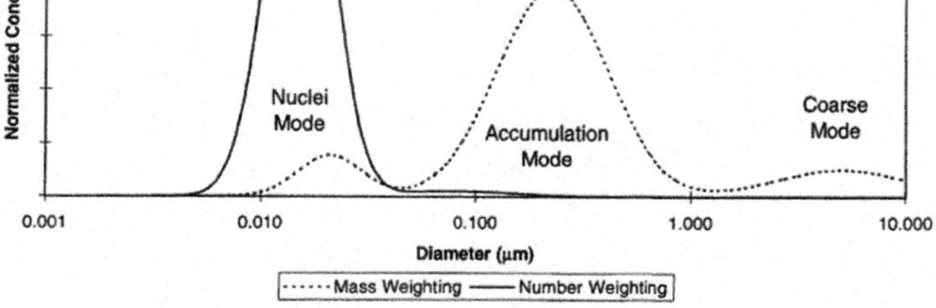

Fig. 3.23 Particle number-size distribution for diesel engines (Kittelson 1998)

Fig. 3.24 Particle number-size distribution of diesel and DME at varying engine speeds (Wei et al. 2014)

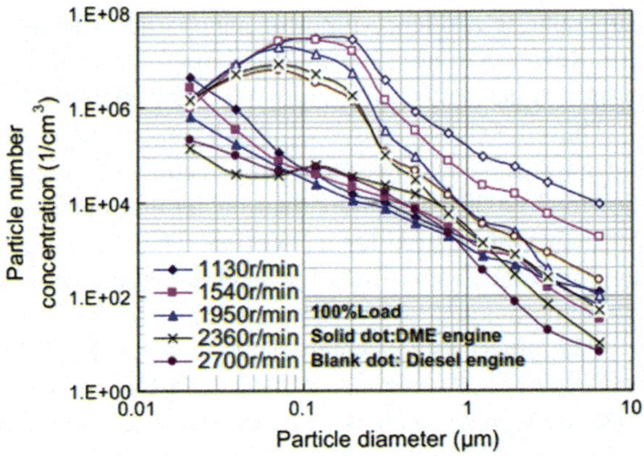

fuel quantity increases the fuel-rich zones and the in-cylinder temperature. Therefore, soot emission from diesel combustion increases with increased engine load. The PN concentration of DME is lower than diesel because of the absence of the C–C bond and higher fuel oxygen in DME, which suppress the formation of soot (Li et al. 2008). The geometric mean diameter of the particles in the accumulation mode ($DGNs_{acc}$) increases with increasing engine load. At high loads, the concentration of soot particles increases. Therefore, many soot particles accumulate and form a chain-like structure. $DGNs_{acc}$ of diesel was 10–30 nm larger than $DGNs_{acc}$ of DME at similar engine operating conditions. Therefore, unlike diesel, small chain-like structures of exhaust particles form in DME engines. This results in lower $DGNs_{acc}$ of exhaust particles of DME than diesel (Li et al. 2008).

3.12.4 *Trace Metals*

The trace metals in the CI engine exhaust pose a serious risk to the environment and human health. Engine friction and wear debris from the combustion chamber, fuel, fuel additives and lubricating oil are primary sources of trace metals in particulates. Agarwal et al. (2010) conducted experiments using straight vegetable oils. They reported that the Iron (Fe), calcium (Ca), magnesium (Mg), and sodium (Na) trace concentrations were higher than nickel (Ni), copper (Cu), zinc (Zn), and lead (Pb), trace concentrations. Table 3.5 discusses the sources of various metallic elements present in soot.

High metal concentrations in the environment can cause serious health problems for humans and animals. Trace metal ingestion can lead to cancer, respiratory, pulmonary and neurotoxicity diseases (Geiger and Cooper 2010). Different trace metals cause various health issues in the human body, which are listed in Table 3.6.

Table 3.5 Source of various trace metals present in soot (Agarwal et al. 2003)

Element	Source component
Sodium (Na)	Coolant, additives
Magnesium (Mg)	Bearings, additives, supercharger, gearbox
Aluminium (Al)	Piston, bearings, dirt, additives, turbochargers
Silicon (Si)	Lubricants, dirt
Potassium (K)	Additives, coolant
Calcium (Ca)	Additives, greases, water
Copper (Cu)	Bronze bushings, bearings
Iron (Fe)	Cylinder liner, piston, piston rings, anti-friction bearings, gears, shafts, rust, crankshaft, valves, valve guide
Zink (Zn)	Bearings, plating, brass components, neoprene seals, additives

Table 3.6 Health effects of various trace metals (Geiger and Cooper 2010)

Element	Health effects
Sodium (Na)	Irritation to skin, eye, nose, and throat
Magnesium (Mg)	Irritate throat, nose, upper respiratory tract or mucous membrane, difficulty in breathing
Calcium (Ca)	Calcium carbonate ($CaCO_3$) dust irritates eyes, nose, mucous membranes, and skin. Contact of ($CaCO_3$) with the eyes causes eyelids redness, pain and inflammation
Copper (Cu)	Eye irritation, irritation of nasal mucous membranes, upper respiratory tract irritation, metal fume fever and nausea. Acute Cu poising can cause methemoglobinemia, liver injury, and haemolytic anaemia
Iron (Fe)	Metal fume fever, coughing, weakness, shortness of breath, fatigue, sweating and pain in muscles, fever, chills, and pain in joints
Zink (Zn)	Large doses inhalation of Zn can cause nausea, stomach cramps, and vomiting. Acute exposure to Zn oxide can result in upper respiratory tract irritation, coughing, chills and fever, substernal pain, rales
Cadmium (Cd)	Short-term effects are lung effects such as bronchial and pulmonary irritation. Long-term effects are Cd accumulation in the kidneys and affect the liver, bone, lung, nervous system and immune system
Chromium (Cr)	Short-term effect of Cr inhalation includes shortness of breath, coughing, wheezing, abdominal pain, vomiting and effects on the respiratory tract. Long-term effects of Cr inhalation are decreased pulmonary function, affected perforations and ulcerations of the septum, bronchitis, Asthma, pneumonia, soreness and nasal itching

3.13 Conclusions

Due to its favourable properties, DME is progressively used in many industries, such as transport and aerosol. DME also poses some challenges to the existing

FIE. However, DME significantly reduces emissions without compromising engine performance.

1. DME has many favourable properties such as high CN, fuel-bound oxygen, absence of C–C bond, superior spray atomisation characteristics, lower self-ignition temperature and ignition delay than mineral diesel.
2. DME also has some disadvantages, such as ultra-low viscosity, minimal self-lubrication properties, high vapour pressure, low critical point, and so on, all of which make it difficult for conventional FIE to manage DME. The lower acoustic velocity caused by DME's high compressibility also causes a long injection delay than diesel. DME is stored at a pressure higher than its vapour pressure to ensure its supply in a liquid phase. All components of the FIE are made of DME-compatible materials to avoid corrosion of parts. Lubricity additives are added to DME to improve its lubricity. Many studies showed a higher BTE of diesel than DME.
3. EGT was lower for DME due to its high latent heat of vaporisation and low in-cylinder temperature. Lower peak pressure of the DME engine was observed than mineral diesel for the same injection volume because DME has a low calorific value (28.43 MJ/kg) than diesel (42.5 MJ/kg). DME has a low HRR than mineral diesel in a premixed combustion phase. A smaller amount of DME is accumulated in the combustion chamber before the SoC due to the small ignition delay of DME. DME's peak fuel line pressure was significantly lower than diesel. DME has low fuel injection velocity and long fuel injection duration than baseline diesel.
4. Some studies indicated that DME generates lower NO_x than diesel-fueled engines due to its high CN, lower heating value, and higher latent heat capacity. Some other studies reported high NO_x emissions from DME combustion because of fast ignition and high peak in-cylinder temperature, increasing the NO_x emissions more than diesel. DME has superior spray atomisation, evaporation characteristics, and a short ignition delay, avoids the over-rich regions due to short mixing time. DME's oxygen concentration also aids in avoiding fuel-rich zones and reducing HC emissions. The use of high EGR rate increases HC emissions. Using larger injector nozzle holes and lower fuel injection pressure (FIP) than diesel resulted in higher CO emissions from DME-fuelled engines since DME combustion required a long fuel injection duration. The split injection strategy reduced the CO emission with advanced main and retarded pilot injections.
5. Soot production is inhibited by the absence of a C–C bond in the molecular structure of DME. Therefore, soot emissions were almost zero in DME combustion. PM emissions are significantly low in DME combustion than diesel due to 35% (w/w) fuel oxygen and the absence of C–C bonds. PM emitted by a DME engine were in the nanoparticle size range. Formaldehyde (HCHO), acetaldehyde (CH_3CHO), and formic acid (HCOOH) were reportedly high in DME than in diesel-fuelled engines.

3.14 Future Scope

Several studies have been conducted on DME utilisation in various engine types. Unexplored areas need to be studied for DME utilisation and improving the engine characteristics to replace diesel. These are as follows:

1. DME SoI timing should be optimised for every engine operating condition to compensate for the retarded SoI timing due to the high compressibility of DME.
2. A high-pressure fuel pump with a large plunger diameter should be explored to supply a higher DME quantity to the engine to match its entire speed-load envelope.
3. An appropriate fuel conditioning system should be developed to eliminate the chances of vapour lock in the fuel lines.
4. Durability studies should be conducted for the FIE components to analyse the DME compatibility of different materials. These results can be used for making a DME-compatible FIE for commercial applications.

References

Agarwal, AK, Bijwe, J, Das, LM (2003) Effect of biodiesel utilization of wear of vital parts in compression ignition engine. J Eng Gas Turbines Power. https://doi.org/10.1115/1.1454114

Agarwal AK, Gupta T, Kothari A (2010) Toxic potential evaluation of particulate matter emitted from a constant speed compression ignition engine: a comparison between straight vegetable oil and mineral diesel. Aerosol Sci Technol 44:724–733. https://doi.org/10.1080/02786826.2010.486386

Agarwal AK, Sharma N, Singh AP (2017) Potential of DME and methanol for locomotive traction in India: opportunities, technology options and challenges. Locomot Rail Road Transp Technol Chall Prospect 129–151. https://doi.org/10.1007/978-981-10-3788-7_7

Arcoumanis C, Bae C, Crookes R, Kinoshita E (2008) The potential of di-methyl ether (DME) as an alternative fuel for compression-ignition engines: a review. Fuel 87:1014–1030. https://doi.org/10.1016/J.FUEL.2007.06.007

Bowman FM, Seinfeld JH (1995) Atmospheric chemistry of alternate fuels and reformulated gasoline components. Prog Energy Combust Sci 21:387–417. https://doi.org/10.1016/0360-1285(95)00008-9

Carbon Dioxide in the Atmosphere is Increasing. https://www.globalchange.gov/browse/indicators/atmospheric-carbon-dioxide. Accessed 15 Aug 2022

Chapman EM, Boehman A, Wain K, Lloyd W, Perez JM, Stiver D et al (2004) Final technical progress report for project entitled, Impact of DME-diesel fuel blend properties on diesel fuel injection systems, pp 1–65. https://doi.org/10.2172/821275

Christensen R, Sorenson SC, Jensen MG, Hansen KF (1997) Engine operation on dimethyl ether in a naturally aspirated, Dl Diesel Engine. SAE Tech Pap. https://doi.org/10.4271/971665

Climate Change Over the Next 100 Years. https://clintonwhitehouse5.archives.gov/Initiatives/Climate/next100.html. Accessed 20 Aug 2022

Contributors to Global Sea Level Rise. https://www.climate.gov/media/12868. Accessed 28 Aug 2022

Dimethyl Ether Market Size & Share and Regional Forecast 2021–2028. https://www.fortunebusinessinsights.com/dimethyl-ether-market-104309. Accessed 04 Sep 2022

European Countries Banning Fossil Fuel Cars and Switching to Electric Cars. https://www.roadtraffic-technology.com/features/european-countries-banning-fossil-fuel-cars/. Accessed 01 Sep 2022

Effects of Global Warming. https://www.livescience.com/37057-global-warming-effects.html. Accessed 17 Aug 2022

Falco MD (2017) Dimethyl ether production from CO_2 rich feedstocks in a one-step process: thermodynamic evaluation and reactor simulation. https://doi.org/10.1016/j.cej.2016.03.009

Formaldehyde Screening Study. Missoula city-county health department. https://www.missoulacounty.us/home/showpublisheddocument/3945/635809567464330000. Accessed 15 Sep 2022

Fundamental aspects of Di-methyl ether. https://afdc.energy.gov/files/pdfs/3608.pdf. Accessed 17 Sep 2022

Geiger A, Cooper J (2010) Overview of airborne metals regulations, exposure limits, health effects, and contemporary research. Draft Environ Anal 3:1–50. https://www3.epa.gov/ttnemc01/prelim/otm31appC.pdf. Accessed 10 Sep 2022

Global Surface Temperature. https://en.wikipedia.org/wiki/Global_surface_temperature. Accessed 22 Aug 2022

Hansen KF, Nielsen L, Hansen JB, Mikkelsen SE, Landälv H, Ristola T et al (2000) Demonstration of a DME (dimethyl ether) fuelled city bus. SAE Tech Pap. https://doi.org/10.4271/2000-01-2005

Hazardous substance fact sheet. New jersey department of health and senior services. https://nj.gov/health/eoh/rtkweb/documents/fs/0758.pdf. Accessed 15 Sep 2022

Huang ZH, Wang HW, Chen HY, Zhou LB, Jiang DM (1999) Study of combustion characteristics of a compression ignition engine fuelled with dimethyl ether. Proc Inst Mech Eng Part D J Automob Eng 213:647–652. https://doi.org/10.1243/0954407991527161

Huang Z, Qiao X, Zhang W, Wu J, Zhang J (2009) Dimethyl ether as alternative fuel for CI engine and vehicle. Front Energy Power Eng China 3:99–108. https://doi.org/10.1007/s11708-009-0013-1

India Planning Commission (2007) The working group report on road transport for the eleventh five year plan, 1–83 http://www.im4change.org/docs/11th_vol3.pdf. Accessed 05 Aug 2022

India's crude oil import bill set to rise 20% to $105 billion in FY19. https://energy.economictimes.indiatimes.com/news/oil-and-gas/indias-crude-oil-import-bill-set-to-rise-20-to-105-billion-in-fy19/63881639. Accessed 07 Aug 2022

Jang J, Yang K, Bae C (2009) The effect of injection location of DME and LPG in a dual fuel HCCI engine. SAE Tech Pap 4970. https://doi.org/10.4271/2009-01-1847

Japar SM, Wallington TJ, Richert JFO, Ball JC (1990) The atmospheric chemistry of Oxygenated fuel additives: t-Butyl alcohol, dimethyl ether, and methylt-butyl ether. Int J Chem Kinet 22:1257–1269. https://doi.org/10.1002/550221205

Jie L, Shenghua L, Yi L, Yanju W, Guangle L, Zan Z (2010) Regulated and nonregulated emissions from a dimethyl ether powered compression ignition engine. Energy Fuels 24:2465–2469. https://doi.org/10.1021/ef9016043

Kanta N, Valera H, Unnithan S, Kumar V, Dhyani V, Mehra S et al (2022) Feasibility study of novel DME fuel injection equipment: Part 1-fuel injection strategies and spray characteristics. Fuel 323:124333. https://doi.org/10.1016/j.fuel.2022.124333

Karpuk ME, Wright JD, Dippo JL, Jantzen DE (1991) Dimethyl ether as an ignition enhancer for methanol-fueled diesel engines. SAE Tech Pap. https://doi.org/10.4271/912420

Kass MD, Daw C (2016) Compatibility of dimethyl ether (DME) and diesel blends with fuel system polymers: a Hansen solubility analysis approach. SAE Int J Fuels Lubr 9:71–79. https://doi.org/10.4271/2016-01-0835

Kim MY, Bang SH, Lee CS (2007a) Experimental investigation of spray and combustion characteristics of dimethyl ether in a common-rail diesel engine. Energy Fuels 21:793–800. https://doi.org/10.1021/060310

Kim MY, Yoon SH, Park KH, Lee CS (2007b) Effect of multiple injection strategies on the emission characteristics of dimethyl ether (DME)-fueled compression ignition engine. Energy Fuels 21:2673–2681. https://doi.org/10.1021/ef0701844

Kim MY, Lee JH, Lee CS (2008a) Combustion characteristics and NOx emissions of a dimethyl-ether-fueled premixed charge compression ignition engine. Energy Fuels 22:4206–4212. https://doi.org/10.1021/ef800221g

Kim MY, Yoon SH, Ryu BW, Lee CS (2008b) Combustion and emission characteristics of DME as an alternative fuel for compression ignition engines with a high pressure injection system. Fuel 87:2779–2786. https://doi.org/10.1016/2008.01.032

Kim HJ, Park SH, Suh HK, Lee CS (2009) Atomization and evaporation characteristics of biodiesel and dimethyl ether compared to diesel fuel in a high-pressure injection system. Energy Fuels 23:1734–1742. https://doi.org/10.1021/800811

Kim HJ, Lee KS, Lee CS (2011) A study on the reduction of exhaust emissions through HCCI combustion by using a narrow spray angle and advanced injection timing in a DME engine. Fuel Process Technol 92:1756–1763. https://doi.org/10.1016/j.fuproc.2011.04.024

Kim HJ, Park SW, Lee CS (2012) Études numériques et expérimentales sur les caractéristiques de combustion et d'émissions d'un éther diméthylique (EDM)- Moteur à auto-allumage rempli de combustible. Oil Gas Sci Technol 67:479–489. https://doi.org/10.2516/ogst/2011130

Kittelson DB (1998) Engines and nanoparticles: a review. J Aerosol Sci 29:575–588. https://doi.org/10.1016/S0021-8502(97)10037-4

Kumar V, Agarwal AK (2021) Material compatibility, technical challenges and modifications required for DME adaptation in compression ignition engines. Energy Environ Sustain 37–57. https://doi.org/10.1007/978-981-16-1513-9_3

Li XL, Huang Z, Wang JS, Wu JH (2008) Characteristics of ultrafine particles emitted from a dimethyl ether (DME) engine. Chinese Sci Bull 53:304–312. https://doi.org/10.1007/s11434-008-0011-4

Longbao Z, Hewu W, Deming J, Zuohua H (1999) Study of performance and combustion characteristics of a DME-fueled light-duty direct-injection diesel engine. SAE Tech Pap. https://doi.org/10.4271/1999-01-3669

MacFarlane IC (1965) NRC publications archive archives des publications du CNRC the consolidation of peat: a literature review. https://doi.org/10.4224/20331668

Mehra S, Agarwal AK (2022) Prospects and challenges of DME fueled low-temperature combustion engine technology. Energy Environ Sustain 261–91. https://doi.org/10.1007/978-981-16-8344-2_10

Methanol Economy. https://www.niti.gov.in/methanol-economy. Accessed 02 Sep 2022

Morsy MH, Ahn DH, Chung SH (2006) Pilot injection of DME for ignition of natural gas at dual fuel engine-like conditions. Int J Automot Technol 7:1–7

Mukherjee NK, Valera H, Unnithan S, Kumar V, Dhyani V, Mehra S et al (2022) Feasibility study of novel DME fuel injection equipment: Part 2-performance, combustion, regulated and unregulated emissions. Fuel 323:124338. https://doi.org/10.1016/J.FUEL.2022.124338

New Climate Institute (2020) Climate action tracker decarbonising the Indian transport sector 2:1–79. https://climateactiontracker.org/documents/832/CAT_2020-12-09_Report_DecarbonisingIndianTransportSector_Dec2020.pdf. Accessed 07 Aug 2022

Oguma M, Goto S, Watanabe T (2004) Engine performance and emission characteristics of DME diesel engine with inline injection pump developed for DME. SAE Tech Pap. https://doi.org/10.4271/2004-01-1863

Oguma M, Shiotani H, Goto S, Suzuki S (2005) Measurement of trace levels of harmful substances emitted from a DME di diesel engine. SAE Tech Pap. https://doi.org/10.4271/2005-01-2202

Oh C, Jang J, Bae C (2010) The effect of LPG composition on combustion and performance in a DME-LPG dual-fuel HCCI engine. SAE Tech Pap. https://doi.org/10.4271/2010-01-0336

Park SH, Lee CS (2013) Combustion performance and emission reduction characteristics of automotive DME engine system. Prog Energy Combust Sci 39:147–168. https://doi.org/10.1016/2012.10.002

Park SH, Lee CS (2014) Applicability of dimethyl ether (DME) in a compression ignition engine as an alternative fuel. Energy Convers Manag 86:848–863. https://doi.org/10.1016/2014.06.051
Park SH, Kim HJ, Lee CS (2010) Macroscopic spray characteristics and breakup performance of dimethyl ether (DME) fuel at high fuel temperatures and ambient conditions. Fuel 89:3001–3011. https://doi.org/10.1016/2010.05.002
Proposal for a "REGULATION OF THE EUROPEAN PARLIAMENT AND OF THE COUNCIL" on the deployment of alternative fuel fuels infrastructure and repealing directive 2014/94/EU of the European Parliament and the Council. https://eur-lex.europa.eu/legal-content/en/TXT/?uri=CELEX%3A52021PC0559. Accessed 01 Sep 2022
Quality of Urban Air Review Group (1996) Airborne Particulate Matter in the United Kingdom. Third Report of the Quality of Urban Air Review Group. https://uk-air.defra.gov.uk/assets/documents/reports/empire/quarg/quarg_11.pdf. Accessed 20 Sep 2022
Sivebaek IM, Sorenson SC, Jakobsen J (2001) Dimethyl ether (DME)-assessment of viscosity using the new volatile fuel viscometer (VFVM). SAE Tech Pap. https://doi.org/10.4271/2001-01-2013
Sorenson SC (2001) Dimethyl ether in diesel engines: progress and perspectives. J Eng Gas Turbines Power 123:652–658. https://doi.org/10.1115/1.1370373
Sorenson SC, Mikkelsen SE (1995) Performance and emissions of a 0.273 liter direct injection diesel engine fuelled with neat dimethyl ether. SAE Tech Pap. https://doi.org/10.4271/950064
Szybist JP, Mclauglin S, Iyer S (2014) Emission and performance benchmarking of a prototype dimehty ether fueled heavy-duty truck. ORNL/TM-2014/59. https://afdc.energy.gov/files/u/publication/ornl_dme_tm-2014-59.pdf. Accessed 13 Sep 2022
Teng H, McCandless JC, Schneyer JB (2001) Thermochemical characteristics of dimethyl ether—An alternative fuel for compression-ignition engines. SAE Tech Pap. https://doi.org/10.4271/2001-01-0154
Thomas G, Feng B, Veeraragavan A, Cleary MJ, Drinnan N (2014) Emissions from DME combustion in diesel engines and their implications on meeting future emission norms: a review. Fuel Process Technol 119:286–304. https://doi.org/10.1016/j.fuproc.2013.10.018
Tripathi A, Park S, Park S, Agarwal AK (2022) Prospects of dual-fuel injection system in compression ignition (CI) engines using Di-Methyl Ether (DME) 223–59. https://doi.org/10.1007/978-981-16-8344-2_9
Tsuchiya T, Sato Y (2006) Development of DME engine for heavy-duty truck. SAE Tech Pap. https://doi.org/10.4271/2006-01-0052
Valera H, Agarwal AK (2019) Methanol as an alternative fuel for diesel engines. Energy Environ Sustain 9–33. https://doi.org/10.1007/978-981-13-3287-6_2
Wei Y, Wang K, Wang W, Liu S, Chen X, Yang Y et al (2014) Comparison study on the emission characteristics of diesel- and dimethyl ether-originated particulate matters. Appl Energy 130:357–369. https://doi.org/10.1016/j.apenergy.2014.05.058
What is Carbon Neutrality and How it Can be Achieved by 2050. https://www.europarl.europa.eu/news/en/headlines/society/20190926STO62270/what-is-carbon-neutrality-and-how-can-it-be-achieved-by-2050. Accessed 25 Aug 2022
World Oil Reserve. https://www.worldometers.info/oil/#:~:text=World%20Oil%20Reserves&text=The%20world%20has%20proven%20reserves,levels%20and%20excluding%20unproven%20reserves. Accessed 12 Aug 2022
Xu L, Wang Y, Liu D (2022) Effects of oxygenated biofuel additives on soot formation: a comprehensive review of laboratory-scale studies. Fuel 313:122635. https://doi.org/10.1016/j.fuel.2021.122635
Yoon SH, Han SC, Lee CS (2013) Effects of high EGR rate on dimethyl ether (DME) combustion and pollutant emission characteristics in a direct injection diesel engine. Energies 6:5157–5167. https://doi.org/10.3390/en6105157
Youn IM, Park SH, Roh HG, Lee CS (2011) Investigation on the fuel spray and emission reduction characteristics for dimethyl ether (DME) fueled multi-cylinder diesel engine with common-rail injection system. Fuel Process Technol 92:1280–1287. https://doi.org/10.1016/j.fuproc.2011.01.018

Zhang Y, Li Y, Liu P, Zhan R, Huang Z, Lin H (2019) Investigation on the chemical effects of dimethyl ether and ethanol additions on PAH formation in laminar premixed ethylene flames. Fuel 256:115809. https://doi.org/10.1016/j.fuel.2019.115809

Zhao Y, Wang Y, Li D, Lei X, Liu S (2014) Combustion and emission characteristics of a DME (dimethyl ether)-diesel dual fuel premixed charge compression ignition engine with EGR (exhaust gas recirculation). Energy 72:608–617. https://doi.org/10.1016/j.energy.2014.05.086

Zhu Z, Li DK, Liu J, Wei YJ, Liu SH (2012) Investigation on the regulated and unregulated emissions of a DME engine under different injection timing. Appl Therm Eng 35:9–14. https://doi.org/10.1016/j.applthermaleng.2011.08.015

Chapter 4
Combustion and Emission Characteristics of Oxygenated Alternative Fuels in Compression Ignition Engines

Tomesh Kumar Sahu and Pravesh Chandra Shukla

Abstract Environmental pollution from petroleum products for energy generation is of grave concern nowadays. Oxygenated alternative fuels like biodiesels, alcohols, etc., have gained much popularity for internal combustion (IC) engines. Due to their inherent oxygen content, these oxygenated alternative fuels possess lower carbon-to-hydrogen (C/H) ratio for same heating value. Supporting the road map towards decarbonization of mobility, there has been a recent uptick in curiosity about the possibility of alcohols in compression ignition (CI) engines. Biodiesel contains ~10% inherent oxygen (*m/m*), and alcohols may contains up to 50% oxygen (*m/m*), affecting the CI engine's combustion, performance, and emission. An overview of CI engines fuelled with oxygenated fuels (biodiesel and alcohol blends) on combustion characteristics, engine performance, and exhaust emissions are presented in this study. Biodiesel and alcohol fuels (methanol, ethanol, and butanol) have been compared to evaluate the impact of different blend ratios, oxygen mass fraction content based on in-cylinder combustion pressure trace, heat release rate, and engine performance as brake thermal efficiency and carbon dioxide (CO_2) emission. Up to 25% lowered CO_2 emission was recorded for oxygenated fuel compared to diesel, with significantly lower particulate emission.

Keywords Oxygenated alternative fuels · Alcohols · Waste cooking oil (WCO) based biodiesel · Compression ignition (CI) engines

Abbreviations

ABE	Acetone-butanol-ethanol
BDE	Biodiesel-diesel-ethanol
BTE	Brake thermal efficiency
CA05	5% of total burn fraction

T. K. Sahu · P. C. Shukla (✉)
Department of Mechanical Engineering, Indian Institute of Technology Bhilai, Sejbahar, India
e-mail: pravesh@iitbhilai.ac.in

P. C. Shukla et al. (eds.), *Renewable Fuels for Sustainable Mobility*, Energy, Environment, and Sustainability, https://doi.org/10.1007/978-981-99-1392-3_4

CA50	50% of total burn fraction
CA90	90% of total burn fraction
CAD	Crank angle degree
CI	Compression ignition
C/H	Carbon to hydrogen ratio
CO	Carbon monoxide
CO_2	Carbon dioxide
EGR	Exhaust gas recirculation
HC	Hydrocarbon
HRR	Heat release rate
IBE	Isopropanol-butanol-ethanol
IC	Internal combustion
ISEC	Indicated specific energy consumption
IMEP	Indicated mean effective pressure
ITE	Indicated thermal efficiency
LTC	Low-temperature combustion
NOx	Nitrogen oxides
PPC	Partially premixed combustion
SI	Spark ignition
TDC	Top dead center
ULSD	Ultra-low sulfur diesel

4.1 Introduction

Global population is likely to reach 9 billion by the year 2040, resulting in a 25% increase in energy demand (Ghadikolaei et al. 2018). Increased energy demand and a growing global urge to minimize carbon emissions impact the usage of conventional fuels and encourage the investigation and implementation of alternative fuels for existing engines (Sahu et al. 2021a, 2022a). Sustainable fuel production processes and utilization have drawn attention to environmental protection by reducing overall carbon dioxide (CO_2) emissions (Ianniello et al. 2021). Biofuels are produced from renewable and sustainable resources, which have proven their use in compression ignition (CI) engines and have been promising in recent years. In this context, biofuels (such as biodiesels, lower and higher alcohols, and their blends with conventional fuels) are extensively explored, showing exciting results regarding emissions reduction and increased efficiency compared with conventional fuels (Ianniello et al. 2021; Shamun et al. 2020; Luca et al. 2022; Sahu et al. 2022c). Innovative technologies for powertrain systems improved the overall engine efficiency and lowered emissions (Belgiorno et al. 2020; Di Blasio et al. 2019). Improved injector nozzle and fuel injection systems (Sequino et al. 2018; Beatrice et al. 2017; Di Blasio et al. 2019), in-cylinder air motion, and EGR systems (Dimitrakopoulos et al. 2019) are being implemented to meet stringent emission regulations and engine performance requirements. Different injection strategies (Sahu et al. 2022d) as well as advanced combustion modes (dual fuel, partially premixed combustion, etc.) (Belgiorno et al. 2019; Dimitrakopoulos et al. 2017; Sahu et al. 2021b) combined with alternative fuels could be an enabler towards clean combustion and reduce the overall CO_2 impact of the transportation sector (Sahu et al. 2022d). It has been observed that oxygenated alternative fuel blends in dual fuel mode lead to the air-fuel mixture in crevice volume showing higher hydrocarbon (HC) emissions (Fraioli et al. 2017; Monsalve-Serrano et al. 2020). Other studies have also reported this behaviour for heavy-duty and light-duty engines (Heuser et al. 2016; Pedrozo et al. 2018). These observations show that HC emissions are higher at lower engine loads and have a similar emission profile to conventional fuel at higher engine loads. Engine operating parameters such as maximum in-cylinder pressure and compression ratio (CR) can affect the air-to-fuel mass trapped in the crevice volume, primarily affecting the HC emission slip in dual fuel combustion (Di Blasio et al. 2015, 2017). Biodiesel-diesel-ethanol (BDE; 78%, 17%, and 5% v/v, diesel, biodiesel, and ethanol, respectively) blends were studied by Krishna et al. (2019) and a marginal decrease in efficiency with reduced carbon monoxide (CO) and comparable CO_2 emissions for BDE has been reported. Saravanan et al. (2020) have tested exhaust gas recirculation (EGR) on ethanol-diesel blends in CI engines and reported that an increase in EGR reduces the efficiency, reduces peak pressure (P_{max}) and heat release rate (HRR) compared to ethanol-diesel fuelled conditions. This is due to a drop in thermal efficiency caused by replacing a large quantity of combustion air with recirculated exhaust gases. Verma et al. (2020) have tested hybrid fuel (ethanol-methanol-diesel-microalgae blends) in a CI engine and observed that the presence of spirulina microalgae shortens the ignition delay while increasing efficiency and decreasing the exhaust gas temperature. Another study shows that EGR is an effective technique to reduce the oxides of nitrogen (NO_x) associated with alcohol fuel where Duan et al. (2021) have observed an advancement in injection timing which advances the HRR and HRR_{max} decreases with increasing the EGR ratio. Li et al. (2019) reported that utilization of isopropanol-butanol-ethanol blend with diesel (IBE30) reduces HRR_{max} value and NO_x emissions. Higher molecular weight alcohols (butanol, pentanol, hexanol, or octanol) blends with Ultra-low sulfur diesel (ULSD) were tested and investigated by Rajesh Kumar et al. (2016). They observed that butanol served promising with the lowest emission levels of smoke, NO_x, and CO; however, it showed longer ignition delay, higher P_{max} and HRR_{max}, and shortest combustion duration among tested alcohols. Neat methanol was tested for varying intake air temperature, resulting in longer ignition delay and lowering the P_{max} with decreasing the intake air temperature (Zincir et al. 2019). Shamun et al. (2018) examined NO_x and Indicated mean effective pressure (IMEP) in a Partially premixed combustion (PPC) engine with methanol and *iso*-octane. They compared the charge-cooling impact of methanol and *iso*-octane using a single injection technique and observed that methanol resulted in lower nitrogen oxide (NO_x) emissions. Higher latent heat of methanol vaporization results in lower NO_x emission (Shamun et al. 2018). It has been observed that a reduction in the fuel injection rate reduced the NO_x and CO_2 emissions and increased the smoke and HC emissions (Sayin et al. 2009). According to the research summarised in the literature survey, it can be noted that biodiesel, methanol, ethanol, and butanol have been tested and compared extensively with varying volume fractions under various injection techniques, blends dual fuel modes, and PPC techniques. Researchers found that the lower carbon-to-hydrogen (C/H) ratio and inherent oxygen content may be the reasons for reduced exhaust emissions. In this framework, this study provides a thorough literature review on comparing different oxygenated alternative fuels (biodiesel, methanol, ethanol, and butanol) blended with diesel, keeping almost constant the oxygen fraction of blends. The analysis includes the comparative study of the main combustion analysis parameters and efficiencies. The first part of the chapter discusses the characteristics of oxygenated alternative fuels. The later part of the chapter discusses the combustion and performance behaviours of oxygenated fuels. This discussion emphasized in-cylinder pressure, heat release rate (HRR), ignition delay, and engine efficiencies.

4.2 Fuel Properties of Oxygenated Alternative Fuels

This section discusses the selected fuel properties of diesel and some of the alternative oxygenated fuels. Table 4.1 reports various fuel properties such as density, kinematic viscosity, heating value, latent heat of vaporization, autoignition temperature, carbon, hydrogen, oxygen content, and cetane number for different alcohols and biodiesels. Alcohols (methanol, ethanol and butanol) show a lower carbon-to-hydrogen ratio (up to 50%) with 50–21% inherent oxygen content (*m/m*). Primarily, alcohols and biodiesel are being used as oxygenated alternative fuels due to the ease of their production, availability, and renewable nature. Alcohol properties significantly change with increasing molecular weight, which can be observed in Table 4.1. For example, the kinematic viscosity of methanol, ethanol, and butanol are 0.59 mm^2/s, 1.2 mm^2/s and 2.51 mm^2/s, respectively. Similarly, heating values are 20.2, 27.1, and 33.1 MJ/kg for methanol, ethanol, and butanol.

Table 4.1 show that the stoichiometric air-fuel ratios for methanol, ethanol, and butanol are 6.6, 9.0, and 11.1 kg of air/kg of fuel, which are significantly lower than diesel, i.e., 14.49 kg of air/kg of fuel. Lower air requirement for chemically complete combustion increases the chances for a higher degree of complete combustion and reduces unburnt hydrocarbon emission. Alcohols show a lower cetane number, higher autoignition temperature, and higher latent heat of evaporation, increasing the ignition delay and decreasing the autoignition characteristics of the fuel.

4.3 Combustion and Performance of Oxygenated Fuels

CI engines are more widespread than SI engines in the farming and heavy-duty transportation sectors owing to their higher durability, performance, and fuel economy. Using oxygenated fuels (mainly alcohols and biodiesels) in CI engines

Table 4.1 Fuel properties of diesel, biodiesel, methanol, ethanol, and butanol (Verhelst et al. 2019; Labeckas et al. 2017a; Pipicelli et al. 2022; Zhang et al. 2011; Alemahdi and Tuner 2020)

Property parameters	Diesel	Methanol	Ethanol	Butanol
Density at 15 °C, kg/m^3	832.7	790	785	810
Kinematic viscosity at 40 °C, mm^2/s	2.13	0.59	1.2	2.51
Net heating value, MJ/kg	43.0	20.2	27.1	33.1
Autoignition temperature, °C	250	363	385	343
Latent heat of vaporization, kJ/kg	270	1178	838	585
Flashpoint, °C	57	12	12	37
Oxygen, max wt%	0	50	34.7	21.6
Carbon, max wt%	86.08	37.5	52.1	64.8
Hydrogen, wt%	12.99	12.5	13.0	13.6
Carbon-to-hydrogen ratio (C/H)	6.5	3.0	4.0	4.7
Stoichiometric air-fuel ratio, kg/kg	14.49	6.6	9.0	11.1
Cetane number	51.4	4	8	25

might offer a more environmentally friendly and viable alternative to conventional diesel (Sahu and Shukla 2022). Despite having renewable nature and superior fuel properties, specific challenges still need to be investigated for alcohols and biodiesels. Since alcohol has a lower cetane number (CN) value than diesel, the autoignition temperature value is higher, resulting in a longer ignition delay in contrast to diesel. In order to avoid this, it is essential to make modifications to the fuel design, and injection strategy or adding ignition improver additives (Sahu and Shukla 2022). Also, hardware modifications like increasing the compression ratio, increasing the intake temperature, and optimizing the EGR rate are ways to make combustion possible for alcohols. In this section, the combustion and performance of oxygenated fuels are discussed and compared with conventional diesel. Jamrozik et al. have tested butanol diesel fuel for oxygen fractions of 0–14.2% (m/m), corresponding to 60% butanol energy fraction (66% in mass fraction) in diesel. Figure 4.1 shows the carbon-hydrogen and oxygen mass fractions of prepared butanol blends. Higher peak pressure and HRR (Fig. 4.2) were observed for butanol blends (till 30% energy fraction of butanol). Further increase in butanol energy fraction showed reduced peak pressure and reduced HRR with retarded location of P_{max} and HRR_{max}. It is attributable primarily to the rising latent heat of vaporization of the blend, which deteriorates the fuel evaporation process, and the low cetane number of n-butanol, which weakens the autoignition characteristics.

Ghadikolaei et al. (2018) investigated CI engines' combustion, performance, and emissions parameters fueled with five distinct alternative fuel mixtures. They used seven different fuels, diesel (D), waste cooking oil (B), methanol (M), ethanol (E), 2-propanol (Pr), n-butanol (Bu), and n-Pentanol (Pe) labeled as DB, DBM, DBE, DBPr, DBBu, and DBPe. These blends were prepared so that the oxygen content of 5% can be retained in the blend and the calorific value remained constant. They

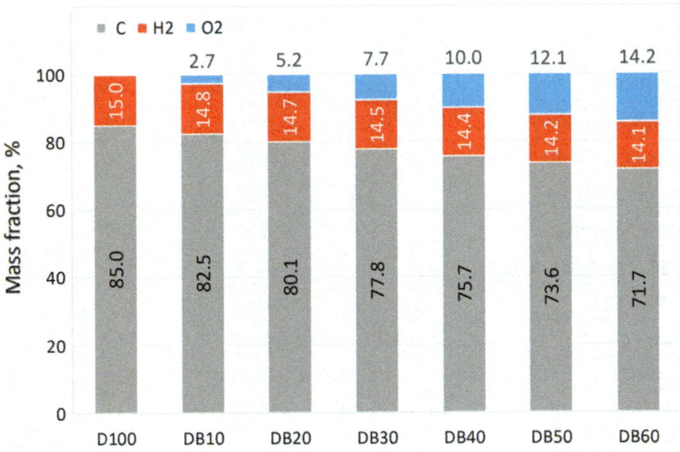

Fig. 4.1 Carbon, hydrogen, and oxygen mass fractions at different diesel-butanol blend ratios (Jamrozik et al. 2021). Copyright © 2021 MDPI. Reprinted with copyright permission

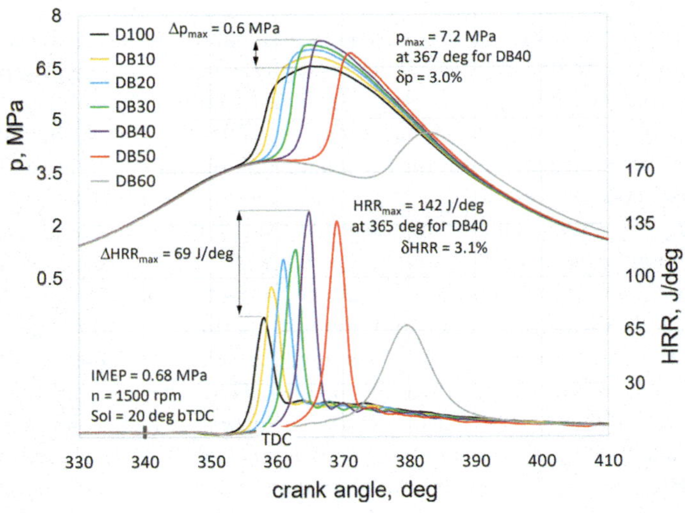

Fig. 4.2 In-cylinder pressure and HRR traces at part-load at different diesel-butanol blend ratios (Jamrozik et al. 2021). Copyright © 2021 MDPI. Reprinted with copyright permission

reported that the peak in-cylinder pressure of all tested blends shows a 1% higher P_{max} than conventional. Peak HRR was elevated by 22.1% for higher alcohols and 14.8% and 5.9% for lower alcohols (DBE and DBM), respectively. Moreover, 5% oxygenated biodiesel showed a lower peak of HRR by 3%. Labeckas et al. (2017a) have studied the variation of ethanol oxygen fraction and biodiesel oxygen fraction for five different oxygen fractions (0, 0.91, 1.81, 2.71, 3.61 and 4.52%). They observed a slightly higher ignition delay for ethanol blends than biodiesel blends when oxygen

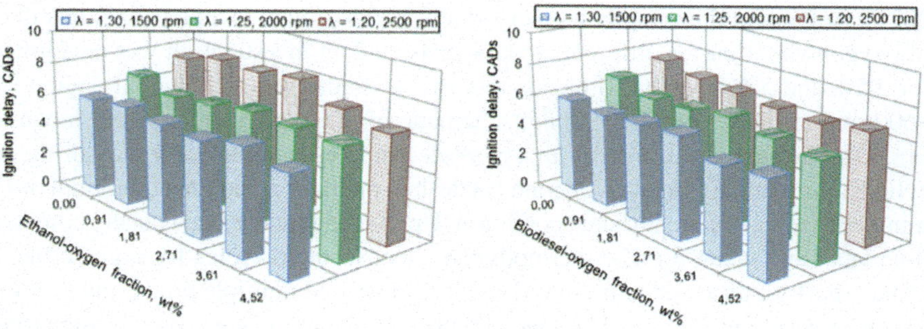

Fig. 4.3 Ignition delay for ethanol and biodiesel oxygen fraction (Labeckas et al. 2017a). Copyright © 2017 Elsevier. Reprinted with copyright permission

fraction (*m/m*) was maintained at 3.61%. However, ethanol showed 1.7 CAD higher ignition delay for 2.71% oxygen at $\lambda = 1.2$ (Fig. 4.3) at 2500 rpm of engine speed.

Nour et al. (2019) tested 10% pentanol/octanol with ethanol—diesel blend (10:10:80) and reduces the peak in-cylinder pressure compared to diesel. Diffusion phase of HRR is enhanced with a diminished premixed phase. Break specific fuel consumption (BSFC) was observed to be higher for pentanol substitution and lower for octanol substitution. Both pentanol and octanol report better stability for blends of ethanol-diesel with significantly lower NO_x and CO_2 emissions. Guedes et al. (Mendes Guedes et al. 2018) have suggested advancing the fuel injection timing by 1 CAD for an increase of every 5% ethanol in the blends of diesel-biodiesel-ethanol, which resulted in a maximum increase of 9% in peak pressure. Prakash et al. (2018) examined the performance of ternary mixtures of bioethanol, diesel, and castor oil. 30% bioethanol demonstrated a longer ignition delay, leading to higher fuel collection and a rapid increase in in-cylinder pressure and HRR. Shamun et al. (2018) tested diesel, biodiesel, and ethanol in a light-duty diesel engine; they underlined the fact that greater oxygen content in the fuel leads to complete combustion and elevates the temperature of the combustion process. Ignition delay was increased when ethanol was present, which resulted in a greater amount of fuel being consumed during the premixed phase of combustion and an increase in the creation of thermal NO_x. Pradelle et al. (2019) experimented with a Euro 3 CI engine and reported that the decreased density and heating value of ethanol caused a rise in specific fuel consumption of around 2% for every 5% increase in that anhydrous ethanol was included in the blend. Ethanol possesses a higher latent heat of vaporization, heat capacity at constant pressure, and a lower cetane number than conventional fuel causes an increase in the ignition delay. Despite these limits, useful thermal energy increased with ethanol content. This improved the engine's efficiency, especially at low loads. Shrivastava et al. (2021) investigated that the ternary blend (Biodiesel 20%—Ethanol 10%-Diesel 70%) delivered superior performance than the conventional diesel. Their study shows a 2% reduction in break thermal efficiency (BTE), a 3% rise in BSFC, a 3% increase in EGT, a 0.86% increase in CO_2, a 0.02% reduction in CO, an 8% reduction in NO_x, and 12 ppm reduction in HC compared to baseline diesel (Shrivastava et al.

2021). Authors of this chapter too performed an experiment to understand the effect of oxygen variation for different alcohol fuels, and an investigation was performed for 2.5% and 5% oxygenated blends in a twin-cylinder water-cooled CI engine for 1600 rpm at 20 Nm engine loading, schematic of the experimental setup shown in Fig. 4.4. The engine was fitted with a piezoelectric pressure sensor for measuring in-cylinder pressure. Engine was operated at desired operating load for a minimum of 15 minutes to reach the steady state condition. Upon reaching the steady state, combustion data were taken for 120 thermodynamic consecutive cycles for a given blend of fuel. Performance and emissions data were measured multiple times, and average data were used for analysis. Methanol, ethanol, and butanol were used for preparing the oxygenated blend, named as M2.5, E2.5, and B2.5 for 2.5% oxygen *m/m* fraction and M5, E5, and B5 for 5% oxygen *m/m* fraction. It contains 5.2% methanol, 7.4% ethanol, and 11.7% butanol respectively for a 2.5% oxygenated blend and 10.3% methanol, 14.8% ethanol, and 23.3% butanol respectively for a 5% oxygenated blend. Figure 4.5 shows the measured in-cylinder pressure and calculated HRR for tested fuels at 20 Nm loading conditions. For 2.5% oxygenated case (Fig. 4.5a), pressure and HRR pattern were similar with slightly higher peak pressure for M2.5, HRR curve was observed to shift away from top dead center (TDC) for all three tested fuels, E2.5 resulted in 2 CAD retarded location of HRR_{max} compared to diesel. Figure 4.5b shows the pressure and HRR pattern for 5% oxygenated fuel. M5 showed the advanced location of peak pressure; however, E5 and B5 showed the retarded location of peak pressure than diesel. E5 and B5 also showed a higher ignition delay and higher HRR_{max} and compared to diesel, the location of HRR_{max} was also retarded by ~3 CAD.

Figure 4.6 shows the value of combustion phasing (CA05, CA50, and CA90 correspond to total heat release of 5, 50, and 90%), combustion duration, and BTE

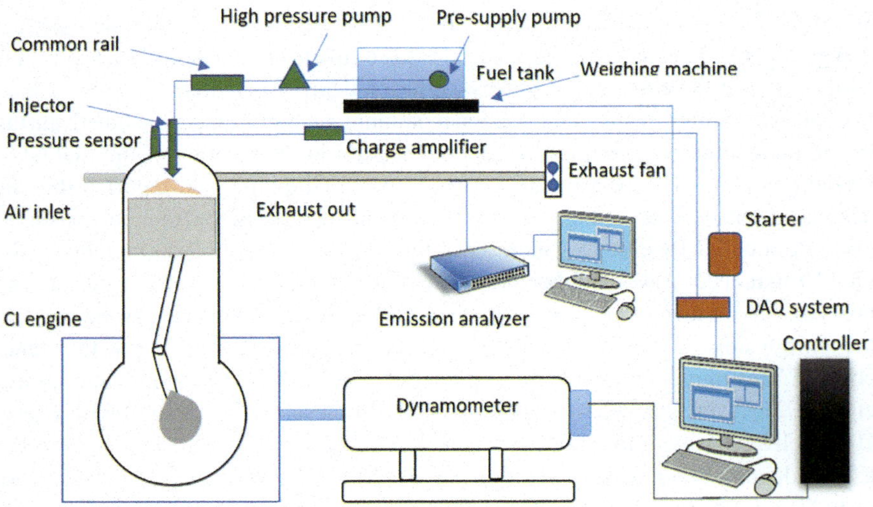

Fig. 4.4 Schematic representation of the experimental setup

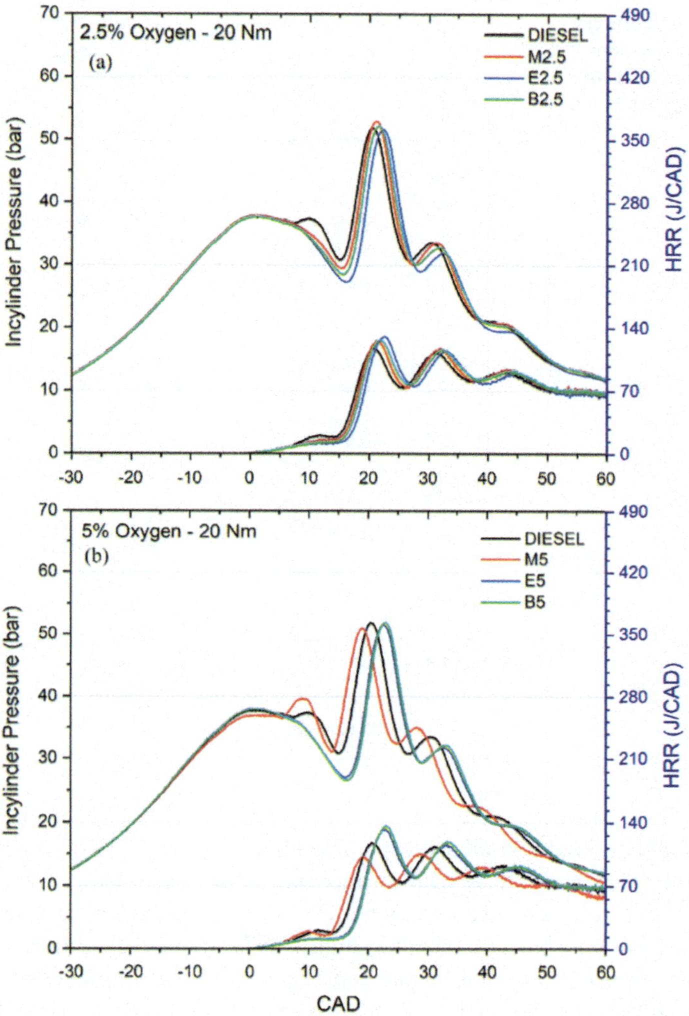

Fig. 4.5 In-cylinder pressure and HRR for 2.5 and 5% oxygenated fuel at 20 Nm

for 2.5 and 5% oxygenated blends at 20 Nm engine loading condition for 1600 rpm. It was noted that all three fuel CA05 locations were advances for 5% oxygenated fuel than 2.5% oxygenated fuel. A similar pattern was also observed for CA50 and CA90 locations. Early CA05 and CA90 of oxygenated fuel resulted in a lower combustion duration (lowered by ~2–3 CAD) for both 2.5 and 5% oxygenated fuel compared to baseline diesel. BTE was observed comparable for both 2.5 and 5% oxygenated fuel, with maximum efficiency of 22% was achieved for M5 and a minimum efficiency of 18.9% was achieved for E2.5 observed for the tested condition (Fig. 4.6).

Labeckas et al. (2017b) reported that BTE is higher after 2.71% oxygen fraction; further, it is intensified with a higher oxygen fraction for ethanol-oxygen mass content (shown in Fig. 4.7) for higher engine rpm of 2000 and 2500 rpm. Biodiesel oxygen

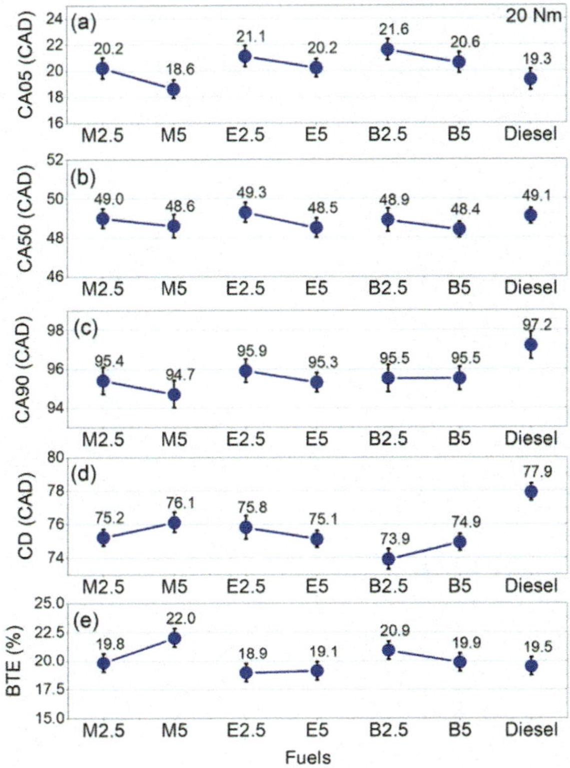

Fig. 4.6 Combustion phasing, combustion duration, and BTE for 2.5 and 5% oxygenated fuel at 20 Nm

mass fraction showed a similar comparable efficiency for 2500 rpm; moreover, it is a slightly increasing trend with an increase in oxygen fraction for 2000 rpm. Compared to both cases, ethanol oxygen fraction is more beneficial than biodiesel oxygen fraction as it showed a relatively higher BTE for tested conditions. Variations of BTE are shown in Fig. 4.8 for the experiment performed by Ghadikolaei et al. (2018) for different fuel blends with a 5% oxygen mass fraction. Figure 4.8 shows that the highest BTE of the tested fuels was observed at 199.5 Nm, resulted in the best engine performance. Low load operations caused lower combustion temperatures, which possibly resulted to incomplete combustion. At high loads, fuel/air ratio is richer, and combustion temperature is higher; however, there is insufficient time for mixing, leading to incomplete combustion and a decrease in BTE. Five loads show that the BTE of DBM (3.5%) and DBBu (1.5%) is higher than that of diesel fuel, while the BTE of other blended fuels is comparable [Ghadikolaei et al. (2018)]. Lower fuel viscosity, improved fuel atomization, and higher oxygen content etc. Improved BTE by enhancing the combustion process, which transforms fuel chemical energy into useful engine work. DBM showed the highest BTE of all the evaluated fuels. Among the investigated alcohols, methanol indicated the shortest chain and lowest molecular weight, which promotes better combustion and the highest BTE (Fig. 4.8). Methanol showed the lowest boiling point, which lowered heat losses and raises BTE.

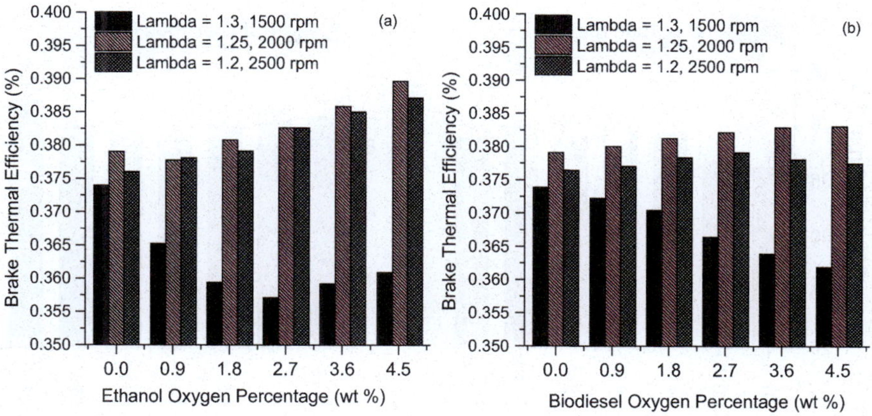

Fig. 4.7 BTE for an ethanol-oxygen mass fraction (left) and biodiesel oxygen mass fraction (right) (Labeckas et al. 2017b). Copyright © 2017 Elsevier. Redrawn with copyright permission

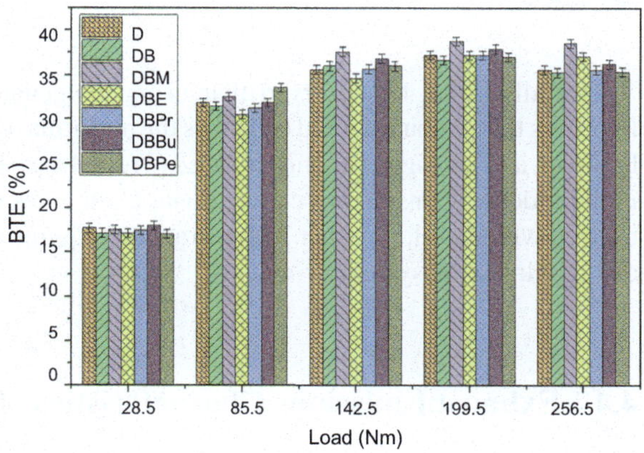

Fig. 4.8 BTE for 5% oxygen mass fraction for different alternative fuels (diesel (D), waste cooking oil biodiesel (B), methanol (M), ethanol (E), 2-propanol (Pr), n-butanol (Bu), and n-pentanol (Pe)) (Ghadikolaei et al. 2018). Copyright © 2018 Elsevier. Redrawn with copyright permission

Jamrozik et al. (2021) reported that DB40 (40% butanol energy fraction in diesel) showed maximum efficiency of 41.3% (Fig. 4.9) with the lowest energy consumption among the tested butanol-diesel blends which is approximately 11% higher than diesel. Higher efficiency is achieved possibly due to the acceleration of chemical oxidation with the higher butanol fraction, however, a further increase in butanol fraction shifted the HRR curve away from TDC resulting in lower useful work and lower efficiency.

Nour et al. (2019) demonstrated that pentanol (10%) increased the BTE by 3% at lower loads for ethanol-diesel blends (10:10:80). Adding octanol to diesel raises BTE by 5% at low and mid loads, whereas at higher loads, diesel is 1% more efficient. The ternary blends' oxygen concentration improved combustion and boosted BTE, especially with octanol.

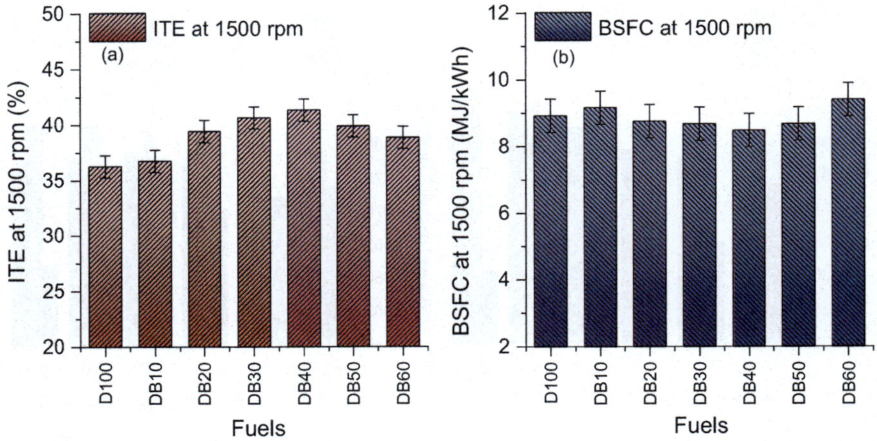

Fig. 4.9 Indicated thermal efficiency (ITE) and indicated specific energy consumption (ISEC) for diesel-butanol blends (Jamrozik et al. 2021). Copyright © 2021 MDPI. Redrawn with copyright permission

Overall, it can be observed that using oxygenated fuel in a limited fraction improves the combustion and enhances the performance characteristics of the engine, however, a much higher fraction also led to retard the peak pressure and offered higher ignition delay. Considering conventional CI engines up to 15% of alcohols can be utilized with diesel, a further higher fraction requires alterations in engine hardware and fuel injection systems.

4.4 Exhaust Emissions Characteristics of Oxygenated Fuels

Locally peak in-cylinder temperature locations and a high degree of pyrolysis lead to NO_x and PM emissions in CI engines. It has been reported that alcohols are beneficial in reducing PM emissions in diesel combustion drastically with almost no/slight change in NO_x emissions. Jamrozik et al. (2021) tested butanol blends with oxygen percentages of 2.7–14.2% (shown in Fig. 4.1, mentioned as DB10 to DB60). They reported NO_x is an increasing trend from DB10 to DB40; moreover, DB60 shows significantly lower NO_x (Fig. 10a) compared to DB50 and similar to diesel D100. This is mainly due to the higher amount of alcohol present in DB50 and DB60 blends, resulting in higher latent heat of vaporization, and reducing the peak cylinder temperature lowers the NO_x formation. CO and CO_2 emissions are in decreasing trend with an increase in butanol fraction; CO shows a sharp decline (Fig. 10b) with an increase in oxygen fraction due to improving combustion quality and lower C/H ratio of overall blends with an increase in butanol.

Kumar et al. (2022) tested the particulate matter (PM) in diesel engines fueled with ethanol-diesel blends and reported that 5–10% of ethanol with diesel fuel reduced

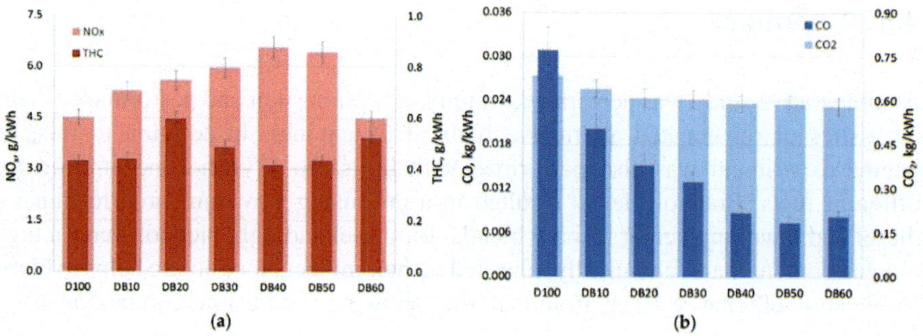

Fig. 4.10 NO_x, THC, CO, and CO_2 emissions for different butanol-diesel blends (Jamrozik et al. 2021). Copyright © 2021 MDPI. Reprinted with copyright permission

significant particulate emissions. 50% reduction in PM was observed using ethanol compared to diesel fuel. This may be due to the use of ethanol fuel with diesel improving the combustion and increasing the combustion temperature which resulted in a reduction in soot formation. Smoke results shown in Fig. 4.11 were performed by authors, which show the emission characteristics of diesel engine fueled with different alcohols with the same oxygen fraction (2.5 and 5% inherent oxygen). Butanol blends with 5% oxygen mass fraction showed a higher reduction compared to diesel and 2.5% butanol oxygenated fuel blend. It was observed that E5 and B5 resulted in significantly higher HRR_{max} and lower combustion duration (Figs. 4.5 and 4.6), which showed that a higher degree of complete combustion resulted in lower smoke emissions.

Fig. 4.11 Smoke emissions for 2.5 and 5% oxygenated alcohol fuel (methanol, ethanol, butanol blends) at 1600 rpm, 20 Nm loading

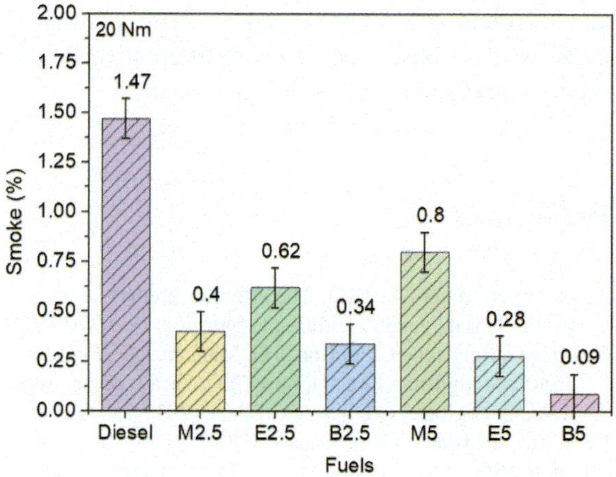

4.5 Summary

A comparative study was performed to study the combustion and performance characteristics of oxygenated alternative fuels in CI engines. In addition to this, an engine experiment was also performed with the same oxygen concentration for different fuels. Butanol blends resulted in a promising way compared to mineral diesel and lower molecular alcohol blends. Biodiesel and light alcohols have a high oxygen content that is chemically inherited in their molecular structure helps to boost combustion efficiency. Apart from that, the following conclusions can be drawn:

- Alcohols (methanol, ethanol, and butanol) with the same oxygen mass fractions showed different results due to different volume fractions of alcohol, ethanol, and butanol. Ethanol and butanol showed higher HRR than methanol blends (keeping the oxygen mass fraction same). Higher volume fraction of ethanol/butanol is required to make the same oxygen mass fraction.
- Methanol blends showed improved efficiency due to the shortest chain and lowest molecular weight among investigated alcohols, leading to better combustion and higher thermal efficiency.
- In general, higher alcohol fraction addition showed increased in-cylinder pressure and higher HRR, which may be damped using ignition improver additives, or higher alcohols such as the use of pentanol or octanol.
- The incorporation of alcohol into diesel fuel led to a considerable decrease in CO and smoke emissions.

The present study indicated that alcohols are promising alternative fuels in terms of PM reduction. Alcohols (especially butanol) indicated significantly higher efficiency compared to diesel due to better combustion. Overall, utilization of these alcohol fuels is also beneficial in terms of total CO_2 reduction in the environment as it is considered as green alternative fuel produced by first-generation or second-generation vegetation/agro products.

References

Alemahdi N, Tuner M (2020) The effect of 2-ethyl-hexyl nitrate on HCCI combustion properties to compensate ethanol addition to gasoline. Fuel 270:117569

Beatrice C, Belgiorno G, Di Blasio G, Mancaruso E, Sequino L, Vaglieco BM (2017) Analysis of a prototype high-pressure "Hollow Cone Spray" diesel injector performance in optical and metal research engines. SAE Technical Paper 2017-24-0073

Belgiorno G, Blasio GD, Beatrice C (2019) Advances of the natural gas/diesel RCCI concept application for light-duty engines: comprehensive analysis of the influence of the design and calibration parameters on performance and emissions. In: Natural gas engines. Springer, pp 251–266

Belgiorno G, Boscolo A, Dileo G, Numidi F, Pesce FC, Vassallo A et al (2020) Experimental study of additive-manufacturing-enabled innovative diesel combustion bowl features for achieving ultra-low emissions and high efficiency. SAE Int J Adv Curr Pract Mob 3(2020–37–0003):672–684

Di Blasio G, Belgiorno G, Beatrice C, Fraioli V, Migliaccio M (2015) Experimental evaluation of compression ratio influence on the performance of a dual-fuel methane-diesel light-duty engine. SAE Int J Engines 8(5):2253–2267

Di Blasio G, Belgiorno G, Beatrice C (2017) Parametric analysis of compression ratio variation effects on thermodynamic, gaseous pollutant and particle emissions of a dual-fuel CH 4-diesel light duty engine. SAE Technical Paper 2017-01-0764

Di Blasio G, Beatrice C, Ianniello R, Pesce FC, Vassallo A, Belgiorno G et al (2019a) Balancing hydraulic flow and fuel injection parameters for low-emission and high-efficiency automotive diesel engines. SAE Int J Adv Curr Pract Mob 2(2019-24-0111):638–652

Di Blasio G, Vassallo A, Pesce FC, Beatrice C, Belgiorno G, Avolio G (2019b) The key role of advanced, flexible fuel injection systems to match the future CO_2 targets in an ultra-light mid-size diesel engine. SAE Int J Engines 12(2):129–144

Di Luca G, Pipicelli M, Ianniello R, Belgiorno G, Di Blasio G (2022) Alcohol fuels in spark ignition engines. In: Di Blasio G, Agarwal AK, Belgiorno G, Shukla PC (eds) Application of clean fuels in combustion engines. Springer Nature Singapore, Singapore, pp 33–54

Dimitrakopoulos N, Belgiorno G, Tuner M, Tunestal P, Di Blasio G, Beatrice C (2017) PPC operation with low ron gasoline fuel. A study on load range on a euro 6 light duty diesel engine. Jpn Soc Mech Eng C308

Dimitrakopoulos N, Belgiorno G, Tunér M, Tunestål P, Di Blasio G (2019) Effect of EGR routing on efficiency and emissions of a PPC engine. Appl Therm Eng 152:742–750

Duan X, Xu Z, Sun X, Deng B, Liu J (2021) Effects of injection timing and EGR on combustion and emissions characteristics of the diesel engine fuelled with acetone–butanol–ethanol/diesel blend fuels. Energy 231:121069

Fraioli V, Beatrice C, Di Blasio G, Belgiorno G, Migliaccio M (2017) Multidimensional simulations of combustion in methane-diesel dual-fuel light-duty engines. SAE Technical Paper 2017-01-0568

Ghadikolaei MA, Cheung CS, Yung K-F (2018) Study of combustion, performance and emissions of diesel engine fueled with diesel/biodiesel/alcohol blends having the same oxygen concentration. Energy 157:258–269

Heuser B, Kremer F, Pischinger S, Rohs H, Holderbaum B, Körfer T (2016) An experimental investigation of dual-fuel combustion in a light duty diesel engine by in-cylinder blending of ethanol and diesel. SAE Int J Engines 9(1):11–25

Ianniello R, Belgiorno G, Di Luca G, Beatrice C, Di Blasio G (2021) Ethanol in dual-fuel and blend fueling modes for advanced combustion in compression ignition engines. In: Alcohol as an alternative fuel for internal combustion engines. Springer, pp 5–27

Jamrozik A, Tutak W, Grab-Rogaliński K (2021) Combustion stability, performance and emission characteristics of a CI engine fueled with diesel/n-butanol blends. Energies 14(10)

Krishna SM, Abdul Salam P, Tongroon M, Chollacoop N (2019) Performance and emission assessment of optimally blended biodiesel-diesel-ethanol in diesel engine generator. Appl Therm Eng 155:525–533

Kumar A, Sahu TK, Shukla PC (2022) Design and fabrication of a partial flow dilution tunnel for particulate mass sampling in an ethanol blended CI engine, 1st edn, vol 1042. IOP Publishing, p 012006

Labeckas G, Slavinskas S, Kanapkienė I (2017a) The individual effects of cetane number, oxygen content or fuel properties on the ignition delay, combustion characteristics, and cyclic variation of a turbocharged CRDI diesel engine—Part 1. Energy Convers Manag 148:1003–1027

Labeckas G, Slavinskas S, Kanapkienė I (2017b) The individual effects of cetane number, oxygen content or fuel properties on performance efficiency, exhaust smoke and emissions of a turbocharged CRDI diesel engine—Part 2. Energy Convers Manag 149:442–466

Li G, Lee TH, Liu Z, Lee CF, Zhang C (2019) Effects of injection strategies on combustion and emission characteristics of a common-rail diesel engine fueled with isopropanol-butanol-ethanol and diesel blends. Renew Energy 130:677–686

Mendes Guedes AD, Leal Braga S, Pradelle F (2018) Performance and combustion characteristics of a compression ignition engine running on diesel-biodiesel-ethanol (DBE) blends—Part 2: optimization of injection timing. Fuel 225:174–183

Monsalve-Serrano J, Belgiorno G, Di Blasio G, Guzmán-Mendoza M (2020) 1D simulation and experimental analysis on the effects of the injection parameters in methane–diesel dual-fuel combustion. Energies 13(14)

Nour M, Attia AMA, Nada SA (2019) Improvement of CI engine combustion and performance running on ternary blends of higher alcohol (pentanol and octanol)/hydrous ethanol/diesel. Fuel 251:10–22

Pedrozo VB, May I, Guan W, Zhao H (2018) High efficiency ethanol-diesel dual-fuel combustion: A comparison against conventional diesel combustion from low to full engine load. Fuel 230:440–451

Pipicelli M, Luca GD, Ianniello R, Gimelli A, Beatrice C (2022) Alcohol fuels in compression ignition engines. In: Application of clean fuels in combustion engines. Springer, pp 9–31

Pradelle F, Leal Braga S, Fonseca de Aguiar Martins AR, Turkovics F, Nohra Chaar Pradelle R (2019) Performance and combustion characteristics of a compression ignition engine running on diesel-biodiesel-ethanol (DBE) blends—Potential as diesel fuel substitute on an Euro III engine. Renew Energy 136:586–598

Prakash T, Geo VE, Martin LJ, Nagalingam B (2018) Effect of ternary blends of bio-ethanol, diesel and castor oil on performance, emission and combustion in a CI engine. Renew Energy 122:301–309

Rajesh Kumar B, Saravanan S, Rana D, Nagendran A (2016) A comparative analysis on combustion and emissions of some next generation higher-alcohol/diesel blends in a direct-injection diesel engine. Energy Convers Manag 119:246–256

Sahu TK, Sarkar S, Shukla PC (2021a) Combustion investigation of waste cooking oil (WCO) with varying compression ratio in a single cylinder CI engine. Fuel 283:119262

Sahu TK, Kshatri R, Shukla PC (2021b) Impact of ethanol on combustion, performance, and emission characteristics of diesel engine. In: Alcohol as an alternative fuel for internal combustion engines. Springer, pp 251–266

Sahu TK, Shukla PC (2022a) Effect of Inherent oxygen mass fraction of alcohol blends with diesel on combustion and emission parameters. Environ Prog Sustain Energy e14030

Sahu TK, Shukla PC (2022b) Combustion and emission characteristics of butanol-diesel blend (B15) doped with diethyl ether, diglyme and ethyl diglyme in a CRDI diesel engine. SAE Technical Paper 2022-01-1073.

Sahu TK, Sahu VK, Mondal A, Shukla PC, Gupta S, Sarkar S (2022a) Investigation of sugar extraction capability from rice paddy straw for potential use of bioethanol production towards energy security. Energy Sour Part A: Recovery Utiliz Environ Effects 44(1):272–286

Sahu TK, Shukla PC, Belgiorno G, Maurya RK (2022b) Alcohols as alternative fuels in compression ignition engines for sustainable transportation: a review. Energy Sour Part A: Recovery Utiliz Environ Effects 44(4):8736–8759

Sahu TK, Kshatri R, Kumar A, Shukla PC (2022c) Combustion stability investigation of ethanol blends (E05, E10) in a twin-cylinder ci engine. SAE Technical Paper 2022-01-0521

Sahu VK, Sahu TK, Shukla PC (2022d) Fuel injection strategies for alcohol utilization in combustion engines. In: Application of clean fuels in combustion engines. Springer, pp 55–69

Saravanan P, Kumar NM, Ettappan M, Dhanagopal R, Vishnupriyan J (2020) Effect of exhaust gas re-circulation on performance, emission and combustion characteristics of ethanol-fueled diesel engine. Case Stud Thermal Eng 20:100643

Sayin C, Ilhan M, Canakci M, Gumus M (2009) Effect of injection timing on the exhaust emissions of a diesel engine using diesel–methanol blends. Renew Energy 34(5):1261–1269

Sequino L, Belgiorno G, Di Blasio G, Mancaruso E, Beatrice C, Vaglieco BM (2018) Assessment of the new features of a prototype high-pressure "Hollow Cone Spray" diesel injector by means of engine performance characterization and spray visualization. SAE Technical Paper 2018-01-1697

Shamun S, Belgiorno G, Di Blasio G, Beatrice C, Tunér M, Tunestål P (2018) Performance and emissions of diesel-biodiesel-ethanol blends in a light duty compression ignition engine. Appl Therm Eng 145:444–452

Shamun S, Belgiorno G, Di Blasio G (2020) Engine parameters assessment for alcohols fuels application in compression ignition engines. In: Singh AP, Sharma YC, Mustafi NN, Agarwal AK (eds) Alternative fuels and their utilization strategies in internal combustion engines. Springer Singapore, Singapore, pp 125–139

Shamun S, Zincir B, Shukla P, Garcia Valladolid P, Verhelst S, Tunér M (2018) Quantification and analysis of the charge cooling effect of methanol in a compression ignition engine utilizing PPC strategy, vol 51982. American Society of Mechanical Engineers, p V001T02A7

Shrivastava K, Thipse SS, Patil ID (2021) Optimization of diesel engine performance and emission parameters of Karanja biodiesel-ethanol-diesel blends at optimized operating conditions. Fuel 293:120451

Verhelst S, Turner JWG, Sileghem L, Vancoillie J (2019) Methanol as a fuel for internal combustion engines. Prog Energy Combust Sci 70:43–88

Verma TN, Nashine P, Chaurasiya PK, Rajak U, Afzal A, Kumar S et al (2020) The effect of ethanol-methanol-diesel-microalgae blends on performance, combustion and emissions of a direct injection diesel engine. Sustain Energy Technol Assess 42:100851

Zhang ZH, Tsang KS, Cheung CS, Chan TL, Yao CD (2011) Effect of fumigation methanol and ethanol on the gaseous and particulate emissions of a direct-injection diesel engine. Atmos Environ 45(11):2001–2008

Zincir B, Shukla P, Shamun S, Tuner M, Deniz C, Johansson B (2019) Investigation of effects of intake temperature on low load limitations of methanol partially premixed combustion. Energy Fuels 33(6):5695–5709

Chapter 5
Functional Use-Based Positioning of Conventional Vehicles in Conjunction with Alternate Low-Emission Fuels

Kumar Saurabh and Rudrodip Majumdar

Abstract India envisages energy diversification and transition to a cleaner fuel mix for the road transport segment, which contributes to about 75% of the country's total CO_2 emissions from the transport sector. In this pursuit, electric vehicles are pitted as a '*one size fits all*' solution to all the problems posed by the current fossil fuel-based transport in the country. The current vehicle fleet running in India is dominated by the Internal Combustion Engine (ICE)-based products (powered by mainly petrol and diesel), and the trend is likely to continue in the near-to-medium term. Therefore, it is necessary to consider alternate fuels for ICE-based vehicles to achieve decarbonization in India's road transport sector. This study discusses the ICE-based alternate fuel options (natural gas, auto-gas and hydrogen) in the light of 28 identified parameters under the 4A framework of energy security, encompassing technical availability, resource availability, infrastructure accessibility, price affordability, social acceptability and environmental acceptability. The 4A framework analysis is carried out to assess the large-scale deployability of the alternate fuel options. The learnings from a few prominent global experiences (compressed natural gas in Argentina, liquefied natural gas in China, auto-gas in Turkey, ethanol in Brazil, biodiesel in Indonesia and hydrogen research across the globe) have been imbibed in the mapping of the possibilities in the Indian context, keeping in mind the diverse functional uses of different vehicle segments within the road transport sector. Biofuel blends are deployable in the short term for all vehicles running on conventional petrol or diesel, whereas the expansion of natural gas usage is constrained by the lack of availability and accessibility beyond a few nodes. The energy transition in the freight segment would need a complete overhaul of the ecosystem since hydrogen appears to be the most prominent alternate fuel in the medium-to-long term.

Keywords Alternate fuels · Internal combustion engine (ICE) · Energy transition · 4A framework

---

K. Saurabh · R. Majumdar (✉)
National Institute of Advanced Studies (NIAS), Indian Institute of Science Campus, Bengaluru, Karnataka 560012, India
e-mail: rudrodip@nias.res.in

K. Saurabh
Manipal Academy of Higher Education (MAHE), Manipal, Karnataka 576104, India

P. C. Shukla et al. (eds.), *Renewable Fuels for Sustainable Mobility*, Energy, Environment, and Sustainability, https://doi.org/10.1007/978-981-99-1392-3_5

Abbreviations

2W	2-Wheeler
3W	3-Wheeler
4W	4-Wheeler
4W-P	4-Wheeler Passenger
4W-NP	4-Wheeler Non-Passenger
APM	Administered Price Mechanism
ARAI	Automotive Research Association of India
BEV	Battery Electric Vehicle
BIS	Bureau of Indian Standards
BPCL	Bharat Petroleum Corporation Limited
BS	Bharat Stage
CBM	Coal Bed Methane
CCUS	Carbon Capture, Utilization and Storage
CNG	Compressed Natural Gas
E&P	Exploration and Production
EBP	Ethanol Blended Petrol
FAME	Faster Adoption and Manufacturing of (Hybrid &) Electric Vehicles
FCEV	Fuel Cell Electric Vehicle
FFSHEV	Flex Fuel Strong Hybrid Electric Vehicle
FFV	Flex Fuel Vehicle
FSRU	Floating Storage Regasification Unit
FV	Freight Vehicle
GAIL	Gas Authority of India Limited
GoI	Government of India
GHG	Green House Gas
GST	Goods and Services Tax
HDT	Heavy Duty Truck
HPCL	Hindustan Petroleum Corporation Limited
HV-A	Heavy Vehicle—Agriculture
HV-B	Heavy Vehicle—Bus
HV-C	Heavy Vehicle—Construction
HV-M	Heavy Vehicle—Municipality
ICD	Inland Container Depot
ICE	Internal Combustion Engine
ICV	Intermediate Commercial Vehicle
IOCL	Indian Oil Corporation Limited
IPP	Import Parity Price
IRENA	International Renewable Energy Agency
IS	Indian Standard
LCV	Light Commercial Vehicle
LNG	Liquefied Natural Gas
LPDI	Liquid Phase Direct Injection

LPG	Liquefied Petroleum Gas
LPI	Liquid Phase Injection
M&HCV	Medium and Heavy Commercial Vehicle
MoPNG	Ministry of Petroleum and Natural Gas
MoRTH	Ministry of Road Transport and Highways
MMT	Million Metric Tonnes
MPV	Multi-Purpose Vehicle
NTPC	National Thermal Power Corporation
OIL	Oil India Limited
ONGC	Oil and Natural Gas Corporation
PLI	Production Linked Incentive
PM	Particulate Matter
PPAC	Petroleum Planning and Analysis Cell
PSU	Public Sector Undertaking
R&D	Research and Development
RBI	Reserve Bank of India
RIL	Reliance Industries Limited
SDG	Sustainable Development Goal
SMR	Steam Methane Reformation
SUV	Sports Utility Vehicle
TERI	The Energy Research Institute
TTW	Tank to Wheel
UT	Union Territory
VRDE	Vehicles Research & Development Establishment
WTT	Well to Tank

Symbols

₹/kg	Rupee per kg
₹/l	Rupee per liter
BCM	Billion Cubic Meter
g/kWh	gram per kilo Watt hour
g/mol	gram per mole
gCO_{2eq}/km	gram CO_2 equivalent per km
GW	Giga Watt
kWh/m^2/year	kiloWatt hour per square meter per year
MJ/kg	Mega Joule per kg
MJ/l	Mega Joule per liter
MJ/m^3	Mega Joule per cubic meter
MMT	Million Metric Tonne
MW	Mega Watt
psi	pound per square inch
TJ	Tera Joule

TMT	Thousand Metric Tonne
USD/liter	US Dollar per liter

5.1 Introduction

The rise of the electric vehicle segment in the Indian automobile industry has initiated the energy transition in the road transport sector of India with an intent to reduce CO_2 emissions. It is in line with Target 7.2 under the aegis of Sustainable Development Goals (SDG) set up in 2015 in Paris that aims to increase the share of renewable energy in the global energy mix (United Nations Environment Program Homepage 2022). Available open-source information suggests that 94% of the global energy requirement for the transport sector was met by liquid fuels (petrol, diesel, fuel oil, aviation jet fuel, etc.) in 2020, followed by natural gas (4%), electricity (1.5%) and Liquefied Petroleum Gas (LPG) (0.5%) (U.S. Energy Information Administration Homepage 2021). In India, among the liquid fuels, diesel and fuel oil contributed to 61% of the energy consumed by the transport sector (U.S. Energy Information Administration Homepage 2021). Gasoline and its ethanol-based blends contributed to 26% of the total energy consumed by the whole Indian transport sector. The other contributors comprise natural gas variants (6%), jet fuel (4%), electricity (2%) and LPG at 1% (U.S. Energy Information Administration Homepage 2021). In the Indian road transport sector, liquid fuels provided 94.5% of the energy required in 2019, followed by natural gas (3.5%), electricity (1.5%) and biomass (0.5%) (International Energy Agency Homepage 2022a).

Since almost all the alternatives to diesel and petrol have to start from a very low base in India, there is a limit to their expansion. Therefore, their potential to replace the incumbent Internal Combustion Engine (ICE)-based fuels (i.e. petrol and diesel) is also limited in the near term (Leach et al. 2020). Hence, there is a need to consider all the probable alternate fuels that can drive ICE-based vehicles in the short-to-medium-term future (by 2040). It is intuitive that the transition from the existing supply chain of diesel and petrol will remain slow due to high system-level inertia comprising the bottlenecks emanating from high investments required, lack of political will, inadequate consumer confidence, and limited incentives and support for the industry. Therefore, it is important to choose the appropriate alternate fuels such that the existing infrastructure can be utilized partially or completely for the transportation, storage, and distribution of such fuels to vehicle tanks. The choice of fuels should also consider the life cycle emissions associated with different stages of handling (Saurabh and Majumdar 2021). This will pave the way towards realizing the much-awaited net-zero carbon emission phase in the long term.

The energy consumed by the road transport sector of India contributed to 84.5% of the total energy consumed by the Indian transport sector (including all modes) in 2018–19, and this sector was responsible for 74% of the CO_2 emissions attributed to the transport sector in India (U.S. Energy Information Administration Homepage

2021; Ministry of Statistics and Program Implementation Homepage 2022; International Energy Agency Homepage 2022b). Further scrutiny of the Indian road transport sector reveals a clear distinction in the emission patterns based on the vehicle segment types. The two and three-wheeler segments in India account for about 80% of the on-road vehicles by number and contribute towards nearly 25% of CO_2 emissions emanating from the Indian road transport sector. Passenger vehicles (cars and buses) account for 15% of the total vehicles by number and are responsible for another 25% share of CO_2 emissions from the road transport sector. It is interesting to note that the rest 5% of the on-road vehicles are light, medium, and heavy freight-carrying commercial vehicles or non-passenger heavy-duty vehicles, yet they are responsible for more than 50% of the CO_2 emissions coming from the road transport sector in India (International Energy Agency Homepage 2022b; Ministry of Road Transport Highways Homepage 2022).

The Ministry of Road Transport and Highways (MoRTH) categorizes the automobiles running on Indian roads based on their functional capability and usage, as detailed in Sect. 5.2 (Ministry of Road Transport Highways Homepage 2021). This segmentation of the vehicles can be used to understand the vehicle usage trends from the viewpoint of the energy transition. It is highly likely that the transition of some of the vehicle categories to the cleaner alternatives would be much faster than the others. One key factor that is likely to drive this transition is the specificity of the functional use. It can be seen that a large share of the electric two-wheeler market in India is dominated by the scooters as compared to motorcycles in the initial transitional phase. Even though motorcycles have nearly two-thirds of the market share, the initial e-transition of scooters is attributed to their specific functional role (intra-city travel and short-distance rides) and specific characteristics such as being gender-neutral, rapidly accelerating, lightweight and convenient (Hardt and Bogenberger 2019). The pace of this transition is not solely guided by consumer choice but also by the enablers of the energy transition. The dimensions enabling the energy transition readiness (as highlighted by the Energy Transition Index 2020) are capital and investment, regulations and political commitment, institutions and governance, infrastructure and innovative business environment, human capital and consumer participation and energy system structure (World Economic Forum Homepage 2020). Therefore, it is necessary to assess the readiness of the alternate fuels vis-à-vis the vehicle categories to embrace each other so that the vehicle–fuel combinations are ready for deployment in the short-to-medium term. Consequently, the focus of this chapter is on the functional use-based positioning of ICE vehicles running on alternate fuels keeping in mind the larger goal of sustainable mobility transition in India.

Apart from petrol and diesel, currently Compressed Natural Gas (CNG) and Liquefied Petroleum Gas (LPG) or auto-gas are used in the ICE-based non-hybrid vehicles in India. Natural gas is also used in the form of Liquefied Natural Gas (LNG) as an alternate transport fuel globally. The first pilot project based on LNG in India happened in 2016. Kerala looks forward to converting its 3000 state-run buses from diesel to LNG in 2022 (CGD India Homepage 2021). Auto-gas is not actively promoted by the Government of India (GoI), but it has found its takers among the three-wheeler auto-rickshaws since the auto-gas driven three-wheelers require less maintenance and the fuel is easily available (Outlook India Homepage 2021). These vehicles are also fuel-efficient and cost-effective as compared to natural gas (Go Gas Homepage 2022). The advocacy group '*Indian Auto LPG Coalition*' considers LPG as the low-hanging alternate fuel for the short-to-medium term which can only be materialized through an active policy push from the GoI (Auto-gas.net Homepage 2022a). Ethanol and biodiesel are used as blends in petrol and diesel, respectively, as alternatives to pure fossil-based fuels, which indicates limited diversification of the energy sources using biomass (a renewable resource). In order to promote the use of biofuels in India, the Government of India (GoI) came up with the National Biofuels Policy in 2018 (Ministry of Petroleum and Natural Gas Homepage 2018). The GoI notified the Green Hydrogen Policy in February 2022 under the National Hydrogen Mission to promote the use of green hydrogen to diversify the energy mix of the country (Ministry of Power Homepage 2022). Big enterprises such as Indian Oil Corporation Limited (IOCL) and Reliance Industries Limited (RIL) have been supportive of the hydrogen-based economy in India by exhibiting a keenness to build hydrogen production facilities and distribution networks (Indian Oil Corporation Limited Homepage 2020a; Reliance Industries Limited Homepage 2021). Other major enterprises that are keen to enter the hydrogen-based energy ecosystem envisaged for the country in the near term include Gas Authority of India Limited (GAIL), Bharat Petroleum Corporation Limited (BPCL), National Thermal Power Corporation, Jindal Steel, Adani Power and Larsen & Toubro (Upstream Online Homepage 2022).

India has also deployed some of the above fuels on a pilot scale in bi-fuel vehicles where two fuels run the same Internal Combustion Engine, but they are stored in two separate fuel tanks and are also supplied to the combustion chamber through different injection lines. The bi-fuel vehicles that run on Indian roads are powered by the fuel combinations such as petrol or diesel in combination with CNG or LPG (for four-wheelers), or diesel in combination with CNG or LNG (mainly for buses), etc. However, hybrid vehicles, which comprise two different power sources (ICE and electric motor), have been introduced into the mainstream of the Indian road transport ecosystem only recently in 2021–22 despite the active push from MoRTH since 2015. The Faster Adoption and Manufacturing of (Hybrid &) Electric Vehicles (FAME) schemes rolled out in 2015 (FAME-I) and 2019 (FAME-II) were aimed at promoting the indigenous manufacturing of hybrid and electric vehicles through Production Linked Incentives (PLI) (Press Information Bureau Homepage 2020). The FAME-II was originally intended to run for 3 years till 31 March 2022, but it was extended for another 2 years as the intended targets were not met largely due to stringent criteria that were not favourable for the local suppliers of raw materials and vehicle parts because of the lack of stability in the placement of orders (Press Information Bureau Homepage 2021a).

Along with the hybrid vehicle architecture, there is yet another vehicle powertrain architecture available in the automotive market that adds to operational flexibility. The Flex Fuel vehicle (FFV) has the technology of a '*flex engine*' where the engine is custom-made to run on 100% petrol or up to 83% blend of ethanol (Alternate Fuels Data Center Homepage 2022a). The fuel in either pure or blended form is stored in a single tank and is delivered to the combustion chamber using the same fuel injection lines. Hybrid vehicles (equipped with both ICE and electric motors) accompanied by flex engine options have emerged as big commercial success in Brazil and China. The success of hybrid flex vehicles has prompted the GoI to issue an advisory to the Indian automotive manufacturers to focus on the indigenous manufacturing of the futuristic Flex Fuel Strong Hybrid Electric Vehicle (FFSHEV), which is essentially a combination of a hybrid vehicle and an FFV (Press Information Bureau Homepage 2021b). The core idea is to create a vehicle that can switch between multiple fuels, viz. petrol, bioethanol, and electric power coming from a battery pack, based on the functional requirements.

Currently, India's automotive industry is dominated by the conventional ICE-based vehicles (gasoline, diesel, and CNG-based) as they accounted for nearly 99.5% of the total vehicles running on Indian roads in FY 2019–20 (VAHAN Dashboard Homepage 2022). An Internal Combustion Engine converts the chemical energy contained in the fuel to mechanical energy necessary for running an automobile. The design of conventional vehicle engines and the associated fuels are selected largely based upon the power and torque requirements of the vehicle (two- or three- or four-wheelers or multi-wheeled trucks or tractors or earth-moving equipment) and the calorific value of the fuel. Based on the vehicle performance requirements, there can be a lot of variability in the vehicle engine design parameters. The variations are usually observed in the number of the rotors and the cylinders in which the pistons move, the cylinder arrangement (vertical or horizontal or V-shaped or radial or opposed), the operating cycles (Spark-Ignition or Compression-Ignition), the strokes (two or four), the valve type and arrangement (T or L or F or I), the valve timing, the charge stratification method (carburetor based or fuel-injection based—air-injection or airless), the combustion chamber design (open or divided), the fuel load-control governing method (throttle-based or quantitative or qualitative), and in the cooling methods (air-cooled or water-cooled) (Indian Institute of Technology Delhi Homepage 2022). The advent of hybrid and bi-fuel flexi vehicles is likely to add further complexities to the design of the vehicles.

This chapter aims to determine the deployability of the alternate fuels that can power the ICE-based vehicles running on Indian roads in the near-to-medium term, through a thorough data-driven analysis. The reliability of the potential alternatives is established through the 4A energy security framework analysis, which has been introduced in Sect. 5.3 of this chapter. Accordingly, the 4A analysis of the prominent ICE alternate fuels has been presented in Sect. 5.5. Further, global case studies around alternate fuels are discussed in detail in Sect. 5.6. Finally, this chapter aims to provide a basis for fuel transition in various vehicle segments which imbibes the learning from international experiences while it is also firmly anchored on the current state of affairs pertinent to India's road transport sector. Such a deep examination based on the 4A analysis of the available alternate fuel options as well as the functional use of different vehicle categories is a novel approach towards addressing the energy transition in the Indian road transport sector.

The introduction section (Sect. 5.1) provides an overview of the available alternate fuels and vehicle designs along with their role in the energy transition of the current Indian road transport system. Section 5.2 provides a detailed account of the vehicle attributes (functional use, fuel type, and status of emission norm compliance) for the vehicles that are running on Indian roads. Section 5.3 discusses the 4A energy security framework (*availability*, *accessibility*, *affordability*, as well as social and environmental *acceptability*) around the alternate fuels. The approach and the methods adopted in the present study are highlighted in Sect. 5.4. An in-depth discussion encompassing the 4A analysis for the three alternate fuels (natural gas, auto-gas, and hydrogen) for ICE-based road transport in India is presented in Sect. 5.5. A few prominent global experiences (CNG in Argentina, LNG in China, LPG in Turkey, ethanol in Brazil, biodiesel in Indonesia, and hydrogen across the globe) have been discussed along the lines of the 4A framework in Sect. 5.6 of this chapter. Based on the data summarized in Sects. 5.2 and 5.5 as well as the examples of the vehicle-fuel combinations across the globe, the possible pairings of the vehicles and the alternate fuels are prescribed and discussed in Sect. 5.7. As India aims for the diversification of its road transport ecosystem by transitioning to alternate fuels and energy sources, the lessons learnt from global experiences have been mapped into the Indian context in Sect. 5.8. The final section (Sect. 5.9) concludes the findings and mentions a possible way forward for India.

5.2 Understanding the Vehicles on Indian Roads

The MoRTH is the nodal ministry responsible for regulating the vehicles running on Indian roads. It divides these vehicles into two main categories; the first one is the '*transport category*' which includes both non-personal passenger and freight vehicles that are used for commercial purposes. While the second one is the '*non-transport category*' that includes personal vehicles that are driven for non-commercial purposes. As per the data from the VAHAN portal, 91% of the vehicles registered in the last 14 years (2009–22) belonged to the non-transport category, while the remaining 9% belonged to the transport category (VAHAN Dashboard Homepage 2022). Two-wheelers (mopeds, scooters, and motorcycles) accounted for 72.9% of the registered vehicles, whereas three-wheelers (autos, e-cargos, e-carts, and e-rickshaws) were only 2.8% of the total vehicles. The four-wheelers (passenger motor cars) had a share of about 16%, while the freight carriers and heavy passenger vehicles (ambulances, inter-city buses, and intra-city buses) contributed to 4.1% and 0.7% of the registered vehicles, respectively. The vehicles for agricultural purposes (tractors, trailers, tillers, and harvesters, etc.) and construction purposes (tractors, trailers, dumpers, excavators, road rollers, and bulldozers, etc.) constituted 2.7% and 0.8% of the total registered vehicles, respectively, during the 14-year period (2009–22). A negligible volume (~0.01%) belonged to the other niche functional use categories such as fire-tenders, quadricycles, cash vans, etc. (VAHAN Dashboard Homepage 2022). Figure 5.1 depicts the share of different vehicle categories among the total vehicles registered in India from 2009 to 2022.

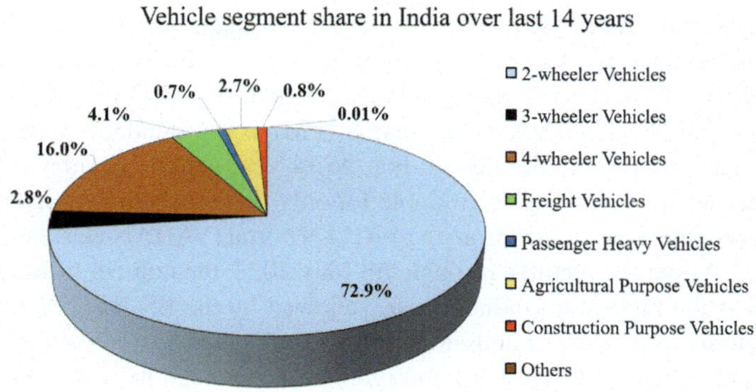

Fig. 5.1 Share of different vehicle categories in the total vehicles registered in India (2009–22). *Data source* VAHAN Portal (2022)

During the aforesaid 14-year period, 81.9% of the total registered vehicles were found to be fuelled by petrol-only, whereas the vehicles running on diesel-only contributed to 14.6% of the total registered vehicles. The share of the registered vehicles running on bi-fuels (petrol in combination with other fuels like ethanol, CNG, LPG, methanol, etc., or diesel in combination with options such as CNG, LNG, etc.) is found to be only 2.25% during the same period. The share of registered vehicles running purely on gaseous fuel (CNG and LPG) were found to be only 0.3%. The share of registered battery-driven electric vehicles (BEVs) (2Ws, 3Ws, and 4Ws combined) was found to be only 0.43% during that period. The share of the registered vehicles based on fuel diversity is depicted in Fig. 5.2.

The emission norms for the fuel and the engines have been designated from 2000 onwards with Bharat Stage (BS)-I and Euro 1. The emission norms were later upgraded to BS-II in 2005, and subsequently to BS-III in 2010. Major changes in the emission standards were mandated by the BS-IV norms that were rolled out in

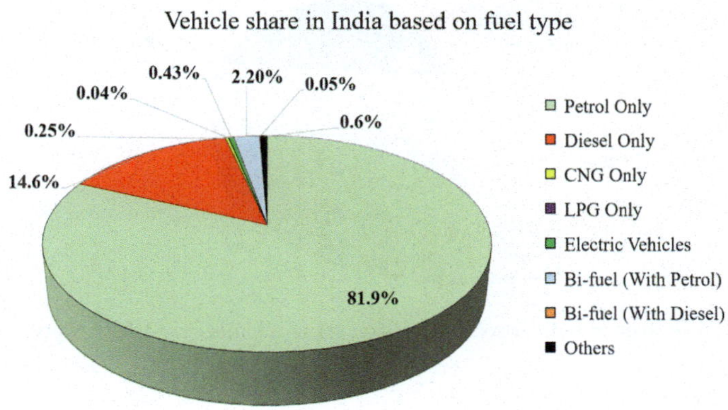

Fig. 5.2 Vehicle share in India based on fuel type. *Data source* VAHAN Portal (2022)

2017 (Pothumsetty et al. 2020). Thereafter, more stringent emission standards were enforced through the BS-VI or Euro 6 norms brought out in 2020 (Vats et al. 2022). The emission norms became progressively stringent for gaseous pollutants (such as CO, HC, and NO_x) and particulate matter (PM) as the standards were gradually upgraded from the BS-I to the BS-VI. For the 14-year period considered, emission norm compliance information is available for only two-thirds of the total registered vehicles as per the data sourced from the VAHAN portal (VAHAN Dashboard Homepage 2022). As per the records available till May 2022, the number of BS-IV (33%) vehicles was the highest on Indian roads, followed by the BS-III (27%) compliant ones. The most recent BS-VI emission norms were abided by 17% of the registered vehicles. The older BS-II and BS-I automobiles contributed to 16% and 7% of the total Indian on-road vehicles, respectively. The relative share of vehicles based on their compliance with the emission norms is shown in Fig. 5.3. Since the missing data on emission norm compliance belong mostly to the earlier years of the chosen duration, the relative share of the BS-VI-compliant vehicles appears to be higher than expected. However, these numbers need to be interpreted in the appropriate context to understand the ground reality of the overall emission compliance of the registered vehicles in India.

The categorization of the vehicle segments by the MoRTH as *transport* and *non-transport* (as explained earlier) needs to be restructured for understanding the vehicle categories from the point of view of the energy transition. Primarily, the vehicles are divided based on the number of wheels, size and shape that dictate their broad functional uses. Further, they are sub-categorized on the basis of their specific functional use. A summary of the granular vehicle categorization based on functional uses is provided below in Table 5.1.

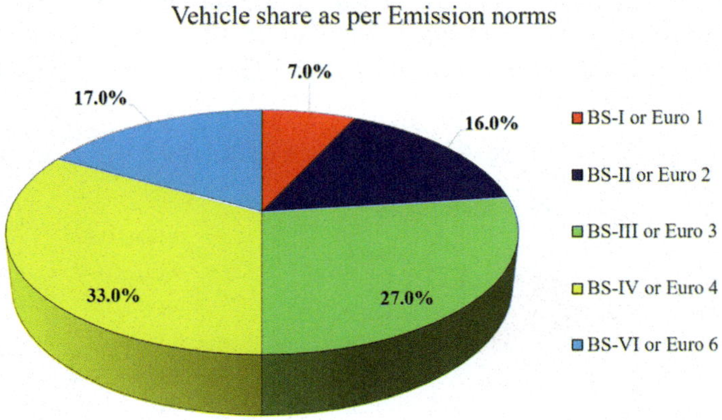

Fig. 5.3 Vehicle share in India based on emission norms. *Data source* VAHAN portal (2022)

Table 5.1 Summary of the granular vehicle categorization based on functional uses

Category	% Share (%)	ICE-based fuels in use	Sub-category
2-Wheeler (2W)	72.90	Petrol-only	Motorized cycle, scooter, moped, motorcycle, performance bike or super-bike, adapted vehicle
3-Wheeler (3W)	2.79	Petrol-only, CNG-only, diesel-only, LPG-only	Passenger-carrying, freight-carrying
4-Wheeler—passenger (4W-P)	15.74	Petrol-only, diesel-only, CNG-only, bi-fuel (Petrol-CNG, Petrol LPG)	Adapted vehicle, hatchback, sedan, SUV, MPV, crossover, van, luxury car, quadricycle, private service vehicle
4-Wheeler non-passenger (4W-NP)	0.29	Petrol-only, diesel-only, CNG-only	Tempo traveler (Maxi Cab), ambulance, camper van, cash van, mobile clinic, library van, mobile workshop, mobile canteen, hearse
Heavy vehicle—bus (HV-B)	0.65	CNG-only, diesel-only	Educational Institution bus, inter-city bus, Intra-city bus, Omnibus
Heavy vehicle—municipality (HV-M)	0.01	Diesel-only	Animal ambulance, auxiliary trailer, breakdown van, fire fighting vehicle, fire tender, fork lift, tower wagon, recovery vehicle, Snorked ladder, tree trimming vehicle, tow-truck
Heavy vehicle—agriculture (HV-A)	2.74	Diesel-only	Tractor, power tiller, harvester, trailer
Heavy vehicle—construction (HV-C)	0.83	Diesel-only	Bulldozer, construction equipment vehicle, crane-mounted vehicle, dumper, earth moving equipment, excavator, modular hydraulic trailer, power tiller, road roller, semi-trailer, tractor, tractor trolley, trailer, vehicles fitted with rig/generator/compressor

(continued)

Table 5.1 (continued)

Category	% Share (%)	ICE-based fuels in use	Sub-category
Freight vehicle (FV)	4.04	Diesel-only, CNG-only	Pick-up (capacity of 1.5 Tonnes or less), LCV (1.5–5.5 Tonnes), ICV (5.5–10.5 Tonnes), M&HCV (10.5 Tonnes or more), articulated vehicles

Source VAHAN Portal

5.3 ICE-Based Fuels for Consideration

The commercial viability of an ICE-based fuel (that is envisaged as an alternative for the Indian ecosystem) depends primarily upon the availability of the fuel. An alternate fuel needs to meet both technical and physico-chemical parameters to be suitable for the Indian environment. Further, there must be an assured and undeterred supply of the fuel to enable meaningful assessment of the overall availability status. The technical viability of any fuel is established by its physico-chemical properties, ignition parameters, and safety attributes in laboratory-based test conditions. The findings for each alternate fuel option for the ICE-based vehicles in India are summarized in Table 5.2 (Saurabh and Majumdar 2022). It can be seen from the table that all fuel options by and large tend to meet the ignition characteristics and safety standards as compared to the incumbent fuels (gasoline and diesel). However, the need for higher storage requirements against the incumbent fuels makes the transition to an alternate fuel a tricky proposition. The safety measures for LPG and Hydrogen need to be tightened through a thorough analysis of test data obtained from the advanced laboratory conditions as well as the continuous operational data available from the pilot-scale projects before these fuels are considered are for a wider vehicle network. Currently, no fuel can be considered a perfect replacement for the prevalent fossil fuel options (i.e. petrol and diesel). Therefore, there is a need to adopt multiple fuel options in the near-to-medium term to ensure a smooth transition to greener mobility through bi-fuels, FFVs (running on blended fuels), and hybrid vehicles.

Once the technical viability of a fuel is ensured, the supply chain management forms the basis for ensuring the reliability of the fuel for large-scale deployment and adoption. The commercial viability of a fuel is established on the basis of the 4As (*Availability, Accessibility, Acceptability, and Affordability*) of energy security (Winzer 2012). The 4A energy security framework analysis for the alternate fuel options is done on the basis of physical raw material *availability*, *accessibility* to the fuel (ready for consumption), *affordability* of the fuel at the consumer end, as well as social and environmental *acceptability* (Asia Pacific Energy Research Centre Homepage 2007). In the present discussion, a total of 28 parameters have been identified for the assessment of an alternate fuel under the 4A energy security framework to understand its commercial viability for deployment in the Indian road transport sector. These 28 parameters are highlighted in Table 5.3. Among the identified attributes,

Table 5.2 A summary of ICE-based fuels based on their technical availability (Saurabh and Majumdar 2022)

Fuel	Ignition	Storage	Safety	Remarks
CNG (Compressed Natural Gas)	Better fuel as calorific value and octane number are higher as compared to petrol ☑	Occupies more space than petrol due to lower volumetric energy density ☒	Narrow range of ignition limits, high auto-ignition temperature and higher diffusion coefficient render it safe ☑	Retrofitting in older gasoline vehicles not effective; good alternate fuel for new vehicles; good alternative over short-to-medium distance
LNG (Liquefied Natural Gas)	Similar to CNG in ignition parameters but differs in the way both are stored ☑	Energy density is one-third of diesel, leading to high storage requirement ☒	Safe fuel as it is non-toxic and noncorrosive; also has a high auto-ignition temperature ☑	Requires well-insulated and optimally pressurized tank to minimize evaporation losses and weathering due to its cryogenic nature
LPG (Liquefied Petroleum Gas)	Slightly superior fuel compared to petrol as it has a similar calorific value but higher RON ☑	Needs more storage space than petrol due to its lower volumetric energy density ☒	Safe in the absence of a spark but in case of fire, it is less safe as it settles on floors due to higher molecular weight ⚠	Burns sustainably as compared to petrol due to its high H-C ratio, but needs robust cylinders for safe storage in the vehicles
Ethanol	Several advantages over petrol, viz. high octane number, high embedded oxygen content, etc. But it has a lower calorific value ☑	More storage needed as its volumetric energy density is less as compared to petrol ☒	Safe as it has a higher flash point and a lower flame temperature in air as compared to petrol ☑	Viable alternative if the flex engine vehicles are facilitated to run on 100% ethanol
Biodiesel	Net calorific value is slightly lower than diesel while kinematic viscosity, specific gravity, and latent heat of vaporization are nearly the same as diesel ☑	Tends to form gels in cold weather because of its low pour point that makes it difficult for blending in diesel in colder regions ⚠	Remains a safe fuel while handling, transporting and storing; flashpoint is higher than regular diesel, but the auto-ignition temperature is relatively lower ☑	Viable alternative to diesel in tropical meteorological conditions (even in its pure form as B100)

(continued)

Table 5.2 (continued)

Fuel	Ignition	Storage	Safety	Remarks
Hydrogen	Superior fuel as calorific value and stoichiometric air-fuel mixture ratio are higher than petrol ☑	Needs higher storage volumes than petrol in liquid form, even larger storage is needed for the gas ☒	Very safe in the absence of a spark but very dangerous once triggered by spark (Self -burning due to low flash point) ⚠	Considered for its utility in both heavy-load ICE-based vehicles and low-to-medium motor-based fuel cell vehicles. Concerns remain about safety and storage

eleven were identified under the *availability* part, whereas six parameters each were identified to assess *affordability* and *acceptability* dimensions. The *accessibility* part was established through five measurable parameters.

5.4 Methods for Assessment

For the assessment of the suitability of a fuel vis-à-vis a vehicle, two approaches have been adopted in this chapter. The first one takes a stock of the viability of the fuel on the basis of the data collected for the parameters enlisted for the 4As (*Availability, Accessibility, Acceptability, and Affordability*) of energy security in Table 5.3. In addition, the case-study approach has also been followed where the success and failure associated with specific alternate fuels have been understood in the light of prominent examples from different parts of the globe. Based on the learnings from both approaches, a final summary is drawn in Tables 5.4 and 5.5 to indicate the likely energy transition pathways for the various categories of vehicles in India (as highlighted in Table 5.1).

5.5 4A-Assessment of ICE-Based Fuels

5.5.1 Natural Gas

Natural gas largely comprises methane, whether it is used as CNG or as LNG. CNG is stored in the gaseous form in the CNG kits at very high pressures (3000–3600 psi). On the other hand, LNG is used as a liquid fuel stored at very low temperatures (−161 °C) and low pressure (20–150 psi). CNG is preferred as a fuel for a wide range of light to medium vehicles due to its moderate energy density (9 MJ/l); while LNG

Table 5.3 Measurable parameters for the 4As of energy security

Availability parameters	**Measured as**
Technical viability	Physico-chemical properties & safety features
Proven resource base	Existence of resources and reserves (in measurable units)
Exploration—domestic & abroad	Involvement of Indian Companies in exploration
Production stability—domestic & abroad	Year-wise volume of production (in measurable units)
Import from abroad	Countries involved and volume imported (as of total consumption)
Foreign policy involved in import	Qualitative analysis of relations with relevant countries
Domestic consumption/essentiality	Sectoral consumptions (in measurable units or as % share)
Consumption in road transport sector	Contribution of road transport sector in total consumption (as %)
Transport medium to refineries/terminals	Capacities of tankers, pipelines & rakes
Storage & distribution	Refining and storage/strategic reserve capacity (in measurable units)
Safety in making the fuel available to the consumer-end	Safety parameters & standards during storage, transportation, and distribution
Affordability parameters	**Measured as**
Upfront price (per unit fuel)	Retail price trends, if available, compared to oil
Subsidies & pricing support	Existing policies and government commitments
Equity & transparency	Transparency in pricing mechanism
International/domestic price volatility	Price range & median price over last decade
International/domestic price stability	Comparison between domestic and international prices
Cost to customer (per km)	Segment-wise cost per km of travel
Accessibility parameters	**Measured as**
Transportation—From refineries/terminals	Capacities of tankers, pipelines, grids
Distribution channels	Govt. & Private companies and total no. of outlets
Distribution channel density	Outlets density (average distance between successive outlets)
Infrastructure—safety and storage	Safety requirements for storage and disbursement facilities
Innovation and prototype development	R&D centres in India; Major activities, and achievements
Acceptability parameters	**Measured as**
Well to tank GHG emissions	Equivalent CO_2 emissions (in measurable units)
Tank to wheels GHG emissions	Equivalent CO_2 emissions (in measurable units)

(continued)

Table 5.3 (continued)

Small/big scale utilisation potential	Relevance for different vehicle segments (based on functional use)
Energy efficiency improvisation potential	Fuel purification standards (enforced by stringent emission norms)
Localization of emissions	On-site or off-site emissions associated with the fuel
Larger social acceptability	Broad perception about the fuel in use and its pre-processing phases (exploration, extraction, refinement, dispatch); Benefits accrued (e.g. employment generation and other economic aspects)

is found to be more suitable for medium to heavy vehicles due to its high energy density (24 MJ/l).

5.5.1.1 Resource Availability

As of 2019, the worldwide total proven reserve of natural gas stands at 198,755 BCM (BP 2020). Russia has the highest share (19%) in the proven global reserves of natural gas, as can be seen in Fig. 5.4 (BP 2020). In contrast, India has only 0.7% of the proven reserves for natural gas of which more than two-thirds come from offshore basins. ONGC, Oil India Limited, and ONGC Videsh are the Indian companies that are involved in the exploration and production (E&P) of natural gas in India and abroad, respectively. Its domestic production has been decreasing over the past decade since 2010–11, while the consumption of natural gas in the country has increased by 4.56% between FY 2014–15 and FY 2019–20 as evident from Fig. 5.5 (Ministry of Petroleum and Natural Gas Homepage 2022). India has been meeting the domestic natural gas demand through imports that have witnessed an increasing trend since 2011–12. India imports nearly 80% of natural gas from Qatar (44%) and the United States (33%), with Russia being the third largest supplier (15%) to the country (Kulkarni and Pimpalkhare 2019).

Nearly 30% of the natural gas consumed in India goes towards manufacturing fertilizers, about 25% is used for industrial purposes (e.g. refining, petro-chemical, sponge iron, etc.), nearly 20% is utilized for power generation, and only 10% is currently used in the transport sector in the form of CNG (Ministry of Petroleum and Natural Gas Homepage 2022). Currently, 6 LNG terminals and 1 Floating Storage Regasification Unit (FSRU) are operating in India. These facilities have a cumulative storage capacity of 64.2 BCM (Global Energy Observatory Homepage 2022). Further, one more FSRU and three more LNG terminals are under construction, which will add a combined capacity of 34.5 BCM to the existing capacity (Ministry of Petroleum and Natural Gas Homepage 2022). The locations and the current status of the LNG terminals and FSRUs in India are shown in Fig. 5.6. The

Table 5.4 Analysis of the status of the incumbent fuel and the adoption of an alternate fuel option by vehicle segments (based on functional use)

	2-Wheeler (2W)	3-Wheeler (3W)	4-Wheeler—passenger (4W-P)	4-Wheeler—non-passenger (4W-NP)	Heavy vehicle—bus (HV-B)	Heavy vehicle—municipality (HV-M)	Heavy vehicle—agriculture (HV-A)	Heavy vehicle – construction (HV-C)	Freight vehicle (FV)
Petrol/Blends	Deployed	Deployed	Deployed	Deployed	Highly Unlikely	Highly Unlikely	Highly Unlikely	Highly Unlikely	Highly Unlikely
Diesel/Blends	Unlikely	Deployed	Deployed	Deployed	Deployed	Deployed	Deployed	Deployed	Deployed
CNG	Unlikely	Deployed	Deployed	Deployed	Deployed	Likely	Unlikely	Unlikely	Likely
LPG	Unlikely	Deployed	Deployed	Unlikely	Unlikely	Unlikely	Unlikely	Unlikely	Unlikely
LNG	Highly unlikely	Highly unlikely	Highly unlikely	Highly unlikely	Likely	Likely	Unlikely	Unlikely	Likely
Ethanol	Highly likely	Highly likely	Highly likely	Highly likely	Likely	Likely	Unlikely	Unlikely	Likely
Biodiesel	Highly unlikely	Highly likely	Highly likely	Highly likely	Highly likely	Highly likely	Highly likely	Highly likely	Highly likely
Hydrogen (ICE)	Highly unlikely	Unlikely	Unlikely	Unlikely	Likely	Unlikely	Unlikely	Unlikely	Highly likely

Table 5.5 Analysis on the other (apart from fuel) factors involved in the adoption of an alternate fuel option by vehicle segments (based on functional use)

	2-Wheeler (2W)	3-Wheeler (3W)	4-Wheeler—passenger (4W-P)	4-Wheeler—non-passenger (4W-NP)	Heavy vehicle—bus (HV-B)	Heavy vehicle—municipality (HV-M)	Heavy vehicle—agriculture (HV-A)	Heavy vehicle—construction (HV-C)	Freight vehicle (FV)
Top priorities while transitioning	Fuel economy, fuel accessibility, low maintenance	Compact storage, fuel economy, vehicle safety	Vehicle comfort, fuel accessibility, power & mileage	Vehicle space, fuel accessibility, performance	Reduced cost of ownership, vehicle safety	Vehicle space, fuel accessibility, performance	Easy fuel storage, fuel accessibility, performance	Easy fuel storage, fuel accessibility, performance	Low emissions, fuel accessibility, easy maintenance
Alternate vehicle architectures	FFVs	FFVs, bi-fuel vehicles	FFVs, bi-fuel vehicles, hybrid vehicles	FFVs, hybrid vehicles	FFVs, hybrid vehicles	Dedicated vehicles (alternate fuel-based)	Dedicated vehicles (biodiesel-based)	Dedicated vehicles (biodiesel-based)	FFVs, bi-fuel vehicles, hybrid vehicles

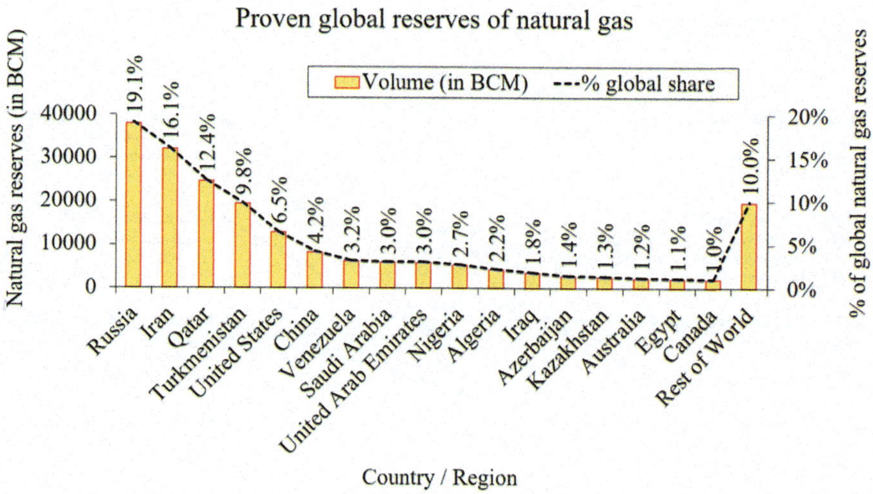

Fig. 5.4 Proven global reserves of natural gas. *Data source* BP Statistical Review (BP 2020)

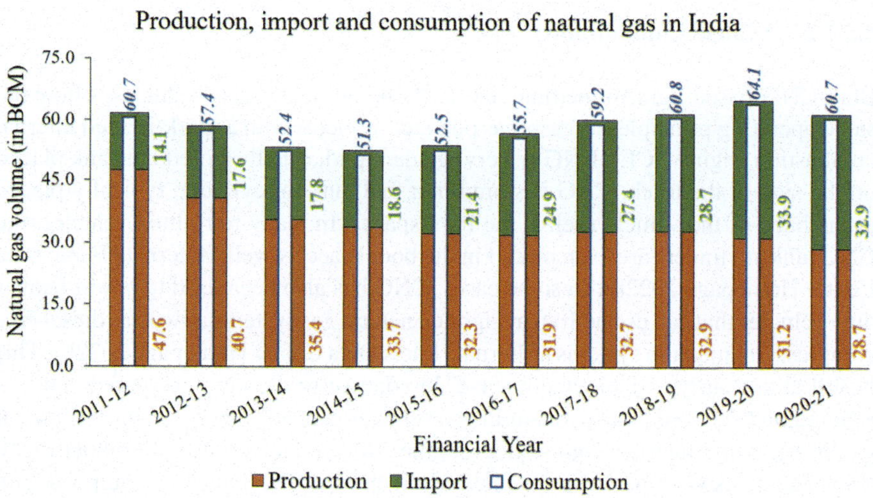

Fig. 5.5 Production, import, and consumption of natural gas in India in the last 10 years. *Data source* MoPNG (2022)

natural gas stored in the abovementioned terminals is distributed across India through the pipelines constructed by GAIL, Reliance Gas Pipeline Limited, Pipelines Infrastructure Limited, and Gujarat State Petronet Limited (Petroleum and Natural Gas Regulatory Board Homepage 2022).

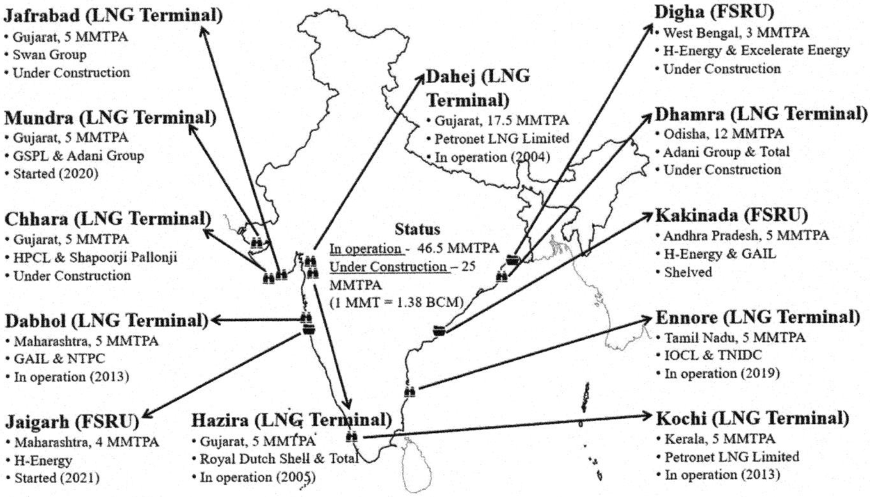

Fig. 5.6 Status of LNG terminals and FSRUs in India. *Data source* MoPNG (2022)

5.5.1.2 Vehicle Compatibility

Both CNG-based light-to-medium and LNG-based heavy-duty vehicles follow the same operating principles as gasoline-powered vehicles with a spark-ignited internal combustion engine (ICE). LNG is a cryogenic liquid, usually stored in a tank placed on the side of the truck. CNG is stored in a fuel tank, or cylinder, typically placed at the back of the vehicle (below the boot space). In many retrofitted vehicles, the CNG tanks/cylinders are even placed in the boot space as well (Alternate Fuels Data Center Homepage 2022b). In some cases, CNG has also been used to power heavy-duty vehicles that can use both spark-ignited natural gas systems as well as diesel-like compression injection systems (Alternate Fuels Data Center Homepage 2022b). This is still at a prototype deployment stage. In contrast to the pure gas-based vehicles running on CNG-only, bi-fuel natural gas vehicles can use either gasoline or natural gas in the same internal combustion engine. Both fuels are stored on-board, and the driver can switch between the fuels while driving. The vehicle is equipped with separate fuel tanks, fuel injection systems, and fuel lines for both fuels (Alternate Fuels Data Center Homepage 2022c).

5.5.1.3 Accessibility

As of 2021, a pipeline network of more than 17,500 km exists in the country for supplying natural gas to end-consumers. About 70% of these pipelines have been constructed by GAIL, 16% by GSPL, 11% by RGPL, and the rest 3% by other enterprises (Petroleum and Natural Gas Regulatory Board Homepage 2022). India envisages a National Gas Grid by 2030 with a total pipeline length of 34,500 km (Press

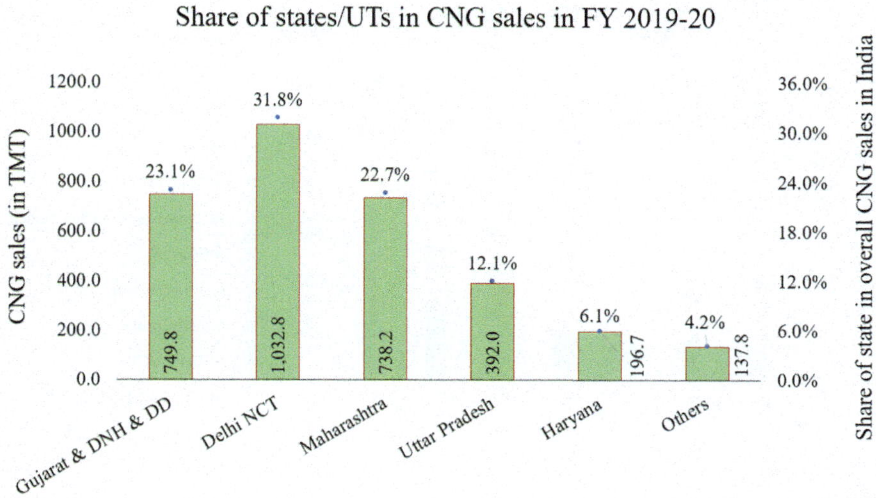

Fig. 5.7 CNG sales in various states of India in FY 2019–20 with their share in consumption. *Data source* MoPNG (2022)

Information Bureau Homepage 2022). This is likely to increase its share in India's energy mix from the current level of 6%. For a long time, the CNG sales in India for the road transport segment pivoted around 3 nodes—Delhi-NCR, Maharashtra, and Gujarat. In FY 2019–20, these nodes contributed 32%, 23%, and 23% of the total sales, respectively, as shown in Fig. 5.7 (Ministry of Petroleum and Natural Gas Homepage 2022). However, the number of CNG stations is increasing across India, and the footprint is even expanding to the States/Union Territories (UTs) where the presence was not there in the past decade. With the growing footprint in other areas, the share of the predominant 3 nodes (Maharashtra, Delhi, and Gujarat) in terms of the number of CNG stations has decreased to nearly 55% in 2020–21 from about 88% in 2014–15 (see Fig. 5.8) (Ministry of Petroleum and Natural Gas Homepage 2022). The CNG distribution outlets were available in 20 states/UTs in 2020–21 (with at least 10 CNG filling stations being present in each State/UT). India follows internationally approved safety standards (IS 15958) with respect to specifications of CNG for automotive purposes, along with other standards such as IS 14504, IS 15641, IS 15130, IS 15126, and IS 15320 with respect to the fuel quality (physico-chemical properties) of CNG at its distribution outlets (Bureau of Indian Standards Homepage 2012).

5.5.1.4 Affordability

The CNG prices in India (Delhi as reference) had largely been stable till mid-FY 2021–22 as seen in Fig. 5.9. However, the price has witnessed a substantial rise thereafter due to global supply shortage, increased demand during post-COVID business

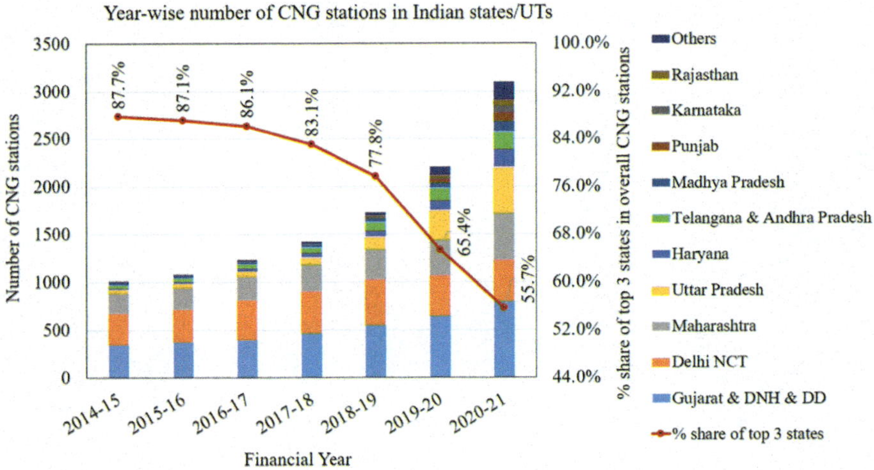

Fig. 5.8 Number of CNG stations in various states/UTs along with the share of top 3 states/UTs. *Data source* MoPNG (2022)

recovery, and the Russia-Ukraine crisis. Over the past 9 months, the CNG prices in Delhi have shot up by more than 70%, from 44.30 ₹/kg in August 2021 to 75.60 ₹/kg in May 2022 (MyPetrolPrice Homepage 2022). However, price stability was observed during normal times due to the subsidies provided by the Government of India, which has led to much lesser price volatility for CNG as compared to gasoline (see Fig. 5.10). Although the international market prices have increased about two-fold over the last 9 months (U.S. Energy Information Administration Homepage 2022), the prices of natural gas for transport purposes have been regulated since it is mainly used in the Indian urban nodes as a daily commuter fuel. Despite H-CNG cars being expensive as compared to their petrol-based counterparts (Kumar and Chakrabarty 2020), they are attractive since the running fuel costs are much lower as compared to gasoline-powered alternatives (Patankar and Patwardhan 2017).

India follows a complex gas pricing system comprising Administered Price Mechanism (APM), and non-APM or free-market gas (Gateway House Homepage 2014). The APM is applicable to domestic gas fields managed by the public sector Oil and Natural Gas Corp. (ONGC) and Oil India Ltd. (OIL), and the price of the domestic produce is determined by the Petroleum Planning and Analysis Cell (PPAC) of the Ministry of Petroleum and Natural Gas (MoPNG). The non-APM gas is sourced either from joint venture fields or through imports (Petroleum Planning Analysis Cell Homepage 2022). The final pricing was devised based on three global benchmarks as per the formula suggested by the six-member panel led by former Reserve Bank of India (RBI) governor C. Rangarajan in December 2012 (Sen 2015). This formula was tweaked in October 2014 through the notification '*New Domestic Natural Gas Pricing Guidelines, 2014*' issued by the central government. The new formula derived the pricing of natural gas in India on the basis of the average rates of liquefied natural gas (LNG) imports into India and the benchmark global gas price (Press Information

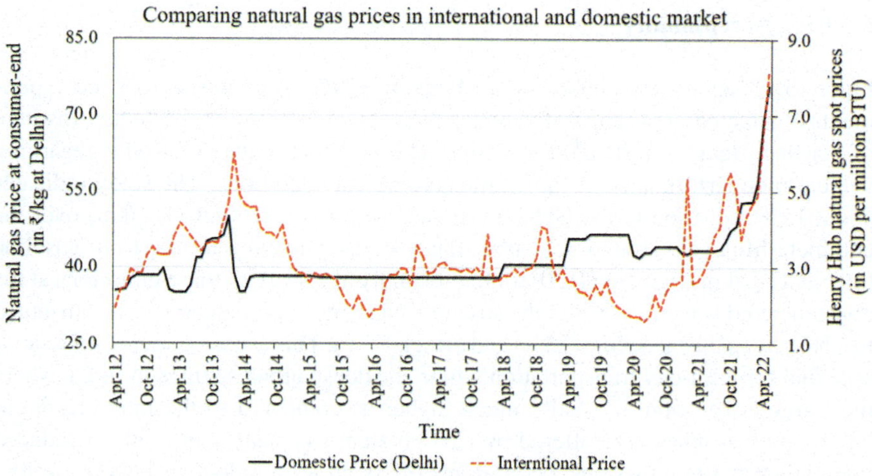

Fig. 5.9 Comparing natural gas prices in the international and domestic markets. *Data source* MyPetrolPrice Homepage (2022), US EIA (2022)

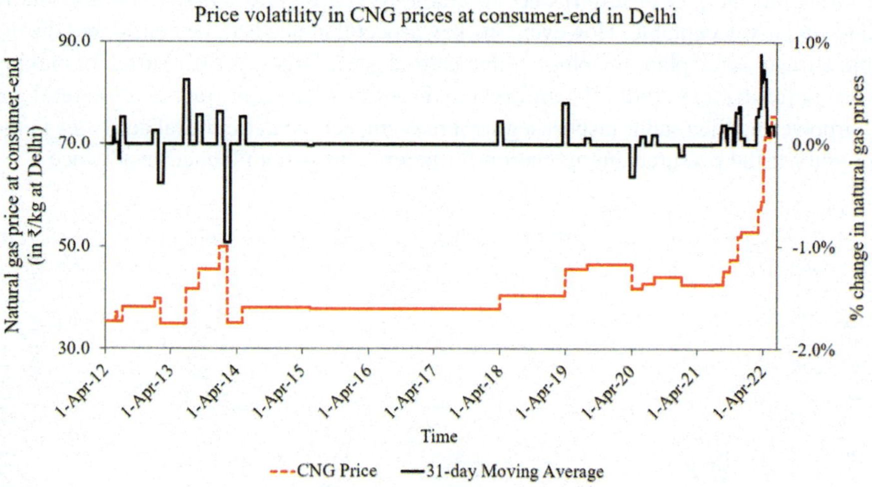

Fig. 5.10 Price volatility in CNG prices at consumer-end in Delhi. *Data source* MyPetrolPrice Homepage (2022)

Bureau Homepage 2019). The pricing was made applicable to all the domestically produced natural gas, irrespective of the source, whether conventional, shale-based or coal-bed methane (CBM) produced by the public sector entities or private sector firms.

5.5.1.5 Acceptability

Figure 5.11 depicts the emissions attributed to CNG, as obtained from the experiments conducted on a range of vehicles by a research group at the Indian Institute of Science, Bengaluru (Patil et al. 2016). This provides valuable insights regarding vehicle-fuel performance in Indian meteorological conditions. The study indicates about 15% reduction in the GHG emissions from CNG (166 gCO_{2eq}/km) over the complete life cycle around vehicular transport, as compared to that from gasoline (192 gCO_{2eq}/km) (Patil et al. 2016). Interestingly, the share of these emissions at the consumer end is much lower in the case of CNG (~60% of total emissions attributed to CNG) as against gasoline (75%) and diesel (73%). This aspect renders CNG a suitable fuel for intra-city transportation within the dense urban settings in India, where the traffic congestion is usually much higher as compared to the inter-city highways. Another advantage offered by CNG is that it is cleaner in terms of reduced non-carbon-based emissions. Natural gas offers an 80% reduction in NO_x, a 90% reduction in Particulate Matter (PM), and a 100% reduction in SO_x emissions, as compared to crude oil (Energy Information Administration Homepage 1999). Such advantages make it a suitable candidate fuel to become prominent in the short term (during the energy transition) since its cradle-to-grave cycle is both environmentally and socially acceptable. However, one key aspect that needs to be considered during the infrastructure planning phase is that natural gas is largely (~30%) used for making urea (a fertilizer) in India. Therefore, the diversion of a major fraction of natural gas for powering the road transport segment may impact the agricultural activities in the country if the resource augmentation for natural gas is not planned in advance.

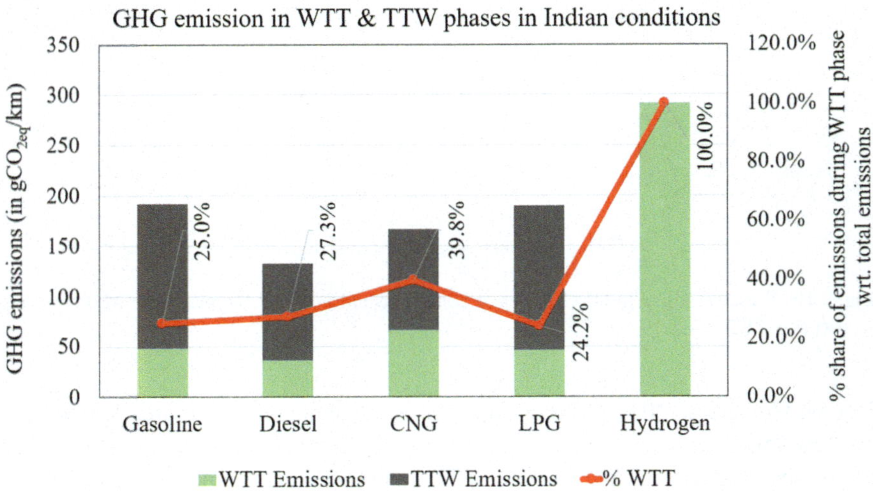

Fig. 5.11 GHG emissions (in gCO_{2eq}/km) in WTT and TTW phases of fuel life cycle. *Data source* Patil et al. (2016)

5.5.2 *Auto-Gas or LPG*

Liquefied Petroleum Gas (LPG) is known as auto-gas when it is used as a transport fuel. LPG comprises a mixture of gases such as propane (24.5 vol%), butane (38.5 vol%), and iso-butane (37 vol%). It exists in the gaseous state at atmospheric pressure and room temperature and its volume reduces by more than 250 times when pressure (30–45 psi) is applied, which converts it into a liquefied state. LPG offers a significantly higher octane number (120) than gasoline (91–100) while its calorific value (45 MJ/kg) is similar to gasoline (43 MJ/kg). However, it has a lower energy content per unit volume of fuel (25.4 MJ/m^3) than gasoline (34.2 MJ/m^3), requiring more space to store. LPG is a safer fuel in the absence of a spark, but it poses a major risk in case it catches fire since it settles near the floor due to its higher molecular weight (54.69 g/mol) as compared to air (28.96 g/mol).

5.5.2.1 Resource Availability

Auto-gas or LPG is a fossil fuel occurring naturally. About 60% of LPG available globally is recovered from the reserves of natural gas and crude oil (wet gas processing), while the remaining 40% is produced during the refining of crude oil (WLPGA Homepage 2022). Recently, a very small amount of it is being produced from renewable vegetable oils and degradable waste materials and is called Bio-LPG. Evidently, the resource availability of LPG can be assessed by looking into the natural gas reserves and the reserves of crude oil. The natural gas reserves have been discussed in Sect. 5.1.1. The fractional distillation of crude oil generates LPG as only 1–4% of its by-products (by volume) (Paczuski et al. 2016). The yield of LPG from crude oil depends on the type of crude available, and the technologies used in the refineries (Energypedia Homepage 2022). The commercial production also depends on the market values of propane and butane as against other by-products. Currently, the share of LPG consumption towards transportation is minuscule (0.65%) in India (Ministry of Petroleum and Natural Gas Homepage 2022). A major share of total LPG available in the country is used for domestic LPG gas distribution (87.5%) and industrial or commercial use (10%) (Ministry of Petroleum and Natural Gas Homepage 2022). Figure 5.12 indicates that the auto-gas consumption in India started to level off during the period between FY 2009–10 and FY 2014–15 (Auto-gas.net Hompage 2020a). Post-2015, there is an increasing trend of auto-gas consumption in India. This can be attributed to the active push from several state governments (especially, Karnataka, Tamil Nadu, Maharashtra, and Gujarat) for using auto-gas as a transport fuel for the three-wheelers running within the respective city limits. In 2019, an estimated 25 lakh vehicles were running on auto-gas in India, of which a large number of vehicles were three-wheelers (Auto-gas.net Hompage 2020a, 2022b). However, the use of auto-gas in the three-wheelers is not likely to increase its demand very sharply, since the average fuel consumption per vehicle is very low (only 300 L per year) (Auto-gas.net Homepage 2022b).

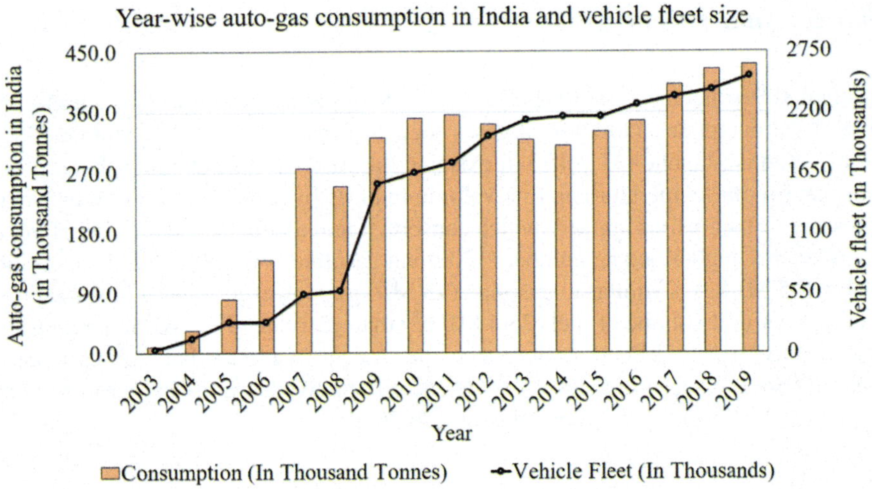

Fig. 5.12 Auto-gas consumption along with vehicle fleet size in India. *Data source* WLPGA (Auto-gas.net Hompage 2020a)

5.5.2.2 Vehicle Compatibility

An LPG vehicle is a vehicle that is auto-gas powered-only, instead of petrol (gasoline) or diesel. LPG vehicles can either be conversions from the existing petrol or diesel-based vehicles or newly manufactured vehicles coming from fresh production lines of the factory. The conversion of a conventional vehicle into an LPG version is done by adding a secondary LPG vehicle fuel system to the petrol- or diesel-powered primary fuel system at a reasonable cost (ELGAS Homepage 2022a). The dual-fuel system allows the vehicle to operate on either auto-gas or petrol (stored in two different tanks), and the manual switching is done from one fuel to another based on the requirement. LPG tanks in modern vehicles are designed donut-shaped to make them compact and save space in the boot by fitting them within the spare wheel wells (ELGAS Homepage 2022a). The most recent and advanced LPG-based fuel system is based on Liquid Phase Direct Injection (LPDI) technology where liquid auto-gas is injected directly into the combustion chamber (ELGAS Homepage 2022b). It has succeeded the Liquid Phase Injection (LPI) technology where liquid auto-gas used to be injected into the intake manifold (ELGAS Homepage 2022b). Before that, technologies were based on the vapor injection method where liquid auto-gas was converted to vapor before it was injected into the intake manifold (ELGAS Homepage 2022c). The LPG-based fuel systems have been commercially deployed on 3-wheelers, cars, buses, motorcycles, tractors, and on light as well as heavy trucks across the globe (ELGAS Homepage 2022c).

5.5.2.3 Accessibility

India has a 3738 km long pipeline network for transporting LPG. The carrying capacity of this pipeline network is 9.54 MMT, and about 90% of this capacity was utilized in FY 2020–21 (Ministry of Petroleum and Natural Gas Homepage 2022). India also has 16 LPG-carrier ships with a cumulative dead weight tonnage capacity of 723 TMT to transport the imported LPG from overseas (Ministry of Petroleum and Natural Gas Homepage 2022). To date, LPG is available in 565 cities (in 23 states/UTs) comprising a network of 1166 stations across India (Indian Auto LPG Coalition Homepage 2022a). All three Public Sector Undertakings (PSUs), Hindustan Petroleum Corporation Limited (HPCL), IOCL, and BPCL, have auto-gas distribution outlets across the states along with big and small private players such as RIL, Aegis Logistics Ltd., Total Oil India Pvt. Ltd., KR Fuels, Go-Gas Auto, Indianoil Petronas Pvt. Ltd., Shiv Energy Pvt. Ltd., Samriddhi Petro Products Pvt. Ltd., Vanaz Engineers Ltd., Essar Oil, and other smaller distributors. Among the major metros, Bangalore, Chennai, Hyderabad, Chandigarh, and Kolkata have effectively brought auto-gas into the mainstream with good availability of filling stations, ensuring accessibility to its largest consumer base, i.e. 3-wheelers (Indian Auto LPG Coalition Homepage 2022b). Figure 5.13 shows that Karnataka, Tamil Nadu, Maharashtra, Gujarat, Kerala, and Andhra Pradesh are the top consumers of auto-gas in India.

Figure 5.13 also shows that the private players have a huge role in the expansion of the distribution network as the top four networked states have nearly 50% of the outlets owned by private players. As of 2018, the private enterprises had more than 40% share in the total number of LPG stations across India. It also establishes the fact that in the regions where the auto-gas distribution network is not robust yet, the PSUs

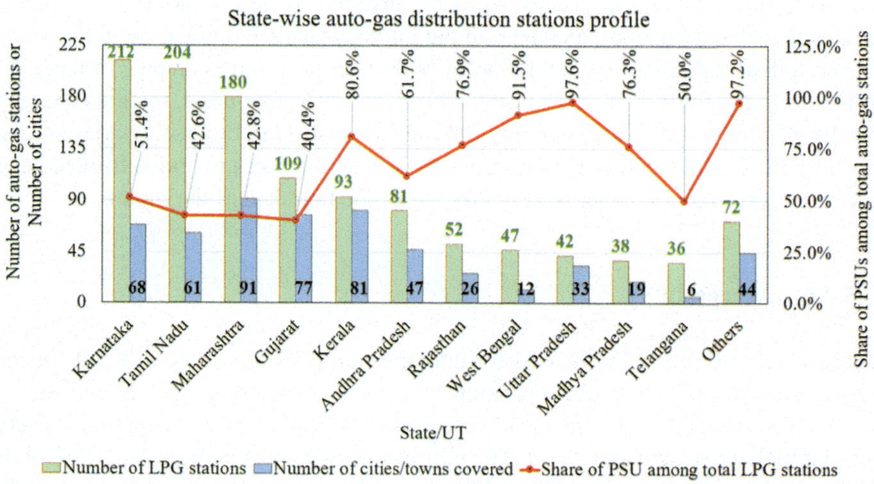

Fig. 5.13 State-wise number of LPG stations in India. *Data source* Indian Auto LPG Coalition (2022a)

need to provide the initial impetus through investments in the new infrastructure to facilitate fuel diversification during the energy transition. This is evident from the 'others' category in Fig. 5.13, which comprise 12 states/UTs. The regions covered under the 'others' category have 97% of the auto-gas outlets owned by the PSUs.

The LPG kits available in the open markets need to be approved by either the Automotive Research Association of India (ARAI), Pune or the Vehicles Research & Development Establishment (VRDE), Ahmed Nagar under Central Motor Vehicles Act, 1989 (Bureau of Indian Standards 2012). In addition, the auto-gas kit and the tank fitted in vehicles must conform to Bureau of Indian Standards (BIS) such as IS 14899:2000 (for LPG containers for automotive use) and IS 15100:2001 (for multi-functional valve assembly on fixed LPG containers) (Bureau of Indian Standards 2012).

5.5.2.4 Affordability

Figure 5.14 presents the average monthly retail price trends for auto-gas in India (averaged over prices in Delhi, Mumbai, Kolkata, and Chennai) (Indian Oil Corporation Limited Homepage 2020b). It can be ascertained from the graph that the price of auto-gas seems to follow a mixed trend of international prices of both crude oil as well as natural gas. Over the past two years (since the advent of the COVID-19 phase), LPG prices have nearly doubled. It had reached 75 ₹/l in May 2022 following a steep rise from the average price of 40 ₹/l prevalent during 2015–20. LPG pricing in India is based on import parity price (IPP) (BusinessLine Homepage 2020). The IPP is determined based on LPG prices in the international market, assuming that the fuel is imported into the country. The IPP, based on Saudi Aramco's LPG price, includes the free-on-board (FOB) price, ocean freight, insurance, customs duties, port dues, etc. This price, quoted in dollars, is then converted to rupees based on the latest exchange rates. Therefore, a change in the international LPG prices or in the value of the Indian rupee against the dollar, or both, translates into a change in the LPG prices in India. Further, the cost of inland freight, marketing costs, margins charged by the oil companies, distributor commission, and taxes (GST), etc., are added to the aforesaid IPP (BusinessLine Homepage 2020). These prices are non-subsidized and are reset every month, generally effective from the first date of the month.

5.5.2.5 Acceptability

Since nearly 90% of LPG is used for cooking purposes, it is very difficult for the Government of India to actively promote auto-gas as an alternate fuel. In addition, it can be seen from Fig. 5.11 that the GHG emission profile for auto-gas in India is very similar to that of petrol and diesel. The figure also shows that auto-gas has 75% of its life-cycle emissions coming in the TTW phase. However, auto-gas emits nearly 80% less Particulate Matter (PM) as compared to petrol and diesel. It also emits about 75% less NO_x as compared to petrol (Saraf et al. 2009). All of these make auto-gas a

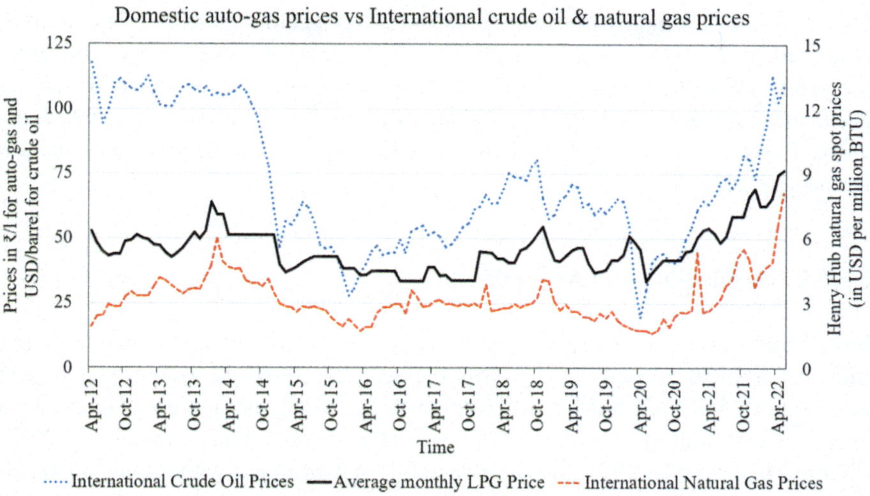

Fig. 5.14 Average monthly domestic auto-gas Prices (averaged over Delhi, Mumbai, Kolkata, and Chennai) versus average monthly international crude oil & natural gas prices. *Data source* EIA (2022), PPAC (2022), Indian Auto LPG Coalition (2022b)

potential alternate fuel for fuel diversification in the short-term during the envisaged energy transition for the road transport sector in the country.

5.5.3 Hydrogen

Hydrogen is deemed a superior fuel for transport due to its very high calorific value (120 MJ/kg). This renders it a suitable alternative to gasoline (44 MJ/kg) (Stanford University Homepage 2022). In addition, its ability to form a better air-fuel mixture (34.3:1) as compared to petrol (14.7:1) leads to more efficient burning (Saurabh and Majumdar 2022). However, it needs high volumes to store, whether used in the liquid or gaseous state, as its energy density in the liquid form (8.5 MJ/l) is considerably lower than that of gasoline (31 MJ/l) (Saurabh and Majumdar 2022). Apart from the storage and safety concerns associated with the flammable nature of hydrogen, the methods of generation of hydrogen cast a shadow on it becoming a potential alternate to existing fuels, i.e. petrol and diesel. The proven techniques for producing hydrogen at a commercial scale include steam methane reformation (SMR), electrolysis, pyrolysis, and gasification (Lamb et al. 2020). The methods that have the potential to generate hydrogen with less GHG emission footprints are low-temperature and high-temperature electrolysis processes (where the electricity is sourced from a green or renewable source), and steam methane reformation where carbon is captured and utilized or stored (Ferrero et al. 2013). Hydrogen is called 'green' if it is generated by electrolysis with electricity sourced from green sources of energy (solar/wind/hydro/tidal/geothermal). As per the hydrogen taxonomy, it is termed 'pink' if it is generated from nuclear energy that has a very less carbon footprint. 'Blue' hydrogen is generated from steam methane reformation or coal gasification processes, where the generated carbon dioxide is captured, stored, and utilized, thereby preventing the release of carbon into the atmosphere (NationalGrid Homepage 2022).

5.5.3.1 Availability and Accessibility

From India's point of view, green hydrogen has attracted the maximum attention as India is a tropical country that is mostly blessed with round-the-year moderate solar insolation (3.5–4.5 kWh/m^2/year) (Kamath et al. 2022), along with adequate wind energy potential along the large coastline. India consumed an estimated 6.7 MMT of 'grey' hydrogen in 2021 that is produced from natural gas without any means for carbon capture, utilization, and storage (CCUS) (LiveMint Homepage 2021a). Nearly 54% (6.6 MMT) of hydrogen is currently consumed by the petro-refining sector while the rest 46% (3.1 MMT) is utilized by the fertilizer industry (LiveMint Homepage 2021a). About 85% of hydrogen is produced and consumed on-site in India since it is safer to transport natural gas (major feedstock for hydrogen) as compared to hydrogen. The transportation of liquid hydrogen needs a huge amount of energy and cost for sustaining very low temperatures (−253 °C) (Hall et al. 2020). Therefore, the transportation of hydrogen to the consumer end may need innovative strategies such as the conversion of hydrogen into liquids such as ammonia or methanol (Hall et al. 2020). However, before envisaging a hydrogen-based clean transport ecosystem, this fuel needs to be consistently produced in a reasonably clean (blue/pink) way for the existing consumers (refining and fertilizer industries). The demand profiles from different sectors and the sustenance of the demand levels should be monitored for a substantial duration in order to reduce investor risk, and to enhance the confidence of the investors and the customers towards adopting the much-desired green hydrogen ecosystem.

5.5.3.2 Vehicle Compatibility

A hydrogen-driven ICE-based vehicle is very similar in architecture to its counterparts that run on either petrol or diesel or CNG. The vehicles coming out of current product lines can be made to run on hydrogen by changing certain components (such as fuel delivery systems and spark plugs) in the existing engine (College of the Desert 2022). Hydrogen can practically power any ICE-based vehicle. However, it needs higher storage capacities (for storing in gaseous form at 350–700 bars) to ensure optimal refills (Energy.gov Homepage 2022). This aspect renders hydrogen-based ICE systems impractical for smaller vehicles (like cars) as they would not be able to accommodate large storage spaces. Vehicles that have significantly higher uptimes than private-use vehicles, such as heavy-duty trucks, buses, and heavy machinery, are a perfect fit for hydrogen-ICE as compared to the battery-based alternative, since they need to operate for a set number of hours, and they cannot afford to stop for long durations to charge their batteries. Such large vehicles are likely to struggle with the excess weights arising from the large lead-acid or Li-ion battery packs. Hydrogen-based heavy trucks would prove to be a boon for areas with low ambient air quality.

5.5.3.3 Affordability

The cost of 'grey' hydrogen in India is estimated to be in the range of 160–200 ₹/kg in 2022, whereas the cost of producing 'green hydrogen' is estimated at 300–320 ₹/kg. Evidently, the shift from grey hydrogen to green appears to be a difficult prospect in the near-to-medium term (Fortune India Homepage 2022). As per a study conducted by The Energy Research Institute (TERI), the possibility of cost reduction through steam methane reforming is not likely to be more than 10% in the coming 20 years. However, one possible way to reduce the levelized cost of production of green hydrogen to below 100 ₹/kg by 2050 is the electrolysis route, which should be powered by the electricity coming from a dedicated renewable energy installation (Hall et al. 2020). The National Hydrogen Policy announced by the GoI in February 2022 aims to boost green energy production in India (Ministry of Power Homepage 2022). It has provisions in place to reduce the green hydrogen production costs by 20–30% to 230–240 ₹/kg on an immediate basis (KPMG 2022). The industry is keen to achieve this target due to some of the progressive policy measures such as open access to solar-based electricity without central surcharge and zero inter-state transmission charges for 25 years, open-market banking of excess power from the self-generated renewable energy, permission to set up bunkers near the ports to store the fuel, and the promotion of indigenous manufacturing of electrolysers, etc., which are proposed in the National Hydrogen Policy released in 2022 (Ahluwalia 2022).

5.5.3.4 Acceptability

A hydrogen-based vehicle is an ultra-low emission option since the vehicle would emit only water vapor following the combustion of hydrogen. Unlike hydrogen-run fuel cell electric vehicle (FCEV), hydrogen-based ICE vehicle is not zero-emission since small amounts of CO_2 (because of the burning of engine oil), and NO_x (significantly lower than petrol/diesel vehicle) are emitted (Digital CCJ Homepage 2022). Figure 5.11 indicates that currently, hydrogen generates higher GHG emissions in India, because of the use of grey hydrogen. However, the emissions in the TTW stage are nearly zero at the consumption end. Hydrogen-powered internal combustion engines can be an excellent near-to-medium term and a near-zero-emission solution that would not require any major change in the vehicle architecture. However, the availability of clean hydrogen needs to be ensured to realize the goal of green mobility via the route of the hydrogen-driven transport sector. With intensive R&D

and investment, hydrogen-based FCEVs would likely emerge as a long-term solution for decarbonizing the long-haul heavy-duty freight transport sector.

5.6 Evaluation of Global Experiences

Since India is aiming for a sustainable transport sector that pivots around prudent diversification of the fuel mix, it is important to take stock of the global experiences regarding the commercial deployment of alternate fuels in ICE-based vehicles for road transport. In this section, the focus is on the deployment of CNG in Argentina, LNG in China, auto-gas or LPG in Turkey, ethanol blended petrol (EBP) in Brazil, biodiesel in Indonesia, and hydrogen (for ICE-based vehicles) research across the globe. In each of the cases, the discussions revolve around the elements of the *4A Framework* (i.e. *Availability, Accessibility, Affordability, and Acceptability*) of energy security as highlighted before in Sect. 5.3. The general meteorological conditions of the countries have been highlighted while delving into the enablers for the aforesaid alternate fuels discussed in the context of the specific countries so that the lessons imbibed from those experiences can be put into proper context while mapping these possibilities for the Indian road transport sector.

5.6.1 CNG in Argentina

Argentina lies mostly in the temperate regions of the southern hemisphere where the average ambient temperature ranges between 5 and 30 °C, across the different seasons in a typical year (WorldData Homepage 2022a). Argentina is one of the most urbanized countries in the world where more than 90% of its population lives in small or big cities (Bolay 2020). Its capital, Buenos Aires, alone holds more than 40% of the total population in Argentina (Bolay 2020). Rapid urbanization has led to a decrease in Argentina's air quality, which forced the government to explore cleaner means of transportation for the urban centres (Gas 1991). The Liquid Fuel Substitution Program (*Plan de Sustitución de Combustibles Líquidos*) was adopted by Argentina in 1984, which aimed to substitute diesel with CNG in the public transportation system to reduce diesel imports (Collantes and Melaina 2011). This program was envisaged after the discovery of the natural gas reservoirs of *Loma de la Lata* in north-western Patagonia (Piragine et al. 2018). The success of this program was realized through the increase in the average total daily consumption of CNG for the next 20 years, as nearly 20% of the natural gas consumption in Argentina was attributed to the road transport sector (International Energy Agency Homepage 2022c).

A. *Success-story of Argentina's gas sector in road transport.* The peak adoption of CNG in Argentina by 2004, as shown in Fig. 5.15, was achieved primarily due to three major steps taken by the government. These measures are discussed below.

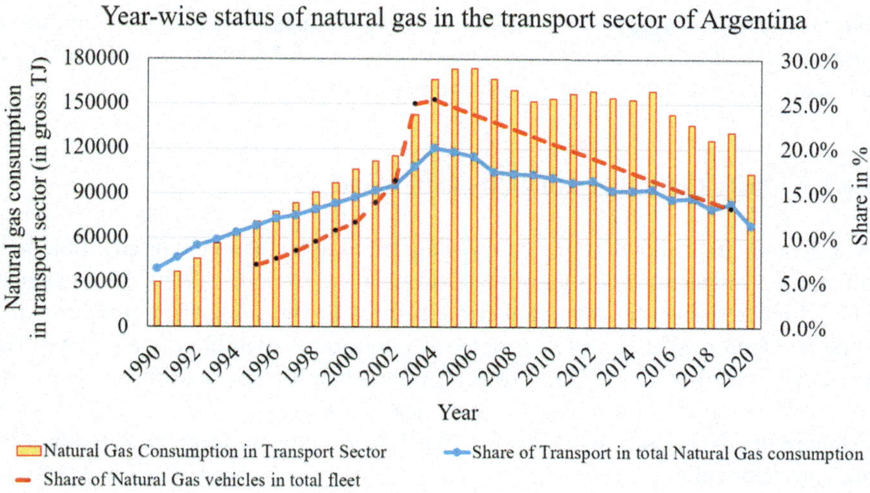

Fig. 5.15 Natural gas use in Argentina's road transport sector from 1990 to 2020. *Data source* Collantes and Melaina (2011)

5.6.1.1 Initial Government Will and Relevant Policies (Availability and Accessibility)

The government policies were targeted to not only promote CNG but also to discourage the use of diesel. The exploration, production, transport, and distribution of domestic natural gas were controlled by the state until 1992 to provide an active thrust towards the adoption of CNG (International Energy Agency Homepage 1999). Thereafter, from 1992 to 1993, the CNG ecosystem was privatized to invite private participation in the different stages of the fuel value chain (i.e. from exploration to distribution) (Vásquez 2016). The regulation of the natural gas ecosystem had a significant impact on capacity building with regard to the refuelling infrastructure.

5.6.1.2 Bringing Together the Stakeholders (Acceptability)

The relevant regulations, norms, and standards were designed and executed for on-board CNG equipment, CNG vehicle modifications, CNG dispensing equipment, and CNG stations (Bolay 2020). This reassured the public about the safety features of CNG fuel and vehicles for road transport. The government withdrew itself from the gas-based transport ecosystem and handed it over to the private sector. The private sector was roped in by assuring them of predictable profits, and this resulted in a robust natural gas transportation and distribution system (Collantes and Melaina 2011). While the government focused on pricing and regulations; the private sector invested towards building the enabling ecosystem. The government lifted the restrictions on the minimum distance between the refuelling stations if the fuel stations offered CNG.

The bureaucratic hurdles were reduced by putting the onus of vehicle certification and inspection on the distribution companies (Collantes and Melaina 2011). It enhanced the sense of ownership among the private players.

5.6.1.3 Price Control and Stability (Affordability)

A wide and quick adoption of CNG was made possible due to the price differential offered in comparison with gasoline (Collantes and Melaina 2011). Further, lines of credit were issued for the initial costs incurred on the conversion of the existing vehicle (Ogunlowo et al. 2015). Due to the domestic availability of natural gas, the prices were kept low by giving subsidies even when the prices of the counterparts increased rapidly (Vásquez 2016). The economic appeal of CNG led to the establishment of CNG-based industries for vehicle conversion to cater to both domestic and international markets.

B. Decline of the gas sector in Argentina's road transport. The measures mentioned above helped in bolstering the position of CNG as an alternate fuel to gasoline and diesel for two decades (1984–2004). In 2004, the demand for CNG in the transport sector of Argentina peaked, and thereafter, the demand started stagnating. The demand came down sharply post-2015 (see Fig. 5.15). The factors contributing to the decline in CNG consumption in Argentina's road transport sector are discussed below.

5.6.1.4 Decreasing Enthusiasm and Preferences of the Consumer

Several factors contributed to the reduced consumer adoption of CNG as a transport fuel since 2004. The primary reason was the decrease in the price difference between CNG and gasoline. As CNG was considered a '*fuel of crisis*' by many people and as the economy of Argentina grew rapidly, the consumers shifted back to gasoline due to more convenience (Collantes and Melaina 2011). CNG vehicles used more trunk space to store fuel, and they needed more refills since CNG provides a lesser riding range owing to a lesser energy density.

5.6.1.5 Lack of Manufacturing Facilities for CNG-Only Vehicles

Argentina mostly relied on the conversion of conventional vehicles by making them capable of running on both gasoline and CNG. Neither the manufacturing nor the rollout of CNG-only vehicles was promoted by the government, which effectively left the option of conversion with the end-users (Collantes and Melaina 2011). The consumers drifted slowly towards gasoline as the price differential between CNG and gasoline reduced with time, and the cost of conversion of the vehicle could not outweigh the savings in fuel costs. The long-term expansion of CNG could not take off since consumer behaviour did not tilt towards the spontaneous adoption of CNG.

The popularity of CNG was mainly attributed to the price differential with respect to gasoline, which caused the continued adoption of converted CNG vehicles for a period of twenty years (1984–2004).

5.6.1.6 Failure to Rope in Heavy-Duty Transportation

The credit for the success of the Liquid Fuel Substitution Program went to privately-owned vehicles (gasoline-based vehicles), not diesel-based heavy-duty vehicles (Collantes and Melaina 2011). The primary reason for the heavy vehicles not adopting CNG was the strong lobbying by the diesel industry and the pricing advantages given to diesel by the government. In addition, the increased cost associated with the conversion of the existing fleet and maintenance of two engines resisted the transition of heavy-duty transport towards CNG.

5.6.2 LNG in China

China is a vast country lying within five time zones geographically and it has a diverse geography that mostly lies in the temperate region of the northern hemisphere. For practical purposes, China can be visualised to be lying on the east side of the Hu Line, since nearly 94% of the Chinese population lives on 43% of the territorial land (Big Think Homepage 2022). The reason lies in the geographical setting of this region, where the land is flatter and wetter making it easier to farm, establish industries and build infrastructure (Sixth Tone Homepage 2022). With the rise of China as an emerging economic power, China used the opportunity as the host of the Beijing Olympics in 2008 to promote cleaner alternatives in the urban centres with the adoption of natural gas in public transport (Higashi 2009). China introduced LNG as a transportation fuel in 1961 (Yuan 2019). However, the major impetus for the same has been evident in the last 5 years (since 2017) (Smajla et al. 2019). Of the total LNG consumed in China, nearly 30% was consumed by the road transport sector in 2019 (Yuan 2019). LNG consumption in road transport in China has been driven by heavy-duty trucks (HDT) and the sales of LNG vehicles have picked up exponentially since 2017 (see Fig. 5.16). As per the reported data, China had about 3.43 lakhs LNG-driven vehicles in 2018, of which the HDTs had a share of about 70% and the rest were buses (Yuan 2019). The number of HDT trucks running on LNG rose to 5.82 lakhs in 2020 (S&P Global Homepage 2021). To cater to the fuel requirements of these vehicles, there is a network of more than 2500 LNG refuelling stations in China (Yuan 2019). The reasons that attributed to the rise of LNG in China are discussed below.

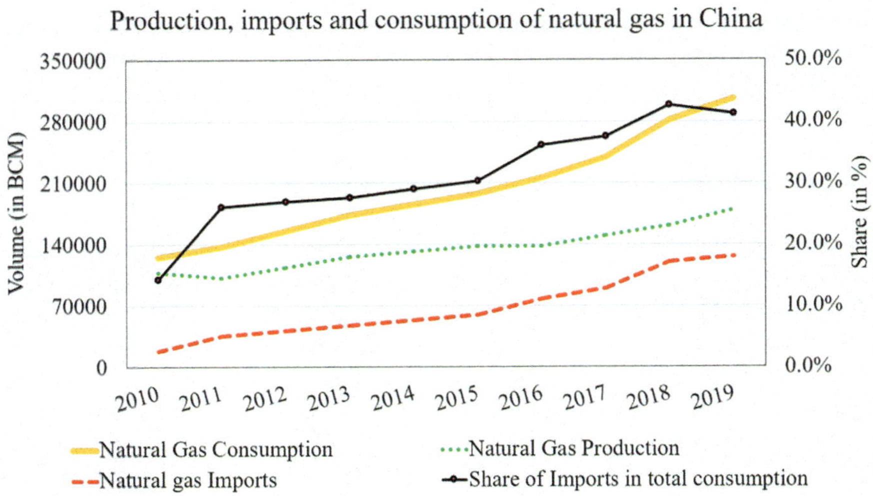

Fig. 5.16 Production, import, and consumption of gas in China and import share in consumption. *Data source* EnerData (EnerData Homepage 2014)

5.6.2.1 R&D, Policy, and Incentives (Acceptability)

R&D, government policy, and incentives act as the first step towards laying the background for the adoption of a new technology. The R&D for LNG utilization in the transport sector can be traced back to the 1960s. However, the research efforts attained maturity and became ready for commercialization in 2008 after the announcement of the National High-Tech R&D Program of China (part of the 863 Program launched in 1986) (Scientific American Homepage 2014). As a part of an active push from the Government, the public-run LNG buses were rolled out first, followed by the LNG adoption by the privately-owned HDTs (Yuan 2019). Fuel subsidies, procurement of LNG vehicles by the State, direct involvement of the State-owned enterprises, and relaxation on the business prerequisites facilitated the growth of LNG infrastructure in China. Stringent regulation enforcement on the fleet of diesel-based trucks motivated the owners to finally shift towards the LNG-based freshly manufactured vehicles (Zhao et al. 2021). The Three-Year Action Plan for Blue Sky Battle was launched in 2018 and it sought to eliminate diesel trucks that did not meet emission requirements, which provided further confidence to the investors in the LNG sector (Carnot-Gandolphe 2019).

5.6.2.2 Promote Cost Competitiveness (Affordability)

Cost-competitiveness acts as the most potent driver for the adoption of a new technological alternative against the prevailing options. In 2013, the cost of running a vehicle on LNG in China was about 20–30% lower than that associated with a diesel-based vehicle (Homepage 2014). This boosted the sales of LNG vehicles in 2013–14. However, the sharp drop in global oil prices in 2015–16 led to a drastic reduction in the price of diesel, followed by a decline in the adoption of LNG vehicles. Since the vehicles sold earlier were in circulation and the robust distribution infrastructure for LNG existed, the share of LNG in the energy mix of the Chinese road transport sector kept on increasing, although at a much slower pace (SP Global Homepage 2022). As shown in Fig. 5.16, LNG consumption in China has seen a steep rise since the global oil prices rose again from 2017 onwards.

5.6.2.3 Fuel Supply and Distribution Assurance (Availability and Accessibility)

Since China uses LNG in other sectors such as residential heating, industries, and city gas purposes, the year-long availability of LNG is indispensable (Columbia University Homepage 2022). As the need for LNG increased with time, China began to import LNG from central Asia through pipelines. In 2018, nearly 40% of LNG consumed in China was imported as reflected in Fig. 5.16. A shortage of LNG was seen in 2017–18 and this situation put a lot of pressure on the utilization of LNG by the transport sector since the available LNG was diverted for high-priority uses such as residential heating. However, the large fleet size of the on-road LNG vehicles and the existing robust LNG refuelling network have a spiralling impact on the growth in the adoption of LNG-based vehicles in China.

5.6.3 Auto-Gas in Turkey

Turkey, located at the confluence of Asia and Europe, lies on the eastern bank of the Mediterranean Sea. Therefore, it experiences a moderate climate throughout the year with average temperatures ranging between 0 and 30 °C (WorldData Homepage 2022b). Turkey also shares the border with one of the top five producers of crude oil (Iraq), as well as a leading producer of natural gas (Iran) in the world. However, Turkey imported nearly 75% of its demand for LPG in 2020 (IHS Markit Homepage 2021). Most of its LPG imports come via the waterway vessels from Algeria, the United States, Kazakhstan, and occasionally from the North Sea (IHS Markit Homepage 2021). The use of LPG in the road transport sector of Turkey started in the 1980s without any government interventions, mainly due to the economic advantages LPG offered to passenger car owners (Karamangil 2007). The domestic LPG tanks were fitted into the vehicles by the local mechanics without much technical and safety analyses. Due to the lack of initial push and promotion from the government's side for using LPG as a motor fuel, auto-gas continued to be used in Turkey's road transport sector in a scattered and sparse manner until 1995. After LPG was approved as a motor vehicle fuel in 1995, the conversion of conventional fuel systems to LPG systems was allowed by the Turkish Ministry of Industry and Commerce (Camadan 2013). As of today, Turkey is the leading nation in terms of using auto-gas in the road transport sector. Turkey accounts for 11% of the global consumption of auto-gas with a network of 10,000 distribution stations across the country (The Economic Times Homepage 2020). The success of auto-gas in Turkey is attributed to the following factors.

5.6.3.1 Government Policies, Certifications, Policies (Acceptability)

Following the approval of LPG as a transport fuel in 1995, the conventional vehicular fuel systems in Turkey underwent conversion to LPG-based vehicles following the technical specifications and international standards for the LPG kit specifications (Karamangil 2007). Later, more institutions from the transport sector were roped in to expedite the conversion of vehicles, and to prevent the illegal conversion of the prevalent vehicles. It led to the sudden spurt of growth of LPG-based vehicles. In the early 2000s, the Turkish government agencies entered the LPG-based automobile ecosystem. The government heavily taxed this sector to reap the benefits without any energy and environmental policy in place. This move had negative repercussions on the adoption of LPG-based vehicles. The tax rate of LPG was raised to 56% in April 2001, further to 104% in June 2001, without distinguishing between the application types (i.e. auto-gas and '*kitchen fuel*') (Karamangil 2007). This *single taxation policy* for both the uses of LPG was withdrawn in August 2002 and the *social protection policy* was kicked in to protect the domestic use of LPG in household kitchens. The higher rates for the auto-gas prompted the LPG distributors to buy LPG from the refineries for its domestic use in the kitchens. However, they diverted the fuel towards the road transport sector for its consumption as auto-gas. Lack of vision and inappropriate measures for the auto-gas sector did not allow the Turkish LPG market to mature and consolidate for quite some time (till 2005).

5.6.3.2 Pricing Advantage Over Petrol and Amortization (Affordability)

For a long time, LPG had been called 'kitchen fuel' in Turkey. Prior to 2000, this fuel enjoyed low taxation from the government, considering its wide use in the kitchens irrespective of household income levels. The price differential between gasoline and the '*kitchen fuel*' was as high as nearly 70%, which triggered the conversion of gasoline-based automobiles to become LPG-based. This trend continued till the late 1990s, and even high-end vehicles were converted to become LPG-based bi-fuel vehicles (Karamangil 2007). The reduced use of gasoline by the passenger cars dented the government's revenue collection from the taxes on gasoline. At the same time, heavy taxes were imposed on the auto-gas sector in the early 2000s since no definitive policy guidelines were available for LPG-based transport in Turkey. This led to chaos in the auto-gas sector resulting in bankruptcies of small-scale enterprises, and a high level of unemployment. However, there was a revision in the policy guidelines in 2005 and the prices and regulations were framed considering both economic as well as environmental aspects. This resulted in the sustainable growth of the LPG-based road transport ecosystem in Turkey (Camadan 2013). Currently, around 4.7 million cars in Turkey are adapted to use auto-gas, which accounts for about 40% of the total car fleet in the country. As per the 2018 data, the total volume of LPG consumed in Turkey was 4.1 million tonnes, of which around 80% could be attributed to the auto-gas used for road transport. In 2018, Turkey emerged as the largest auto-gas market in the world followed by South Korea (Autogas.net Homepage 2020b).

5.6.3.3 Easy Availability and Accessibility (Availability and Accessibility)

Although the Turkish government approved the use of LPG as a transport fuel in 1995 through an executive decision, a lack of legislation limited the expansion of the certified converters for the legal conversion of gasoline-based vehicles to the LPG-based version. This resulted in rampant illegal conversion of automobiles since LPG was substantially cheaper than gasoline. The enactment of the Liquefied Petroleum Gases Market Law in 2005 reformed the auto-gas market in Turkey by introducing new guidelines and pertinent regulations. Under this law, the distribution and transportation of auto-gas as well as the handling of LPG cylinders throughout their life cycle were regulated by a single independent authority (Camadan 2013). Currently, Turkey has the largest auto-gas refuelling network in the world, measured in terms of the number of dispensing sites (Autogas.net Homepage 2022c). In 1997, the auto-gas (LPG meant for the transport sector) business was handled by LPG distribution companies, which supplied LPG to households in cylinders, and in bulk to the industry. During the early years of market expansion (1995–97), the sizeable demand could not be met easily through early-stage infrastructural support. However, with the organized efforts of the LPG distributors to meet the growing demand, auto-gas consumption increased enormously within three years and reached nearly 1.3 million tonnes in 2000 (Karamangil 2007). This success eventually prompted the gasoline distributors (who did not keep auto-gas earlier) to diversify the product mix by adding auto-gas to their stations. For the gasoline distributors, this involved relatively smaller fresh investments in their existing refuelling stations, yielding large additional profits, which also greatly increased their shares in the auto-gas market. During the period of rapid market expansion (2005 onwards), the availability of LPG at the distribution stations increased significantly. However, increased competition among multiple distributors led to reduced profit margins for the distributors, despite the increase in overall auto-gas consumption (Autogas.net Homepage 2022c). Turkey has the second highest number of auto-gas dispensing sites globally (~5000) after Poland (~6000) (Indian Auto LPG Coalition Homepage 2022c).

5.6.4 Ethanol Blended Petrol (EBP)/Biofuels in Brazil

The rise of biofuels in Brazil is directly linked to the first oil crisis of 1973 when the Arab-Israel Yom Kippur war led to an unprecedented rise in the prices of oil (as high as 4 times) (Gee and McMeekin 2011). This prompted Brazil to reduce its dependency on oil since 80% of the demand used to be met by imports. At the same time, Brazil had a surplus stock of sugarcane owing to the fall in its exports. In order to tackle both the problems using a common thread of solution, the Government launched the Ethanol program, also known as '*ProAlcool*', aimed mainly at crude oil import substitution (Goodrich 1981). Sugarcane is grown in most of the Brazilian states. However, it is heavily grown in the south-central region, which has favourable soil and climate conditions for its cultivation. The state of São Paulo contained more than 55% of the total sugarcane harvested area of Brazil in 2018 (Zheng et al. 2022). Apart from favourable natural conditions, the availability of abundant land and low-cost labor boosted sugarcane production. The sugarcane and ethanol industry has grown to huge proportions in recent times, and Brazil accounted for nearly 38.6% of the world's total sugarcane production and 29.5% of the world's total ethanol production in 2019 (Zheng et al. 2022; Alternate Fuels Data Center Homepage 2022d).

5.6.4.1 Availability of Raw Material and Flex Vehicles Coupled with Innovation (Availability)

In the mid-1970s, ethanol was blended with petrol at 5–10% by volume to promote its use and consumption in the road transport sector of Brazil. The Brazilian government mandated the blending of anhydrous ethanol into gasoline for gasoline-powered light vehicles at 15% by volume in 1982, 22% in 1992, and 27% in 2015 (Abel et al. 2021). With the availability of E27 and E100 in the Brazilian market, the average blend of ethanol in gasoline has reached 48% in 2019 (The Indian Express Homepage 2020). The availability of ethanol increased from 158 million gallons in 1975 to 3330 million gallons in 2002 to 8570 million gallons in 2019, aided by government investments in R&D and capacity building for higher production (Alternate Fuels Data Center Homepage 2022d; Meyer et al. 2014). Subsequently, the prices came down substantially and became comparable to that of oil, which resulted in the removal of subsidies provided to it (Goldemberg et al. 2004). Over the two decades (1985–2003), the demand for ethanol-only vehicles dropped heavily as the prices of ethanol increased due to a reduction in domestic output and increased imports. The introduction of flex-fuel engines in 2003 resulted in a further push for ethanol-only fuel as a major transport fuel (Freitas and Kaneko 2011). In 2021, more than 80% of the vehicle sales in Brazil belonged to the flex-fuel vehicle category (Statista Homepage 2022).

5.6.4.2 Policy Stability and the Role of Institutions with a Vision (Acceptability & Accessibility)

Brazil persisted with the use of ethanol in road transport even when the prices of gasoline came down during the period 1985–2000. This policy stability resulted in the continued expansion of the ethanol-based ecosystem. The government, the private sector manufacturers, the cooperative of sugar producers, the regulatory bodies, and the research institutions joined hands together at different junctures to create a market for both fuel and vehicles (Meyer et al. 2014). Intensive research efforts encompassing the fuel and the vehicles continued in close coordination with the stakeholders. The prices and supply chains were managed for ethanol vis-à-vis gasoline, and the distribution channels were created under state supervision (Carvalho 2013). *Petrobras*, a public enterprise, built the distribution and pump infrastructure for pure ethanol, as opposed to ethanol-blended gasoline. The coordination and cooperation among the various institutions resulted in economies of scale for ethanol as a dominant transport fuel in Brazil.

5.6.4.3 Price Optimization (Affordability)

Brazil adopted a preferential tax regime that ensured the attractiveness of ethanol as a fuel. Parallelly, innovation and investments towards developing ethanol as a stand-alone transport fuel (devoid of any blending) boosted the confidence among consumers about the long-term prospects of this fuel as well as the reliability of the flex-fuel engines (Carvalho 2013). The ethanol pump price was fixed at 59% of the price of gasoline, and the ethanol-only vehicles were subjected to lower sales tax and licensing fees as compared to the gasoline-based vehicles in the 1980s (Lovins and Datta 2005). Since ethanol has a lower specific energy content than gasoline, ethanol pump prices had to be lower than that of gasoline to remain competitive. Once the economies of scale were achieved, the price volatility of ethanol became much lesser. Since the commencement of the Brazilian Ethanol Program, the cost of production for ethanol came down by more than two-thirds over 30 years (from 1 USD/liter in 1975 to 0.3 USD/liter in 2005) (Meyer et al. 2014). Because of the present-day competitiveness of ethanol, its prices are no longer regulated to keep it artificially low.

5.6.5 Biodiesel from Indonesia

Indonesia is an equatorial country, where the annual average rainfall in the range of 2,000–3,000 mm with more than 200 rainy days per year, relative humidity ranging between 75 and 80%, and ambient temperature in the range of 22–32 °C provide the ideal conditions for the optimal growth of palm trees (Ahmed et al. 2021). Indonesia launched a massive deforestation drive for large-scale commercial plantations of palm trees, and today it accounts for nearly 60% of the global supply of palm oil (Mongabay Homepage 2022). Indonesia started considering alternate sources of fuel (especially, biofuels as blends) in the mid-2000s when its crude oil production started decreasing and was not adequate for meeting the domestic demand (which forced Indonesia to commence import of oil in 2004) (Silalahi et al. 2020). However, the use of B20 (i.e., the blend of 20% biodiesel by volume with conventional diesel) for all the sectors was formally authorized by regulation *41/2018* released in 2018 (Rahmanulloh 2018). Since early 2020, Indonesia has mandated the use of B30 (which is the highest biodiesel blending percentage in the world), to reduce its import dependence on diesel (Reuters Homepage 2022). Despite the high economic and environmental costs, the country's persistence with biodiesel as an alternate fuel has been possible because of the following enablers.

5.6.5.1 Availability of the Raw Material (Availability)

Since Indonesia is the largest producer of palm oil (a first-generation source of biodiesel) globally since the mid-2000s, the raw material needed to produce biodiesel is abundantly available. Since the first-generation biofuels have triggered an intense debate in Indonesia regarding fuel competing with food and resource sustainability (Yasinta and Karuniasa 2021), the country needs to intensify research efforts towards developing biodiesels based on the second and the third-generation feedstocks for sustaining the commercial availability of biodiesel in the transport sector.

5.6.5.2 Positive Impact on Society and Government Perseverance (Acceptability)

With the increase in palm oil production, the income of small farm holders has also increased (Euler et al. 2017). In 2005, presidential regulations (Nos. 1 and 5) were rolled out to promote and mandate the use of biodiesel blends in the energy mix (Silalahi et al. 2020). The coordination between the different stakeholders (government, biodiesel producers, and distribution companies) has ensured a continued rise in the biodiesel blending percentage since 2009.

5.6.5.3 Appropriate Pricing (Affordability)

The government collects a special fee called 'Palm Oil Estate Fund' (under presidential regulation no. 61) on the export of palm oil, which is used to reduce the gap between the prices of diesel and biodiesel and to promote R&D on the feedstocks of next-generation biodiesels (Rochmyaningsih 2019). This fund is not utilized to compensate for the lower energy content of biodiesel (33 MJ/l) as compared to that of petro-diesel (36 MJ/l) in terms of prices at the pump.

5.6.6 Hydrogen (in ICE-Based Vehicle Commercial Pilot Projects) from the World

Hydrogen-based ICE vehicles are among the most thoroughly researched alternate fuel-driven options for the future to achieve deep decarbonization in the transport sector. They are at a very advanced stage of research and pilot projects have come up, although the commercial deployment has not yet taken place. Japan-based Toyota is one of the leading enterprises engaged in cutting-edge research on hydrogen-based ICE vehicles (Green Car Reports Homepage 2021). The collaborative efforts among the leading Japanese automotive manufacturers Subaru, Mazda, Toyota, Kawasaki, and Yamaha are aimed at enhancing research and development initiatives for hydrogen combustion engine-based projects (Green Car Reports Homepage 2021). These vehicles are intended to provide an equivalent level of drivability, range, and safety as compared to conventional fuel vehicles.

The immediate availability of hydrogen combustion engines coupled with extensive knowledge regarding engine production, durability, and maintenance as well as the capability of combustion engines to run on both hydrogen as well as conventional fuels (gasoline) provide us with a bridging technology that can promote the growth of full-fledged hydrogen infrastructure in the medium-to-long term. The factors that render hydrogen a high-potential combustion engine alternate fuel are detailed in Sect. 5.3 of this chapter. However, there are several identifiable roadblocks that have restricted the desired growth of the hydrogen-ICE vehicles, despite the proven mettle of such engines for the converted gasoline-based vehicles, bi-fuel vehicles (hydrogen with gasoline and hydrogen with diesel), and dedicated hydrogen vehicles.

5.6.6.1 Emissions and Efficiency (Acceptability)

The generation of hydrogen from clean sources is essential for it to be a viable alternate fuel source in the transport sector with respect to the existing fossil fuels. Since its biggest advantage lies in the near-zero TTW emissions, this fuel needs to be produced from electrolysis with electricity generated from green sources. Currently, almost all the hydrogen generated on the planet for energy purposes is generated from fossil fuels (natural gas or coal gasification) (Rocky Mountain Institute Homepage 2019). Figure 5.11 shows that in India, where currently grey hydrogen is the only hydrogen option that is produced and used, the GHG emissions from hydrogen-based vehicles (during its WTW emission cycle) would turn out to be 50% higher as compared to the gasoline-based counterpart. As per a study done in India, NO_x emissions associated with both gasoline- and hydrogen-based ICE vehicles are low, exhibiting similar values (<5 g/kWh) at engine speeds higher than 2000 rpm (Natarajan et al. 2013). At lower rpm values, NO_x emissions turn out to be very high (>90 g/kWh) for gasoline-only vehicles, whereas for hydrogen it remains lower than 5 g/kWh even at lower engine speeds (Natarajan et al. 2013). Most of the hydrogen-ICE vehicles use spark-ignition with port-fuel injection technique, and NO_x emissions can be easily minimized by using a standard 3-way catalytic converter (Verhelst and Wallner 2009). NO_x emissions can also be mitigated by maintaining optimal stoichiometric air-fuel ratios for the hydrogen fuel in the combustion chamber (lean operation) (Das 2016). However, the drawback associated with the port-fuel injection (PFI) method is 'backfiring' due to a very low quenching gap associated with hydrogen (0.064 cm) (Das 2016). This can be overcome by using the direct injection technique (Verhelst and Wallner 2009). The direct injection (DI) technique is in the advanced stages of development and is still some distance away from being commercially available. This technology is also likely to be more expensive as compared to the port-fuel injection counterpart. The market insights suggest that the PFI technology will arrive by 2023, whereas the DI is expected to reach the market by 2025 (SAE International Homepage 2022).

The low efficiency delivered at the wheels by the hydrogen-based combustion engine (~20–25%), coupled with higher storage requirements (4 times more in the case of liquid hydrogen and 19 times more in the case of gaseous form as compared to gasoline) render it a weak competitor against hydrogen-based fuel cell alternative (efficiency of 30–50% at wheels) and battery electric vehicles (efficiency of 39–76% at wheels) (Albatayneh et al. 2020). Therefore, the hydrogen-ICEs would prove to be beneficial only for heavy-duty applications that are difficult to electrify, e.g. trucks with a daily haul of 250 miles (400 km) or more.

5.6.6.2 Creation of Distribution Networks (Accessibility)

The number of hydrogen fuel stations has been limited in the countries where hydrogen fuel cell vehicles have found some takers. This substantially restricts the movement of a hydrogen vehicle beyond a certain range. The estimated capital cost of setting up an average hydrogen station is around $1.9 million (U.S. Department of Energy Homepage 2020). This prohibitively high cost tends to discourage the proprietors to promote the creation of hydrogen infrastructure. Further, while filling up a hydrogen vehicle fuel tank takes only a few minutes, the consumers would have to wait at the filling stations since sufficient pressure needs to build in the storage tank at the distribution outlets to completely fill the vehicle tank (FirstPost Homepage 2021). In the case of mass adoption of hydrogen ICE vehicles, congestion at the distribution stations is highly likely. Safety also remains an area of concern for hydrogen storage facilities at the distribution stations. High-intensity explosions at hydrogen refuelling and storage facilities in Norway and South Korea in the recent past have raised questions about the safety of hydrogen (which is highly flammable) for mass uptake (FirstPost Homepage 2021). These incidents have prompted the local communities to oppose the establishment of new hydrogen refuelling stations and production facilities in their vicinities.

5.6.6.3 Clean and Green Raw Material (Availability and Affordability)

Currently, green hydrogen is 2–3 times more expensive as compared to blue hydrogen, and further cost reductions are needed to ensure mass uptake. The cost of renewable electricity needed to power the electrolyser unit is the largest single-head cost for the on-site production of green hydrogen (IRENA 2020). This renders the production of green hydrogen more expensive than blue/grey hydrogen, regardless of the cost of the electrolyser. Therefore, the low energy price for renewable electricity is a key enabler for producing green hydrogen at competitive market prices.

Low electricity cost alone is not enough for competitive green hydrogen production, and further reductions in the cost of electrolysis facilities are also needed. The cost of an electrolyser is the second largest single-head cost component associated with green hydrogen production (IRENA 2020). This calls for a need to reduce investment costs for electrolysis plants by 40% as compared to the current level in the short term, and by up to 80% in the long-term (IRENA 2020). Increasing the electrolyser plant module size from the current typical sizes of 1–10 MW to larger sizes (20 MW or more), automated electrolyser-stack production envisaging large requirements coming from the GW-scale manufacturing facilities, efficiency and flexibility in operations, and diversification of the use of hydrogen for promoting economies of scale can help in bringing the prices down. However, as per the study by the International Renewable Energy Agency (IRENA), the cost reductions in electrolysers cannot compensate for the high prices of 'green electricity' (IRENA 2020). Therefore, the right policy directions are necessary for any country envisaging a hydrogen-driven economy, and, in particular, a hydrogen-based road transport sector.

5.7 Pairing of Fuels and Vehicles

Following Sect. 5.5 of this chapter, vehicle-alternate fuel pairing is done by considering the functional uses of the vehicles and the essence of the 4A framework analysis for the ICE-based fuels. Tables 5.4 and 5.5 summarize the suitable pairings between the vehicle segments (elaborated in Sect. 5.2) and the alternate fuel options for the road transport sector in India, by taking into consideration the technical availability, resource availability, fuel accessibility, cost of ownership (affordability), and social as well as environmental acceptability.

5.7.1 Two-Wheeler (2W) Segment

The 2-Wheeler segment in India largely comprises motorcycles and scooters and nearly all of the ICE-based 2Ws are petrol-only vehicles. Therefore, ethanol can be a highly likely alternate fuel option to petrol for the 2Ws in India with the introduction of a flexible engine. Since India's green energy aspirations pertinent to the transport sector largely revolve around electric motor-based green mobility, the incentives for the energy diversification in the ICE-based conventional 2-wheelers would possibly remain limited in the foreseeable future due to the rapid adoption of e-mobility by the 2W segment in India. This indicates that, although ICE-based 2W vehicles can potentially run on several fuel options, such as diesel, CNG, LPG, and hydrogen, it is unlikely that these fuels will be commercially deployed in the 2-Wheelers in India. One of the most important considerations for the 2W vehicle owners in India is the fuel economy provided by the vehicle (Bansal et al. 2021). In India, the ownership of a 2W vehicle (scooter or motorcycle) is directly proportional to income levels (ThePrint Homepage 2022). It makes the adoption of a completely new technology slow in case of a large price differential from the existing technology. Further, the 2W vehicles have a large presence across the country, including the deeper pockets of India (ThePrint Homepage 2022). Therefore, the accessibility of the distribution outlets for commercially deployed fuel is a very crucial factor for 2W vehicle users. Looking at the aforesaid aspects, an ethanol-based FFV is the most suitable ICE-based alternative to petrol-only 2Ws in India in the near-to-medium term since the infrastructure requirement, local availability of ethanol feedstock, apt pricing with respect to gasoline, and lower emission levels make the social acceptance of ethanol very high.

5.7.2 Three-Wheeler (3W) Segment

The 3-Wheeler segment in India largely comprises the commercial vehicles that move both passengers (~75% of total 3Ws) and goods (~25%). It is mostly used for the last mile connectivity in India by the common public in both rural as well as urban areas (The EarthBound Report Homepage 2019). Therefore, the cost of travel (on a per km basis) remains paramount in this mode of transportation. Since this segment generates a wide range of employment opportunities (drivers, auto-mechanics, and people associated with supply chains), the 3Ws have witnessed a wide array of experimentation regarding the commercial deployment of diverse energy sources for the ICEs. In India, particularly, the 3W segment has been at the forefront of the adoption of new technology, be it CNG-only, LPG-only, bi-fuels (petrol-CNG, petrol-LPG) or e-3Ws (BEVs). The key considerations for the 3W vehicle segment are the total cost of ownership, compactness of fuel storage tanks, safety of the vehicle, and accessibility of fuelling stations (WRI India Homepage 2020). Since the storage requirements and initial infrastructure costs are high for both hydrogen and LNG, the deployment of these alternate fuels in India's 3W segment is highly unlikely. Ethanol and biodiesel can be easily adopted by the 3Ws in India as they can be easily blended with conventional gasoline, and petro-diesel, respectively. Hybrid vehicles (running on both an electric motor and a combustion engine) are less likely to make a significant mark as well, primarily because of the high initial costs involved as compared to dedicated-fuel-based vehicles.

5.7.3 Four-Wheeler Passenger (4W-P) Segment

The 4-Wheeler passenger segment in India is the most diverse vehicle segment in India in terms of functional use, initial cost involved, vehicular technology, as well as personal choices and purchasing capacities of the consumers. The vehicles in this segment are used for both personal as well as commercial purposes, such as cabs or taxis. Along the lines of fuel diversification in the 3-Wheeler segment, the taxis in the 4W-P segment are also much diversified in terms of fuel used. Among the 4W-P vehicles, there are various sub-categories based on their functional roles, such as hatchback, sedan, sports utility vehicle (SUV), multi-purpose vehicle (MPV), cross-overs, and luxury cars, etc. (Karunakar 2017) The hatchback buyers are generally first-time 4-wheeler vehicle owners who prioritize mileage over any other parameter. The sedan owners tend to give importance to cabin comfort during the inter-city and intra-city rides. Therefore, among cab owners and customers, a sedan is the most preferred sub-category in the 4W-P segment. The SUVs are designed to perform better on off-roads, and therefore, they are expected to offer better performance in terms of engine power and torque. The MPVs are meant to perform multiple functions (running under different conditions, carrying heavier loads, etc.), and therefore, the carrying capacity (especially, boot space) and engine performance are of paramount importance.

The accessibility of the fuel remains important for all the sub-categories since the 4W-P owners are likely to prefer comfort over cost (Chugh et al. 2011). Therefore, this segment can be even targeted by those alternate fuel options in the short term for which the cost of infrastructure is higher. It makes this segment (4W-P) the most sought-after category by the electric motor-based automobile manufacturers (battery-based as well as fuel-cell based). Hydrogen and LNG are not likely to be potential alternatives for the combustion engine in this segment due to the large storage volume required by the fuel tanks. Furthermore, they would require more frequent refills as compared to the incumbent fuels (gasoline and diesel), which is likely to demotivate the private owners due to the 'range anxiety'. The bi-fuel vehicles from the 4W-P segment (such as Tata Tiago, Hyundai Aura, Maruti Suzuki Dzire, etc.) running on both petrol and CNG are already commercially deployed in India. The automobile industry is looking forward to expand the usage of multi-fuel vehicles (Petrol-CNG vehicles), FFVs, and hybrid vehicles (hybrid Honda City and MG Hector launched in 2022) in the near-to-medium term.

5.7.4 Four-Wheeler Non-Passenger (4W-NP) Segment

The 4-Wheeler non-passenger segment in India is a niche segment from the point of view of the vehicle chassis framework, i.e. a monocoque chassis (Force Motors Limited Homepage 2022). All the vehicles in this segment have similar drivetrain characteristics but they are divided into sub-categories based on the specific functional use. The maxi-cab or tempo-traveler has the largest share in this segment, while the other sub-categories such as ambulances, and camper vans are also used frequently. Akin to the MPVs (in the 4W-P segment), safety, vehicle performance, and interior space remain the most important criterion for the 4W-NP vehicle segment. This renders the deployment of dedicated gas-based alternatives such as LPG, LNG or hydrogen unlikely in the near-to-medium term. Currently, the 4W-NP vehicles run on a dedicated fuel system (petrol or diesel or CNG) based on the daily travel distances. Flex-fuel and hybrid vehicles can be deployed in the 4W-NP segment since they have the potential to fulfil the functional requirements over a long period.

5.7.5 Heavy Vehicle—Bus (HV-B) Segment

Buses are heavy vehicles that are widely used for mass mobility in India. They cater especially to intra-city transit, as well as long-distance inter-city travel. They can be state-owned (public transport) as well as privately owned (by educational institutions, industries, and large enterprises). Since the buses form a crucial part of the multi-modal public transport system (comprising sub-urban railways, metro, and even mono-rail services) and a large fraction of the buses running in India are state-run (LiveMint Homepage 2021b), the HV-B segment attracts major attention from the administration towards decarbonization of the public transport sector. The omnibuses in the HV-B segment are slated for inter-city travel, especially in the areas where cheaper rail connectivity is either limited or unavailable (The Economic Times Homepage 2022). The adoption of alternate fuels for different sub-categories of buses will largely depend upon the functional requirements over a long period. The city buses in India mostly run either on diesel or CNG as the single fuel in the vehicle. The current use of single fuel is largely due to the limited daily distance covered by the city buses as well as the availability of the fuel at the local bus depots. Therefore, depending on the creation of the required infrastructure and availability of the fuels, it is possible to transform the city bus sub-segment towards adopting alternate fuel options such as LNG, ethanol, and hydrogen. In contrast, long-distance inter-city buses are more likely to adopt hybrid and flex-fuel vehicles to minimize emissions as well as running costs. In this pursuit, ethanol, biodiesel, and blended fuels can be adopted in the near term. As the infrastructural network for alternate fuels widens, even hydrogen and LNG can be adopted for low-emission long-distance travel in the long term.

5.7.6 Heavy Vehicle—Municipality (HV-M) Segment

The heavy vehicle-municipality segment in India consists of those vehicles that are owned by the local governing bodies for the upkeep of municipal areas. Therefore, the overall count of these vehicles is very less in the country. However, the quest for improved municipal services in progressive urban settings will lead to some growth in this segment. Therefore, this segment cannot be overlooked as India embraces rapid urbanization. HV-M vehicles are strictly used within a limited area, and they perform specific functions such as firefighting, towing, and recovery of breakdown vehicles, etc. Almost all of these vehicles run on diesel due to the heavy-duty performances required from them. The likely alternate fuel option for this segment can be biodiesel which can be used in the existing powertrains. Dedicated-fuel system based on CNG or LNG or ethanol or hydrogen can also emerge based on the availability of a particular fuel within the municipal limits.

5.7.7 Heavy Vehicle—Agriculture (HV-A) Segment

The heavy vehicle—agriculture segment is a large segment in India comprising tractors, trailers, and harvesters. These vehicles mostly use conventional diesel. As these vehicles run in those areas that are often inaccessible through roads, the easy storage of additional fuel in separate containers near the vehicles is a must. It rules out the use of gas-based alternate fuels for running the combustion engine in the HV-A segment, since the storage requirements for all the gas-based fuels are stringent in terms of safety attributes. Therefore, biodiesel is the only viable non-fossil-based alternate fuel for this segment.

5.7.8 Heavy Vehicle—Construction (HV-C) Segment

The heavy vehicle—construction segment is similar to the HV-A segment in terms of its use at remote locations with limited fuel accessibility, and engine power requirements. Therefore, nearly all construction vehicles are currently powered by conventional diesel since it is easy to store in containers without many safety concerns. Therefore, the only plausible alternate option for the HV-C segment also is biodiesel.

5.7.9 Freight Vehicle (FV) Segment

The freight vehicle segment is the trickiest segment towards achieving the dual objective of decarbonization and energy diversification. This segment is the largest contributor to the harmful emissions in the road transport sector which necessitates immediate adoption of clean and green energy sources. However, the comparatively low number of vehicles in the FV segment (currently accounts for only ~4% of the total vehicle fleet in the country) tends to attract a lukewarm response from the manufacturers regarding any transition (Ministry of Road Transport & Highways

Homepage 2022). The vehicles from this segment are used solely for commercial purposes starting from intra-city load transportation to long-haul cross-country goods transport. With increasing travel distance, the load-carrying capacities of the vehicles tend to be higher to reduce the cost per km. The cost of transporting the goods tends to be higher for both small freight vehicles going over long distances and the heavy freight vehicles being used for short-distance transportation (Aljohani and Thompson 2020). Considering the load-carrying capacities of the vehicles, the FV segment can be divided into four broad categories, viz., pick-up (payload carrying capacity of 1.5 tonnes or less), light commercial vehicle (payload of 1.5–5.5 tonnes), intermediate commercial vehicle (payload of 5.5–10.5 tonnes), and medium and heavy commercial vehicle (load capacity of 10.5 tonnes or higher). Since this road transport segment is particularly polluting, the long-term vision towards the de-carbonization of this segment must encompass other cheaper and greener modes of transport such as railways (fully electrified Dedicated Freight Corridors).

Considering the technical availability of the fuel options discussed before, and the existing vehicle technologies, all the gas-based alternate fuel options remain plausible for the FV segment. However, in order to shift towards a new alternate fuel option, it is indispensable to ensure fuel availability at the fuel stations, and fuel accessibility on the expressways, national highways, as well as state highways across the country. This would require a massive investment to create the clean fuel infrastructure, and a robust long-term planning will be needed to deploy the same in a phased manner such that any undue disruption can be avoided. A completely new infrastructure for LNG and hydrogen will have to be implemented in a top-down manner with the medium and heavy commercial vehicles embracing the transition first. In parallel, CNG or ethanol can be adopted by the pick-ups and light commercial vehicles as the alternate fuel options in the near-to-medium term since these vehicles will be largely transporting the goods to the towns and cities that lie at a considerable distance from the expressways and national highways. Since CNG and ethanol have ready takers in the form of 3-Wheeler vehicles, 4-Wheeler passenger cars, and buses within the city range, fuel diversification will be much easier for pick-ups and light commercial vehicles. The feasibility of such a system will require the dry container port such as Inland Container Depots (ICDs) along the expressways and the national highways where the containers can be transferred from a medium/heavy commercial vehicle to a light one (pick-up or light commercial vehicle) and vice versa. Such a supply chain network would require heavy investments and intensive logistical support. However, the country should actively push for such a transition since it has the potential to not only generate huge employment but also result in a substantial reduction in India's emission profile attributed to heavy-duty transport.

5.8 Lessons for India from the Global Experiences

All the case studies discussed in Sect. 5.6 of this chapter point towards a specific set of factors that have led to the complete or partial success in the deployment of alternate fuel in the road transport sector. Some of the lessons can be imbibed by Indian policymakers as the Indian road transport sector is on the cusp of transitioning towards a greener energy mix.

5.8.1 Long-Term Availability of the Resources for Alternate Fuel

The domestic availability of the raw materials for an alternate fuel would provide the necessary confidence to the government towards devising a long-term plan around that fuel. From the global experiences discussed in Sect. 5.6, the natural gas availability in Argentina, sugarcane production in Brazil ensuring availability of ethanol, and palm oil plantation in Indonesia ensuring biodiesel feedstock availability indicate the significance of domestic raw material availability. Domestic availability also leads to price stability of the fuel, sustained R&D initiatives, and capacity building for the fuel ecosystem. Such enablers would help in withstanding the possible volatilities in the demand for the alternate fuel driven by the sudden changes (global reduction in the crude oil prices) with regard to the incumbent fossil fuels (petrol and diesel) in road transport.

5.8.2 Impetus from the Government for Adoption of Alternate Fuels

Since a new fuel ecosystem requires heavy investment for building infrastructure, the initial push for the adoption of an alternate fuel should come from the government. Both long-term vision and short-term plans are important while the regulations and guidelines are framed. Extensive stakeholder consultation needs to take place prior to the formulation of a strategy for the large-scale adoption of any new or alternate fuel. Further, the government needs to ensure that the bureaucratic hurdles are minimized to facilitate ease of doing business. The role of the government is very much evident from the CNG adoption in Argentina, LNG promotion in China, and ethanol deployment in Brazil where the respective governments played a pivotal role in integrating the efforts of various institutions for building an ecosystem. The role of the State is indispensable during the initial years of commercial deployment where the state-owned public companies would ensure the demand generation (as seen from Chinese initiatives regarding LNG) and strengthen the supply chain using the state-owned logistics and distribution outlets (as evident from *Petrobras* activities in Brazil). On the contrary, an absence of active involvement from the government side had resulted in an unorganized auto-gas ecosystem in Turkey, where a long time was taken to embrace the necessary reforms, which were acquired through multiple setbacks in the market and subsequent incremental corrective measures between 1995 and 2005.

5.8.3 *Vital Role of Regulations, Policy Guidelines, Standards, and Institutions*

The creation of a commercial deployment ecosystem for any fuel involves substantial planning, investment, and effective coordination between the government and the industry. Therefore, it is necessary to create independent bodies with administrative and financial autonomy to facilitate informed decision-making in consultation with the stakeholders. Government representatives may be appointed in such institutions to oversee the state of affairs. However, the direct involvement of the government departments in the daily operations and decision-making processes of such bodies may be avoided to get rid of bureaucratic bottlenecks. The international standards for fuel generation, fuel dispensation, fuel storage, and fuel use in vehicles should be reviewed carefully to craft the appropriate guidelines specific to India so that the safety aspect is not compromised. This would ensure confidence in the consumers and the investors towards sustained adoption of an alternate fuel option. The experiences of CNG adoption in Argentina, LNG in China, ethanol in Brazil, and biodiesel in Indonesia indicate that substantial time and effort are needed to create the enabling atmosphere for transitioning to a new fuel. For example, China has persisted with the scrapping of the old heavy-duty trucks which did not meet the emission requirements, while the initiatives were taken towards making LNG adoption attractive through other means.

5.8.4 *Sustained Promotion of the Ecosystem in Place in the Face of Challenges*

A strong presence of incumbent fuels (petrol and diesel) in the energy mix of the road transport sector of a country makes it difficult to switch to a new option abruptly. The continued use of the existing fuel is attributed to system inertia as well as consumer behaviour. Therefore, it is important for the government to lay out the new ecosystem in a gradual manner with a long-term plan, while the incumbent fuel can be slowly phased out through negative incentives. During the transition phase, the private sector enterprises should be actively involved in capacity building and infrastructure creation. The promotion for the adoption of the new or alternate fuel should continue until it reaches a critical level that enables self-sustenance. The successful adoption of alternate fuels has been experienced globally through persistent government support in the face of acute adversity. Brazil sticking with ethanol even when the prices of crude oil dropped heavily during the period 1985–2000, and China ensuring the availability of LNG to the road transport sector in the face of supply shortage in 2017–18 shows the importance of continued support from the government.

5.8.5 Substantial and Continued Investment in the Research and Development (R&D)

The evolution of the transport industry of any advanced industrial power pivots around the continued research and development (R&D) efforts directed towards innovations regarding fuel and vehicle technologies for reduced emission levels and better performance. There is also a need for investment towards developing a commercial deployment strategy. Further, investments are also needed to upgrade the traffic and road management systems. The insights from the R&D should provide regular inputs to the policymakers to help in formulating progressive policies for the road transport sector. The aspirational trajectory of a nation changes dynamically over a period of time. Therefore, the national targets have to be revisited periodically, and realism needs to be incorporated into any national transition plan. The importance of R&D is best realized from the induction of Flex Fuel Vehicles (FFVs) in Brazil in the early 2000s. The use of ethanol stagnated during 1985–2000 due to the reduction in crude oil prices and the limited riding range of ethanol-only vehicles introduced in the late 1970s. With the induction of FFVs, the current uptake of ethanol in the road transport sector of Brazil is at an all-time high.

5.8.6 Creation of 'Economies of Scale' Through Product Diversification and Privatization

Any alternate fuel is essentially an energy source that can also be used in industries and crude oil refineries, as well as for residential heating, power generation, etc. Such diversified use leads to a huge demand which is not necessarily contingent upon the usage in any particular segment. Diversified use of fuel favours the '*economies of scale*' and makes the production and distribution profitable for the involved enterprises. After the initial push from the government-owned enterprises towards the adoption of the new fuel option in the transport sector, the private players must be offered relaxed business prerequisites to build up on the base created by the government and expand the footprint. The privatization in Argentina (for CNG), China (for LNG), and Brazil (for ethanol) have shown the potential for the expansion of the alternate fuel ecosystems by the private sector once the government provided the initial push in the right direction. However, a proper regulatory mechanism should be devised to ensure higher accountability on behalf of the private sector. One possible way is to transfer the ownership of the ecosystem to the private players, while the adherence to the regulations should be monitored by the government agencies. In the case of LNG expansion in China, the private sector was given the onus of inspection and implementation of standards to ensure safety during fuel consumption in China's road transport sector.

5.8.7 Price Parity to Encourage the Spontaneous Adoption by the Consumer

The final success of new fuel adoption in the transport ecosystem largely depends upon the affinity of the end user towards the alternate option. The affordability of the alternate fuel (i.e. reduced cost of ownership over a specific period) acts as the primary driver during the initial period. The global experiences show that the respective governments adopted different methods to ensure price parity for the end-user. In Argentina, price differentials were offered to promote CNG, and lines of credit were issued for converting the existing gasoline-based vehicles. China offered subsidies on the prices of LNG, while Brazil adopted a preferential tax regime to ensure the competitive cost of ethanol against gasoline. Indonesia had established a separate fund from the sales of edible palm oil to subsidize biodiesel and support R&D.

5.8.8 Changing Consumer Behaviour with Respect to Long-Term Adoption of New Eco-System

With the proven track record of the incumbent fuels (petrol and diesel) with respect to their safety, performance, and accessibility through dispensation outlets, it is always difficult to transform consumer behaviour. The price differential alone is not enough, and it is important to make the customer aware of the long-term value (in terms of reliability) of the new fuel and vehicle technologies. There have been instances where the consumers have gone back to the conventional option based on their experiences. The return to gasoline from CNG in Argentina was triggered by the boot space constraints, whereas consumers went to back gasoline from auto-gas in Turkey in the early 2000s due to the price volatility of the fuel in place. In contrast, Brazil witnessed the re-adoption of ethanol after the introduction of the FFVs (a novel vehicle technology) in its road transport ecosystem. Evidently, R&D plays a very crucial role in influencing consumer behaviour by continuously providing value addition to the end consumer experience.

5.9 Conclusion

India is a diverse country with sub-continental features in terms of geographical attributes and meteorological parameters, economic features in terms of infrastructure development and connectivity and consumer behaviour in terms of purchasing power and technology awareness. Envisaging energy transition in the Indian road transport sector is therefore an ambitious and herculean task. Since the fuels behave differently in different topography and climate, it is important to create a flexible ecosystem based on the regional opportunities and constraints with respect to the adoption of alternate fuels. Therefore, it is essential to have pilot projects on a regional basis in India before the complete roll-out of a uniform plan across the country. As the over-arching goal of transforming the existing ecosystem system would revolve around sustainability with a focus on emission reduction, the new transport fuel ecosystem must consider all 4A factors (availability, accessibility, affordability and acceptability) in its ambit for '*just transition*'.

Currently, India has a robust infrastructure in place for handling crude oil encompassing the transportation through oil tankers for fractional distillation at refineries, transportation of the refined products to the distribution outlets and dispensing to the vehicle tanks. However, the crucial limitation in the life cycle of crude oil in India lies in its huge dependence on imports of crude oil, as is also the case for natural gas. Therefore, the need for reducing the import dependence together with the availability of the alternate energy source would dictate the future trajectory of the energy transition in the Indian road transport sector. Concerns regarding the domestic availability of natural gas restrict its potential in the long-term clean energy mix. However, considering the existing natural gas transportation infrastructure (pipelines) and the vehicular technological choices (e.g. bi-fuel vehicles) available for urban settings, natural gas remains a relevant component in the scheme of energy diversification in the near-to-medium term. Biofuels (ethanol and biodiesel) are already in use in India as a blend with conventional fossil fuels (petrol and diesel). However, the adoption of dedicated vehicles running on pure biofuels in the country remains highly unlikely in the near-to-medium term since India has barely achieved a 10% ethanol blend in petrol and 0.1% biodiesel blend in diesel. Moreover, India has only 2.4% of the global land available and sustains 18% of the global population which makes the food crops a priority over the fuel crops (to be used as feedstocks for biofuels) in the context of land usage. Enhanced research towards the utilization of third-generation raw materials would pave the way forward for sustainable biofuels in India. The production of green hydrogen in India remains a possibility due to its tropical location, which ensures round-the-year availability of insolation for renewable electricity generation. However, transitioning to green hydrogen from the currently prevalent grey hydrogen would need a massive overhaul in the infrastructure and a sustained demand driven by a diversified end-user portfolio in the country.

It is important to note that a new ecosystem aiming for the sustainable road transport sector in India cannot be envisaged without integrating other modes of transport (metros rails, railways, airways and waterways) into the overall framework. Various modes of transport are complementary to each other in terms of passenger and goods transport. Therefore, the creation of a new ecosystem support for any alternate fuel needs a detailed look from an integrated perspective. The role of the public transport system is paramount, especially in the urban centers across India. Along with the prominent mass-transit modes, particular emphasis should be given to the vehicle segment catering to the last-mile connectivity for reducing emissions. Consumer behaviour can be transformed towards cleaner fuel and vehicles by upgrading the same keeping in mind the consumer experience. Any futuristic ecosystem should not be designed and deployed in an abrupt manner. Instead, the country's preparedness in terms of resources (availability of raw material and technology), the robustness of the existing infrastructure and its flexibility to transition to alternate fuel infrastructure (accessibility), costs associated with alternate fuels and novel vehicular technologies (affordability), and consumer awareness, as well as market experiences of the vehicle manufacturers and distributors (acceptability) should be taken into account during the energy diversification and energy transition in India's road transport sector.

References

Abel RC, Coney K, Johnson C, Thornton MJ, Zigler BT, McCormick RL (2021) Global ethanol-blended-fuel vehicle compatibility study (No. NREL/TP-5400-81252). National Renewable Energy Lab (NREL), Golden, CO (United States). https://doi.org/10.2172/1832216

Ahluwalia S (2022) India's green hydrogen policy: tentative beginnings'. Obs Res Found 23. https://www.orfonline.org/expert-speak/indias-green-hydrogen-policy/. Accessed 23 Aug 2022

Ahmed A, Ishak MY, Uddin MK, Abd Samad MY, Mukhtar S, Danhassan SS (2021) Effects of some weather parameters on oil palm production in the Peninsular Malaysia. https://doi.org/10.20944/preprints202106.0456.v1

Albatayneh A, Assaf MN, Alterman D, Jaradat M (2020) Comparison of the overall energy efficiency for internal combustion engine vehicles and electric vehicles. Rigas Tehniskas Universitates Zinatniskie Raksti 24(1):669–680. https://doi.org/10.2478/rtuect-2020-0041

Aljohani K, Thompson RG (2020) An examination of last mile delivery practices of freight carriers servicing business receivers in inner-city areas. Sustainability 12(7):2837. https://doi.org/10.3390/su12072837

Alternate Fuels Data Center Homepage (2022a) How do flexible fuel cars work using ethanol?. https://afdc.energy.gov/vehicles/how-do-flexible-fuel-cars-work. Accessed 23 Aug 2022a

Alternate Fuels Data Center Homepage (2022b) Natural gas vehicles. https://afdc.energy.gov/vehicles/natural_gas.html. Accessed 23 Aug 2022b

Alternate Fuels Data Center Homepage (2022c) How do natural gas vehicles work?. https://afdc.energy.gov/vehicles/how-do-natural-gas-cars-work. Accessed 23 Aug 2022c

Alternate Fuels Data Center Homepage (2022d) Global ethanol production by country or Region. https://afdc.energy.gov/data/10331. Accessed 23 Aug 2022d

Asia Pacific Energy Research Centre Homepage (2007) A quest for Energy Security in the 21st Century: resources and Constraints. Institute of Energy Economics, Japan. https://aperc.or.jp/file/2010/9/26/APERC_2007_A_Quest_for_Energy_Security.pdf. Accessed 23 Aug 2022

Autogas.net Hompage (2020a) Auto-gas Incentive Policies—2020 update. https://auto-gas.net/wp-content/uploads/2020/10/WLPGA-Autogas_Incentives_Policies-2020-REPORT.pdf. Accessed 23 Aug 2022

Autogas.net Homepage (2020b) Turkey, the world's largest Autogas market. https://auto-gas.net/mediaroom/turkey-the-worlds-largest-autogas-market/. Accessed 23 Aug 2022

Auto-gas.net Homepage (2022a) Indian autogas industry urges government to include LPG in gas projects. https://auto-gas.net/mediaroom/indian-autogas-industry-urges-government-to-include-lpg-in-gas-projects/. Accessed 23 Aug 2022

Auto-gas.net Homepage (2022b) Autogas market trends. https://auto-gas.net/government-policies/autogas-incentive-policies/india/. Accessed 23 Aug 2022

Autogas.net Homepage (2022c) How did Turkey's Autogas refuelling network get to be the world's biggest?. https://auto-gas.net/mediaroom/how-did-turkeys-autogas-refuelling-network-get-to-be-the-worlds-biggest/. Accessed 23 Aug 2022

Bansal P, Dua R, Krueger R, Graham DJ (2021) Fuel economy valuation and preferences of Indian two-wheeler buyers. J Clean Prod 294:126328. https://doi.org/10.1016/j.jclepro.2021.126328

Big Think Homepage (2022) China's most important border is imaginary: the Hu Line. https://bigthink.com/strange-maps/hu-line/. Accessed 23 Aug 2022

Bolay JC (2020) Urban dynamics and regional development in Argentina. In: Urban planning against poverty. Springer, Cham, pp 167–202. https://doi.org/10.1007/978-3-030-28419-0_6

BP (2020) Statistical review of world energy 2020, 69th Edn. https://on.bp.com/3yw7RNv. Accessed 23 Aug 2022

Bureau of Indian Standards (2012) Liquefied petroleum gas (LPG) containers for automotive use—Specifications. https://law.resource.org/pub/in/bis/S08/is.14899.2000.pdf. Accessed 23 Aug 2022

Bureau of Indian Standards Homepage (2012) Compressed natural gas (CNG) for automotive purposes—Specifications. https://law.resource.org/pub/in/bis/S11/is.15958.2012.pdf. Accessed 23 Aug 2022

BusinessLine Homepage (2020) All you wanted to know about LPG pricing formula. https://www.thehindubusinessline.com/opinion/columns/slate/all-you-wanted-to-know-about-lpg-pricing-formula/article30496150.ece. Accessed 23 Aug 2022

Camadan E (2013) The competition level in the Turkish LPG market. Energy Sour Part B 8(2):154–161. https://doi.org/10.1080/15567240903330400

Carnot-Gandolphe S (2019) China's quest for blue skies: the astonishing transformation of the domestic gas market. Études de ľIfri,, Ifri. https://doi.org/20.500.12592/3jrvpc

Carvalho PN (2013) Government Intervention to strengthen the ethanol sector: lessons from Brazil. Evidence and Lessons from Latin America, Policy Brief. http://ella.practicalaction.org/wp-content/uploads/files/130320_ENV_BraEthPro_BRIEF3.pdf. Accessed 23 Aug 2022

CGD India Homepage (2021) Petronet LNG to collaborate with KSRTC to set up natural gas dispensing stations in Kerala. https://cgdindia.net/petronet-lng-to-collaborate-with-ksrtc-to-set-up-natural-gas-dispensing-stations-in-kerala/. Accessed 23 Aug 2022

Chugh R, Cropper M, Narain U (2011) The cost of fuel economy in the Indian passenger vehicle market. Energy Policy 39(11):7174–7183. https://doi.org/10.1016/j.enpol.2011.08.037

Collantes G, Melaina MW (2011) The co-evolution of alternative fuel infrastructure and vehicles: a study of the experience of Argentina with compressed natural gas. Energy Policy 39(2):664–675. https://doi.org/10.1016/j.enpol.2010.10.039

College of the Desert (2022) Module 3—Hydrogen use in internal combustion engines. https://www1.eere.energy.gov/hydrogenandfuelcells/tech_validation/pdfs/fcm03r0.pdf. Accessed 23 Aug 2022

Columbia University Homepage (2022) Guide to Chinese Climate Policy. https://chineseclimatepolicy.energypolicy.columbia.edu/en/natural-gas. Accessed 23 Aug 2022

Das LM (2016) Hydrogen-fueled internal combustion engines. In: Compendium of hydrogen energy. Woodhead Publishing, pp 177–217. https://doi.org/10.1016/B978-1-78242-363-8.00007-4

De Freitas LC, Kaneko S (2011) Ethanol demand under the flex-fuel technology regime in Brazil. Energy Econ 33(6):1146–1154. https://doi.org/10.1016/j.eneco.2011.03.011
Digital CCJ Homepage (2022) Cummins, other hydrogen ICE proponents hoping for zero-emission status. https://www.ccjdigital.com/alternative-power/article/15293434/cummins-hoping-hydrogen-ice-gets-zeroemission-status. Accessed 23 Aug 2022
Energy Information Administration Homepage (1999) Natural gas 1998—Issues and trends. https://www.eia.gov/naturalgas/archive/056098.pdf. Accessed 23 Aug 2022
EnerData Homepage (2014) Effect of price reforms on the demand of LNG in transport in China. https://www.enerdata.net/publications/executive-briefing/china-lng-price-reforms-effets.html. Accessed 23 Aug 2022
Energypedia Homepage (2022) Liquefied petroleum gas. https://energypedia.info/wiki/Liquefied_Petroleum_Gas_(LPG). Accessed 23 Aug 2022
ELGAS Homepage (2022a) LPG conversion—LPG conversion price—LPG versus petrol (gas or gasoline) conversion myths. https://www.elgas.com.au/blog/690-7-lpg-conversions-myths-autogas/. Accessed 23 Aug 2022
ELGAS Homepage (2022b) Liquid direct injection LPG conversion—LPG conversion—Gas conversion. https://www.elgas.com.au/blog/681-liquid-injection-lpg-vs-vapour-injection-conversions/. Accessed 23 Aug 2022
ELGAS Homepage (2022c) LPG engines for LPG vehicles—How does an LPG engine work? How does LPG work. https://www.elgas.com.au/blog/2207-how-does-an-lpg-engine-work/. Accessed 23 Aug 2022
Energy.gov Homepage (2022) Hydrogen storage. https://www.energy.gov/eere/fuelcells/hydrogen-storage. Accessed 23 Aug 2022
Euler M, Krishna V, Schwarze S, Siregar H, Qaim M (2017) Oil palm adoption, household welfare, and nutrition among smallholder farmers in Indonesia. World Dev 93:219–235. https://doi.org/10.1016/j.worlddev.2016.12.019
Ferrero D, Lanzini A, Santarelli M, Leone P (2013) A comparative assessment on hydrogen production from low-and high-temperature electrolysis. Int J Hydrog Energy 38(9):3523–3536. https://doi.org/10.1016/j.ijhydene.2013.01.065
FirstPost Homepage (2021) Explained: what is a hydrogen internal combustion engine, and can it be a real alternative to battery EVs?. https://www.firstpost.com/tech/auto/explained-what-is-a-hydrogen-internal-combustion-engine-and-can-it-be-a-real-alternative-to-battery-evs-10116721.html. Accessed 23 Aug 2022
Force Motors Limited Homepage (2022) Corporate Presentation. https://connect.forcemotors.com/Employee_Information/User_data/CorporateFMLPresentation.pdf. Accessed 23 Aug 2022
Fortune India Homepage (2022) Green hydrogen costs to halve to ₹160–170 per kg by 2030. https://www.fortuneindia.com/macro/green-hydrogen-costs-to-halve-to-160-170-per-kg-by-2030/107549. Accessed 23 Aug 2022
Gas DN (1991) Assessment of costs and benefits of flexible and alternative fuel use in the US transportation sector. United States. https://www.osti.gov/servlets/purl/6360489
Gateway House Homepage (2014) Decoding natural gas pricing in India. https://www.gatewayhouse.in/decoding-natural-gas-pricing-in-india/. Accessed 23 Aug 2022
Gee S, McMeekin A (2011) Eco-innovation systems and problem sequences: the contrasting cases of US and Brazilian biofuels. Ind Innov 18(03):301–315. https://doi.org/10.1080/13662716.2011.561029
Global Energy Observatory Homepage (2022). http://globalenergyobservatory.org/list.php?db=Transmission&type=LNG_Ports. Accessed 23 Aug 2022
Go Gas Homepage (2022) Auto LPG dispensing stations. https://gogas.co/auto_lpg_station. Accessed 23 Aug 2022
Goldemberg J, Coelho ST, Nastari PM, Lucon O (2004) Ethanol learning curve—The Brazilian experience. Biomass Bioenerg 26(3):301–304. https://doi.org/10.1016/S0961-9534(03)00125-9
Goodrich RS (1981) Brazil enters the biofuels era. South Res Inst Bull (United States) 33(2). https://www.osti.gov/biblio/5464641

Green Car Reports Homepage (2021) Several Japanese vehicle makers expand the use of hydrogen internal combustion engines. https://www.greencarreports.com/news/1134191_japanese-expand-hydrogen-internal-combustion-engines. Accessed 23 Aug 2022

Hall W, Spencer T, Renjith G, Dayal S (2020) The potential role of hydrogen in India: a pathway for scaling-up low carbon hydrogen across the economy. In: The energy and resources institute (TERI), New Delhi, India. https://www.teriin.org/event/potential-role-hydrogen-india-harnessing-hype-teri-report-launch-event. Accessed 23 Aug 2022

Hardt C, Bogenberger K (2019) Usage of e-scooters in urban environments. Transp Res Procedia 37:155–162. https://doi.org/10.1016/j.trpro.2018.12.178

Higashi N (2009) Natural gas in China: Market evolution and Strategy. In: IEA energy markets and security working papers. https://citeseerx.ist.psu.edu/viewdoc/download?doi=10.1.1.378.3252&rep=rep1&type=pdf. Accessed 23 Aug 2022

IHS Markit Homepage (2021) Turkey's NGL production is set to increase after new gas field discovery, but will not be sufficient to greatly reduce dependence on imports. https://ihsmarkit.com/research-analysis/turkeys-gas-processing.html. Accessed 23 Aug 2022

Indian Auto LPG Coalition Homepage (2022a) Where to fill?. http://www.iac.org.in/alds-stations. Accessed 23 Aug 2022a

Indian Auto LPG Coalition Homepage (2022b) Auto LPG in India. http://www.iac.org.in/auto-lpg-in-india. Accessed 23 Aug 2022b

Indian Auto LPG Coalition Homepage (2022c) Auto LPG Worldwide. https://www.iac.org.in/auto-lpg-worldwide. Accessed 23 Aug 2022c

Indian Institute of Technology Delhi Homepage (2022) Course: MEL345-ICE Engines—Classification of internal combustion engines. https://web.iitd.ac.in/~ravimr/courses/mel345/classification.pdf. Accessed 23 Aug 2022

Indian Oil Corporation Limited Homepage (2020a) Indian oil backs hydrogen fuel cell vehicles. https://www.iocl.com/MediaDetails/55152. Accessed 23 Aug 2022

Indian Oil Corporation Limited Homepage (2020b) Previous price of AutoGas. https://iocl.com/auto-lpg-previous-price. Accessed 23 Aug 2022

International Energy Agency Homepage (1999) Regulatory reform in Argentina's Natural Gas Sector. https://iea.blob.core.windows.net/assets/97faadca-0bab-4738-9a4e-1efb2fd7ee22/RegulatoryReforminArgentinasNaturalGasSector.pdf. Accessed 23 Aug 2022

International Energy Agency Homepage (2022a) Data and statistics. https://www.iea.org/data-and-statistics. Accessed 23 Aug 2022a

International Energy Agency Homepage (2022b) India Energy Outlook 2021. https://www.iea.org/reports/india-energy-outlook-2021. Accessed 23 Aug 2022b

International Energy Agency Homepage (2022c) Data and Statistics–TFC by sector: Argentina. https://www.iea.org/data-and-statistics. Accessed 23 Aug 2022c

IRENA (2020) Green hydrogen cost reduction: scaling up electrolysers to meet the 1.5 °C climate goal. International Renewable Energy Agency, Abu Dhabi, ISBN: 978-92-9260-295-6. https://gawg.info/files/papers/IRENA20a.pdf. Accessed 23 Aug 2022

Kamath HG, Majumdar R, Krishnan AV, Srikanth R (2022) Cost and environmental benefits of coal-concentrated solar power (CSP) hybridization in India. Energy 240:122805. https://doi.org/10.1016/j.energy.2021.122805

Karamangil MI (2007) Development of the auto gas and LPG-powered vehicle sector in Turkey: a statistical case study of the sector for Bursa. Energy Policy 35(1):640–649. https://doi.org/10.1016/j.enpol.2006.01.004

Karunakar B (2017) Indian passenger vehicle industry: strategic analysis with focus on the big four firms. NMIMS J Econ Publ Policy 2(1):47–72. http://epp-journal.nmims.edu/wp-content/uploads/2017/april/indian-passenger-vehicle-industry-strategic-analysis-with-focus-on-the-big-four-firms-karunakar-b.pdf. Accessed 23 Aug 2022

KPMG (2022) India's green hydrogen ambition. https://assets.kpmg/content/dam/kpmg/in/pdf/2022/03/Indias-green-hydrogen-ambition.pdf. Accessed 23 Aug 2022

Kulkarni SS, Pimpalkhare A (2019) India's import diversification strategy for natural gas: an analysis of geopolitical implications. Obs Res Found Issue Brief (330). https://www.orfonline.org/wp-content/uploads/2019/12/ORF_IssueBrief_330_NaturalGasImports_NEW.pdf. Accessed 23 Aug 2022

Kumar P, Chakrabarty S (2020) Total cost of ownership analysis of the impact of vehicle usage on the economic viability of electric vehicles in India. Transp Res Rec 2674(11):563–572. https://doi.org/10.1177/361198120947089

Lamb JJ, Hillestad M, Rytter E, Bock R, Nordgård AS, Lien KM, Burheim OS, Pollet BG (2020) Traditional routes for hydrogen production and carbon conversion. In: Hydrogen, biomass and bioenergy. Academic Press, pp 21–53. https://doi.org/10.1016/B978-0-08-102629-8.00003-7

Leach F, Kalghatgi G, Stone R, Miles P (2020) The scope for improving the efficiency and environmental impact of internal combustion engines. Transp Eng 1:100005. https://doi.org/10.1016/j.treng.2020.100005

LiveMint Homepage (2021a) Govt charts course for usage of new-age fuel. https://www.livemint.com/industry/energy/govt-charts-course-for-usage-of-new-age-fuel-11625078901655.html. Accessed 23 Aug 2022

LiveMint Homepage (2021b) The changing face of public-private transport. https://www.livemint.com/brand-post/the-changing-face-of-public-private-transport-11611059501406.html. Accessed 23 Aug 2022

Lovins AB, Datta EK (2005) Winning the oil endgame. Innovations for profits, jobs and security. https://doi.org/10.1093/jwelb/jwp019

Meyer D, Mytelka L, Press R, Dall'Oglio EL, de Sousa Jr PT, Grubler A (2014) Brazilian ethanol: unpacking a success story of energy technology innovation. Energy Technol Innov 275. https://doi.org/10.1017/CBO9781139150880.027

MyPetrolPrice Homepage (2022) CNG Price graph in Delhi. https://www.mypetrolprice.com/2/CNG-price-in-Delhi. Accessed 23 Aug 2022

Mongabay Homepage (2022) With ban on palm oil exports, Indonesia reaps condemnation and praise. https://news.mongabay.com/2022/04/with-ban-on-palm-oil-exports-indonesia-reaps-condemnation-and-praise/. Accessed 23 Aug 2022

Ministry of Petroleum and Natural Gas Homepage (2018) National Policy on Biofuels 2018. https://mopng.gov.in/files/uploads/NATIONAL_POLICY_ON_BIOFUELS-2018.pdf. Accessed 23 Aug 2022

Ministry of Petroleum and Natural Gas Homepage (2022) Indian Petroleum and Natural Gas Statistics 2020–21. https://mopng.gov.in/files/TableManagements/Indian-Petroleum--Natural-Gas_2020-21.pdf. Accessed 23 Aug 2022

Ministry of Power Homepage (2022) Green Hydrogen Policy. https://powermin.gov.in/sites/default/files/Green_Hydrogen_Policy.pdf. Accessed 23 Aug 2022

Ministry of Road Transport & Highways Homepage (2021) Order specifying types of motor vehicles on the basis of its usage. https://morth.nic.in/sites/default/files/notifications_document/Draft%20So%20-%20vehicle%20classification%20on%20usage_0001.pdf. Accessed 23 Aug 2022

Ministry of Road Transport & Highways Homepage (2022) Road Transport Year Book 2016–17. https://morth.nic.in/sites/default/files/Road%20Transport%20Year%20Book%202016-17.pdf. Accessed 23 Aug 2022

Ministry of Statistics and Program Implementation Homepage (2022) Energy Statistics 2019. https://mospi.gov.in/documents/213904/301563//Energy%20Statistics%202019-final1602098943619.pdf/b7e6e6b8-12da-7207-8f6a-b31ec54289a0. Accessed 23 Aug 2022

Natarajan S, Abraham M, Rajesh M, Subash GP, Kunal R, Das L (2013) Delhy 3w-hydrogen fuelled Hy-Alfa three wheeler (No. 2013-01-0224). SAE Technical Paper. https://doi.org/10.4271/2013-01-0224

NationalGrid Homepage (2022). https://ngrid.com/3ygOwP1. Accessed 23 Aug 2022

Ogunlowo OO, Bristow AL, Sohail M (2015) Developing compressed natural gas as an automotive fuel in Nigeria: lessons from international markets. Energy Policy 76:7–17. https://doi.org/10.1016/j.enpol.2014.10.025

Outlook India Homepage (2021) Auto LPG still finding favour in India, Why Govt not helping to boost it?. https://www.outlookindia.com/website/story/business-news-auto-lpg-still-finding-favour-in-india/392169. Accessed 23 Aug 2022

Paczuski M, Marchwiany M, Puławski R, Pankowski A, Kurpiel K, Przedlacki M (2016) Liquefied petroleum gas (LPG) as a fuel for internal combustion engines. In: Alternative fuels, technical and environmental conditions, vol 13. https://doi.org/10.5772/61736

Patankar M, Patwardhan A (2017) The switch to CNG in two urban areas in India: how was this achieved?. In: The business of sustainable mobility—From vision to reality. Routledge, pp 171–186. https://doi.org/10.4324/9781351280969-13

Patil V, Shastry V, Himabindu M, Ravikrishna RV (2016) Life-cycle analysis of energy and greenhouse gas emissions of automotive fuels in India: part 2-well-to-wheels analysis. Energy 96:699–712. https://doi.org/10.1016/j.energy.2015.11.076

Petroleum and Natural Gas Regulatory Board Homepage (2022) Natural Gas Pipelines Network in India—As on 30.09.2021. https://www.pngrb.gov.in/data-bank/NGPL-14122021.pdf. Accessed 23 Aug 2022

Petroleum Planning & Analysis Cell Homepage (2022) Natural gas price. http://ppac.org.in/WriteReadData/userfiles/file/NG_Price.pdf. Accessed 23 Aug 2022

Piragine MA, Shah S, Coutroubis A (2018) Study of potentials and challenges in unconventional oil and gas industry: an argentinian case study. In MATEC web of conferences, vol 210. EDP Sciences, p 02007. https://doi.org/10.1051/matecconf/201821002007

Pothumsetty R, Viswam N, Thomas MR (2020) Bharat stage IV–VI-challenges and strategies. Int J Recent Technol Eng 8(5):2614–2623. https://doi.org/10.35940/ijrte.e5850.018520

Press Information Bureau Homepage (2019) Pricing of natural gas. https://pib.gov.in/Pressreleaseshare.aspx?PRID=1579718. Accessed 23 Aug 2022

Press Information Bureau Homepage (2020) Performance of Fame India Scheme. https://pib.gov.in/newsite/PrintRelease.aspx?relid=198872. Accessed 23 Aug 2022

Press Information Bureau Homepage (2021a) Year-end review of Ministry of Heavy Industries. https://pib.gov.in/PressReleasePage.aspx?PRID=1784161. Accessed 23 Aug 2022

Press Information Bureau Homepage (2021b) Shri Nitin Gadkari calls for manufacturing of Flex Fuel Vehicles (FFV) and Flex Fuel Strong Hybrid Electric Vehicles (FFV-SHEV) complying with BS-6 Norms in a time bound manner within a period of six months. https://pib.gov.in/PressReleaseIframePage.aspx?PRID=1785557. Accessed 23 Aug 2022

Press Information Bureau Homepage (2022) Target to raise the share of natural gas in energy mix to 15% in 2030. https://pib.gov.in/PressReleaseIframePage.aspx?PRID=1779741. Accessed 23 Aug 2022

Rahmanulloh A (2018) Indonesia expands biodiesel mandate. U.S. Foreign Agricultural Service, GAIN Report Number: ID1826. https://apps.fas.usda.gov/newgainapi/api/report/downloadreportbyfilename?filename=Indonesia%20Expands%20Biodiesel%20Mandate_Jakarta_Indonesia_9-18-2018.pdf. Accessed 23 Aug 2022

Reliance Industries Limited Homepage (2021) Chairman's Statement—44th annual general meeting. https://www.ril.com/Downloadfiles/ChairmanCommunications/RIL-AGM-44.pdf. Accessed 23 Aug 2022

Reuters Homepage (2022) Indonesia, Malaysia commit to biodiesel mandates despite higher prices. https://reut.rs/3HXHkeV. Accessed 23 Aug 2022

Rochmyaningsih D (2019) Making peace with oil palm. Science 365(6449):112–115. https://doi.org/10.1126/science.365.6449.112

Rocky Mountain Institute Homepage (2019) The truth about hydrogen. https://rmi.org/the-truth-about-hydrogen/. Accessed 23 Aug 2022

SAE International Homepage (2022) Hitting the gas on hydrogen tech for commercial trucks. https://www.sae.org/news/2022/05/hydrogen-technology-commercial-trucks. Accessed 23 Aug 2022

Saraf RR, Thipse SS, Saxena PK (2009) Comparative emission analysis of gasoline/LPG automotive bifuel engine. Int J Civil Environ Eng 1(4):199–202. https://doi.org/10.5281/zenodo.1331891

Saurabh K, Majumdar R (2021) An approach towards comprehensive life cycle assessment (LCA) for surface transport automotive fuels. In: International conference on sustainable technology and Development 2021. Southern University of Science and Technology (SUSTech) and Elsevier, Shenzhen China. https://www.researchgate.net/publication/356510333. Accessed 23 Aug 2022

Saurabh K, Majumdar R (2022) Fuels for sustainable transport in India. In: Clean fuels for mobility. Springer, Singapore, pp 27–55. https://doi.org/10.1007/978-981-16-8747-1_3

Scientific American Homepage (2014) What a transportation revolution in China looks like. https://www.scientificamerican.com/article/transportation-revolution-in-china/. Accessed 23 Aug 2022

Sen A (2015) Gas pricing reform in India—Implications for the Indian gas landscape. The Oxford Institute for Energy Studies Paper, NG 96. https://ora.ox.ac.uk/objects/uuid:eaee0d6f-f0e7-4ec9-94f8-921b76d9a01b. Accessed 23 Aug 2022

Silalahi FTR, Simatupang TM, Siallagan MP (2020) Biodiesel produced from palm oil in Indonesia: Current status and opportunities. AIMS Energy 8(1):81–101. https://doi.org/10.3934/energy.2020.1.81

Sixth Tone Homepage (2022) Hu Line: China's forgotten frontier. https://www.sixthtone.com/news/1000459/hu-line-chinas-forgotten-frontier. Accessed 23 Aug 2022

Smajla I, Karasalihović Sedlar D, Drljača B, Jukić L (2019) Fuel switch to LNG in heavy truck traffic. Energies 12(3):515. https://doi.org/10.3390/en12030515

S&P Global Homepage (2021) Analysis: high LNG prices trigger gas demand destruction in China's downstream sectors. https://www.spglobal.com/commodityinsights/en/market-insights/latest-news/lng/090221-analysis-high-lng-prices-trigger-gas-demand-destruction-in-chinas-downstream-sectors. Accessed 23 Aug 2022

S&P Global Homepage (2022). Commodities 2022: China's natural gas demand, LNG import growth to slow. https://www.spglobal.com/commodityinsights/en/market-insights/latest-news/lng/011222-commodities-2022-chinas-natural-gas-demand-lng-import-growth-to-slow. Accessed 23 Aug 2022

Stanford University Homepage (2022) Course: PH240—Two hydrogen cars. https://stanford.io/3QQi9yN. Accessed 23 Aug 2022

Statista Homepage (2022) Number of registered new light vehicles in Brazil from 2014 to 2021, by fuel type. https://www.statista.com/statistics/712070/number-of-registered-automobiles-in-brazil/. Accessed 23 Aug 2022

The EarthBound Report Homepage (2019) The role of three-wheelers in sustainable urban transport. https://earthbound.report/2019/03/12/the-role-of-three-wheelers-in-sustainable-urban-transport/. Accessed 23 Aug 2022

The Economic Times Homepage (2020) Increasing auto LPG usage for a cleaner environment—What India could learn from Global Example. https://auto.economictimes.indiatimes.com/news/oil-and-lubes/opinion-increasing-auto-lpg-usage-for-a-cleaner-environment-what-india-could-learn-from-global-example/74588861. Accessed 23 Aug 2022

The Economic Times Homepage (2022) Tech and data will drive the future of Indian inter-city bus travel sector. https://economictimes.indiatimes.com/small-biz/sme-sector/tech-and-data-will-drive-the-future-of-indian-inter-city-bus-travel-sector/articleshow/88706545.cms. Accessed 23 Aug 2022

The Indian Express Homepage (2020) Brazil wants India to step up ethanol production to 'balance' world sugar market. https://indianexpress.com/article/business/brazil-puts-onus-on-india-to-balance-world-sugar-market-6430649/. Accessed 23 Aug 2022

ThePrint Homepage (2022) Only 8% Indian families own cars, NFHS finds. Over 50% still use bicycles, bikes & scooters. https://theprint.in/india/only-8-indian-families-own-cars-nfhs-finds-over-50-still-use-bicycles-bikes-scooters/971413/. Accessed 23 Aug 2022

U.S. Department of Energy Homepage (2020) Hydrogen fuelling stations cost. https://www.hydrogen.energy.gov/pdfs/21002-hydrogen-fueling-station-cost.pdf. Accessed 23 Aug 2022

U.S. Energy Information Administration Homepage (2021) International Energy Outlook 2021. https://www.eia.gov/outlooks/ieo/pdf/IEO2021_Narrative.pdf. Accessed 23 Aug 2022

U.S. Energy Information Administration Homepage (2022) Natural gas. https://www.eia.gov/dnav/ng/hist/rngwhhdm.htm. Accessed 23 Aug 2022

United Nations Environment Program Homepage (2022) GOAL 7: affordable and clean energy. https://www.unep.org/explore-topics/sustainable-development-goals/why-do-sustainable-development-goals-matter/goal-7. Accessed 23 Aug 2022

Upstream Online Homepage (2022) India's Gail awards contract for upscaled green hydrogen electrolyzer. https://www.upstreamonline.com/energy-transition/india-s-gail-awards-contract-for-upscaled-green-hydrogen-electrolyser/2-1-1218594. Accessed 23 Aug 2022

VAHAN Dashboard Homepage (2022). https://vahan.parivahan.gov.in/vahan4dashboard/. Accessed 23 Aug 2022

Vásquez PI (2016) Argentina's oil and gas sector. Wilson Center, Washington, DC, USA. https://www.wilsoncenter.org/sites/default/files/media/documents/publication/argentinasoilgas.vasquez.pdf. Accessed 23 Aug 2022

Vats I, Singhal D, Tripathy S, Jena SK (2022) The transition from BS4 to BS6 compliant vehicles for eco-friendly mobility in India: an empirical study on switching intention. Res Transp Econ 91:101131. https://doi.org/10.1016/j.retrec.2021.101131

Verhelst S, Wallner T (2009) Hydrogen-fueled internal combustion engines. Prog Energy Combust Sci 35(6):490–527. https://doi.org/10.1016/j.pecs.2009.08.001

Winzer C (2012) Conceptualizing energy security. Energy Policy 46:36–48. https://doi.org/10.1016/j.enpol.2012.02.067

WLPGA Homepage (2022) Where does LPG come from?. https://www.wlpga.org/about-lpg/what-is-lpg/where-does-lpg-come-from/. Accessed 23 Aug 2022

World Economic Forum Homepage (2020) Fostering effective energy transition 2020 edition. https://www3.weforum.org/docs/WEF_Fostering_Effective_Energy_Transition_2020_Edition.pdf. Accessed 23 Aug 2022

WorldData Homepage (2022a) The climate in Argentina. https://www.worlddata.info/america/argentina/climate.php. Accessed 23 Aug 2022a

WorldData Homepage (2022b) The climate in Turkey. https://www.worlddata.info/asia/turkey/climate.php. Accessed 23 Aug 2022b

WRI India Homepage (2020) Busting the cost barrier: why electric three-wheelers make business sense. https://wri-india.org/blog/busting-cost-barrier-why-electric-three-wheelers-make-business-sense. Accessed 23 Aug 2022

Yasinta T, Karuniasa M (2021) Palm oil-based biofuels and sustainability in Indonesia: assess social, environmental and economic aspects. IOP Conf Ser: Earth Environ Sci 716(1):012113 (IOP Publishing). https://doi.org/10.1088/1755-1315/716/1/012113

Yuan Y (2019) China—The 'move' to LNG: examining the primary drivers for uptake of LNG as a transport fuel in China. In: 19th international conference and exhibition on liquefied natural gas. https://www.gti.energy/wp-content/uploads/2019/10/22-LNG19-03April2019-Yuan-Yuan-paper.pdf. Accessed 23 Aug 2022

Zhao Q, Huang W, Hu M, Xu X, Wu W (2021) Characterizing the economic and environmental benefits of LNG heavy-duty trucks: a case study in Shenzhen, China. Sustainability 13(24):13522. https://doi.org/10.3390/su132413522

Zheng Y, dos Santos Luciano AC, Dong J, Yuan W (2022) High-resolution map of sugarcane cultivation in Brazil using a phenology-based method. Earth Syst Sci Data 14(4):2065–2080. https://doi.org/10.5194/essd-14-2065-2022

Chapter 6
Strategies for Efficient Utilization of Methanol in Compression Ignition Engines

Sharad Pardhe, Javed Ahamad, Inderpal Singh, Parmod Kumar, and Atul Dhar

Abstract The energy crisis, global warming, and air pollution are the major problems of the modern world. Over the last couple of decades, millions of people are suffering due to worsening air pollution. The transportation sector mainly relies on fossil-fueled internal combustion engines, which significantly contribute to GHG emissions. Alternative fuel technologies are promising options for mitigating vehicular emissions. Among the many alternative fuels, methanol is the best contender as dual fuel or dedicated fuel in internal combustion engines. This review chapter presents the efficient use of methanol in CI engines. Many experimental and numerical studies have been conducted using methanol in CI engines. Methanol fueling improves engine performance, and combustion characteristics and simultaneously reduces emissions due to less carbon and higher oxygen content in methanol. Formaldehyde emissions from methanol fueling is a major concern. Although the cetane number of methanol is quite low, chemical properties of methanol make it possible to use it at higher compression ratios, which can be achieved in CI engines, and lead to higher thermal efficiency. The high self-ignition temperature of methanol requires high pressure and temperature for combustion, which can be achieved in a CI engine running at a higher compression ratio. Due to the high LHV of methanol, it can be used in CI engines as a blend with diesel. Methanol acts as the primary fuel in blending, and diesel acts as the pilot fuel for burning the methanol. Without blending, methanol can be introduced with hot boosted air or methanol can be preheated to deal with the cold start problems. There are various methods through which methanol can be introduced in CI engines such as port injection, direct injection, and direct mixture with diesel. In this chapter, we have discussed the effect of methanol on performance, combustion, and emissions (mainly NO_x, SO_x, and carbon-based emissions) in CI engines along with dual-fuel technology.

Keywords Methanol · Thermal efficiency · Compression ratio · Dual-fuel combustion

---

S. Pardhe · J. Ahamad · I. Singh · P. Kumar · A. Dhar (✉)
School of Engineering, Indian Institute of Technology Mandi, Himachal Pradesh, Mandi 175075, India
e-mail: add@iitmandi.ac.in

P. C. Shukla et al. (eds.), *Renewable Fuels for Sustainable Mobility*, Energy, Environment, and Sustainability, https://doi.org/10.1007/978-981-99-1392-3_6

Abbreviations

aTDC	After Top Dead Center
BDC	Bottom Dead Center
BP	Brake Power
BSFC	Brake-Specific Fuel Consumption
BTE	Brake Thermal Efficiency
CA50	Crank Angle at 50% fuel mass burned
CAD	Crank Angle Duration
CFD	Computational Fluid Dynamics
CH_3OH	Methanol
CI	Compression Ignition
CN	Cetane Number
CO	Carbon monoxide
CO_2	Carbon Monoxide
COP	Conference of Parties
CRDI	Common Rail Direct Injection
D	Diesel
DICI	Direct Injection Compressed Ignition
DMCC	Diesel Methanol Compound Combustion
EGR	Exhaust Gas Recirculation
FIP	Fuel Injection Pressure
GHGs	Green House Gases
GIE	Gross Indicated Efficiency
H/C	Hydrogen/Carbon
HCCI	Homogeneous Charge Compression Ignition
HRF	High Reacting Fuel
HRR	Heat Rejection Rate
ICE	Internal Combustion Engine
IMEP	Indicated Mean Effective Pressure
LCV	Lower Calorific Value
LHR	Low Heat Rejection
LHV	Latent Heat of Vaporisation
LRF	Low Reactive Fuel
M	Methanol
M(X)	X % of Methanol
M/D	Methanol/Diesel
M/D/M	Methanol/Diesel/Methanol
NITI	National Institute of Transforming India
NOx	Nitrogen Dioxide
PAH	Polycyclic Aromatic Hydrocarbons
PFI	Port Fuel Injection
PPC	Partially Premixed Combustion Mode
RCCI	Reactive Controlled Compression Ignition

r_p	Premixed Ratio
SFC	Specific Fuel Consumption
SN	Smoke Number
SOF	Soluble Organic Fraction
SOI	Start of Ignition
SOx	Sulfur Dioxide
TBC	Thermal Barrier Coatings
TDC	Top Dead Center
UHC	Unburned Hydrocarbon Carbon
YSZ	Yttria Stabilized Zirconia
λ	Fuel–Air Ratio

6.1 Introduction

Transportation, which acts as an enabler and booster for economic development and growth, highly depends on engine technology. This application turns out to be the leading user of internal combustion engines. Use of fossil fuel-driven ICEs further leads to the over-exploitation of fossil fuels contributing to global warming and environmental degradation. The effect of CO_2 emission on global warming is a very challenging issue for the twenty-first century. Decarbonization progress is not keeping pace with the target set by the 2015 Paris Agreement. In November 2020, at COP26 in Glasgow, countries pledged to regulate their CO_2 emissions target by 2030. Much attention was focused on GHG emissions, emitted by the energy and transportation sector. For decarbonization in the transportation sector, the use of alcohol-fueled engines is a promising solution (Zhen and Wang 2015).

Methanol (CH_3OH) has turned out to be a promising solution as an alternative fuel for internal combustion engines. Methanol has gained more interest due to its availability, ease of production, and its availability as a liquid fuel that makes it easy to store and carry. In India, the Methanol Economy program is started by NITI Aayog (https://www.niti.gov.in/methanol-economy) to reduce dependency on fossil fuels and reduce crude oil imports. Under this program, methanol production plants will be set up to produce 20 metric million tons per annum. Also, methanol will be used from the transport sector to the energy sector, which will further reduce the cost associated with methanol due to the economy of scale and help develop methanol fuel distribution infrastructure in India. In the transportation sector, methanol blending can help to improve urban air quality due to lower NO_x, SO_x, and particulate matter emissions from pure methanol and methanol blended with diesel and gasoline in ICE (Reitz et al. 2020). Methanol can also be produced from renewable waste as well as by reacting hydrogen with carbon captured from the atmosphere or industrial exhaust (Shamsul et al. 2014). It can provide a method to use carbon-based fuels in a carbon-neutral manner. Methanol can be used in pure form or as a blend in ICEs. Combustion of methanol blends in engines is not very challenging. When used as a blend, the poor auto-ignitability characteristic of pure methanol is taken care of by the other fuel, which is easily auto-ignitable also known as "ignition improver" (Aakko-Saksa et al. 2020). The most challenging part lies in operating engines with pure methanol. It is well known that methanol has a lower cetane number, higher latent heat, and high auto-ignition temperature, thus making it difficult to use as a fuel in CI engines (Verhelst et al. 2019). It requires considerable modifications in the conventional CI engines. One can rely on several concepts to produce an alcohol-fueled engine. One feasible option is the use of LHR. LHR requires the use of TBCs. The main motive behind this is to thermally insulate the interior of the engine to tap the heat to attain higher temperatures for the combustion of methanol. The second option is the port-fuel injection of methanol. This strategy requires proper timing of the valves to incorporate the start of the injection of methanol. The variable valve timing concept can be useful to keep the high-temperature exhaust gases well inside the combustion chamber thus maintaining an appropriate ignitable temperature for the next cycle (Kamo et al. 1989). The third option is the reforming of methanol fuel with the use of glow plug ignition. This solution can be used for running the compression ignition engine purely with methanol. This method helps us to overcome the use of an ignition improver to reach the appropriate ignitable conditions for methanol. Lastly, comes the method which has gained a lot of interest from researchers over the past few years, i.e., to run engines on methanol in HCCI mode and PPC mode. These concepts depend on overcoming the mixing and evaporation delays, i.e., sufficient time should be provided for mixing before auto-ignition. The unfavorable part of this method is the rising peak pressure at higher loads than the safe operating and noise range. To counter this challenge, preheating the intake air is helpful. This leads to the use of EGR and non-inter-cooled turbocharged engines to attain an appropriate peak air temperature. Extensive research is in progress to incorporate methanol as a fuel for the ICEs. Significant modifications are needed in the engine to utilize methanol as fuel as per the methods mentioned above. The following sections discuss in more detail the strategies to use methanol in CI engines along with the corresponding impact on the performance, combustion, and emissions characteristics of the engines.

6.2 Methanol Injection Strategy in CI Engines

There are three popular ways through which methanol can be utilized in CI engines namely, direct mixture, fumigation, and dual injection. These methods are described in the below subsections.

6.2.1 Direct Mixture

This is the easiest and most economical method to run CI engine fueled with methanol. In this method, instead of using 100% diesel, the methanol–diesel blend is injected into the cylinder. This method doesn't require much hardware modification. In this technique blending of methanol and diesel takes place inside the fuel tank. But methanol solubility in diesel is very poor and in case of higher moisture content, phase separation in the blends takes place. Thus, for proper mixing of methanol–diesel and to prevent phase separation before injection, little number of additives or stabilizers is added. Also, methanol solubility in diesel can be enhanced by mixing at higher temperatures (Saxena et al. 2021).

Phase separation problem in the methanol–diesel mixture can be dealt with two possible methods—by mixing cosolvent and emulsifier. The cosolvent prepares homogenous mixing of methanol–diesel by acting as a bridge between diesel and methanol molecules and the emulsifier reacts as a suspending agent, which suspends the methanol droplet in the diesel. In the emulsification process, for producing proper blending, a mixer, and heating step is required. However, in cosolvent, there is no heating step involved, it simplifies the mixing by allowing the fuel to be splash blended. This is a reason why cosolvent is chosen rather than emulsification to make stable blending of methanol–diesel. Co-solvent can be added up to a certain quantity in methanol–diesel blend; otherwise, it will hamper the spray characteristics and degrade the intake energy content. Therefore, for preventing the phase separation co-solvent such as dodecanol, oleic acid, iso-octanol, isooctyl nitrates, acetone, n-butanol, and n-octanol can be added to methanol–diesel blend (Shi et al. 2005; Pang et al. 2006; https://www.jstor.org/stable/44468053?seq=1; https://hero.epa.gov/hero/index.cfm/reference/details/reference_id/1249169).

Huang et al. (2004a) used both oleic acid (10–15%) and iso-butanol (1–2% by volume) in their investigation to inhibit the MD phase separation. Their result shows that increasing methanol quantity in diesel–methanol blend leads to longer ignition delay, that resulted in increased HRR peak and shorter combustion duration. Agarwal et al. (2019) added a little amount of dodecanol (by 1% volume) in the MD mixture to stabilize mixture and found that there is stable combustion of the MD blend as compared to baseline diesel.

6.2.2 Fumigation

In this technique, air and vaporized methanol are injected via intake manifold and diesel is directly injected from high-pressure injection system. This method requires an additional vaporizer along with the extra line and tank for the methanol. In fumigation, as discussed earlier, methanol will be vaporized and mixed with the air which reduces intake charge temperature that increases the density. Thus, more amount of air can be introduced into the cylinder and more power can be obtained. Although this approach increases the engine weight, it can replace around 50% of diesel with methanol (Imran et al. 2013). Fumigation of methanol in the CI engine prevents the formation of PAHs (polycyclic aromatic hydrocarbons). Experimental results show that methanol fumigation in diesel engines lead to lower emission without affecting engine performance. Odeca et al. (1992) used the methanol fumigation method in the DICI engine for optimizing the NOx emissions and smoke concentrations. Their result shows an increase in methanol fraction, improved fuel consumption along with reduced smoke concentration. Heisey and Lestz (1981) replaced around 55% of diesel through fumigated alcohol (methanol and ethanol) in a DICI engine. They observed that thermal efficiency is improved at heavy and moderate loads by fumigated alcohol in the diesel engine. Baranescu (1986) investigated the fumigation of alcohol in the multi-cylinder diesel engine. The alcohol was fumigated in the cross-over pipe between turbocharger and intake port. The result showed a significant increase in peak pressure rise rate in fumigated methanol as compared to baseline diesel leading to knocking.

6.2.3 Direct Injection

For tackling the solubility issue of methanol in diesel, a dual injection approach is the best option. In this method, methanol and diesel are directly introduced into the combustion chamber by different fuel streams. Commonly, in the direct injection technique, low reactive/low cetane number fuel (viz., methanol and gasoline) is injected while the piston is moving from BDC to TDC, and high reactive fuel/high cetane fuel is injected when piston reaches TDC. This method has two approaches, port injection and direct injection. In port injection (Fig. 6.1), methanol is introduced through port via intake manifold, and diesel is injected through the main injection system at high pressure. In direct injection (Fig. 6.2), methanol and diesel are injected from the same injection system but separated orifice and at different angles and timing. In both techniques, methanol acts as a primary fuel, and diesel acts as a pilot fuel, which helps methanol to burn. For port fuel injection, larger modification is needed in the engine which leads to increased cost and complexity of engine geometry, which restricts the usage of port fuel injection. However, in direct injection, due to the absence of surface contact of methanol before the injection leads to high momentum of methanol while injecting. The high momentum of methanol stream results in proper homogeneous mixture of MD, which results in proper combustion of fuel and low pollutant emission. Therefore, the direct injection strategy has great future research prospects.

6.3 Modifications in CI Engines

Methanol properties like low CN, LCV, and high LHV lead to many challenges in IC engines. For dealing with such problems, modifications need to be done in the IC engines when methanol is used as fuel. Moreover, methanol polarity leads to material compatibility issues, which can be overcome by using methanol-compatible materials only. The required modifications in different engine parts are as follows.

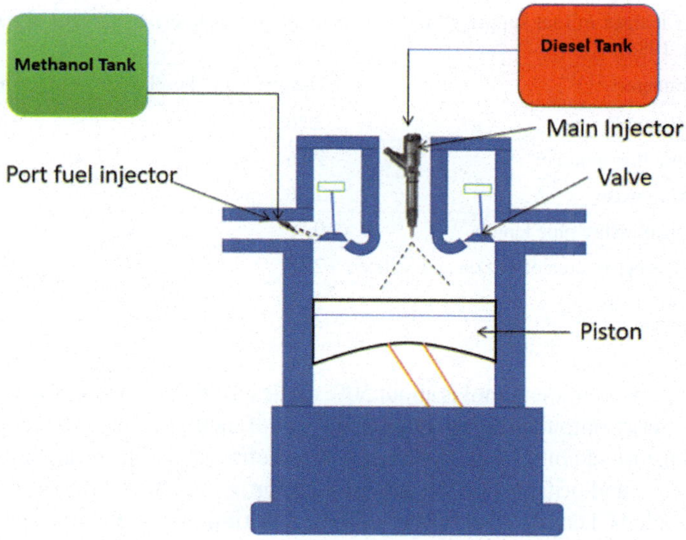

Fig. 6.1 Schematic diagram of port injection technology in CI engine (Saxena et al. 2021)

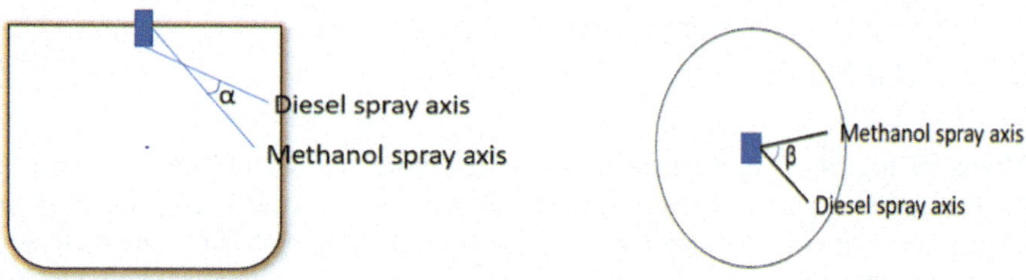

Fig. 6.2 Schematic presentation of diesel and methanol spray axis in direct injection technique (Li et al. 2020)

6.3.1 Fuel-Injection System

Energy density of methanol is about 50% of diesel and gasoline energy density. Therefore, for obtaining the same energy output as diesel and gasoline engines, double amount of methanol needs to be introduced. It can be achieved by modifying the fuel supply system, and by increasing the flow capacity. On fuel supply systems Aigal et al. (1986) have done an experimental study. In this study, they discussed modifications in the fuel supply system when methanol is utilized as fuel for the CI engine instead of diesel. According to their findings, the atomization and injection characteristics of methanol were different from the diesel. The droplet size of methanol after injection is smaller than diesel droplet. Methanol viscosity is more than diesel viscosity which may cause leakage problems in the engine. Table 6.1 consists of important changes in the fuel supply system when methanol substitutes for diesel in the CI engine.

Table 6.1 Changes in fuel supply system for methanol direct injection in comparison to diesel (Aigal et al. 1986)

S.N	Parameters	Change w.r.t. diesel-fueled operation (times)
1	Fuel supply rate	2.26
2	Refraction volume	2.26
3	Spring mass	2.26
4	High-pressure pipe length	0.70
5	The effective area of nozzle and orifice	2.18
6	Spring stiffness	2.26

Energy density of methanol is about 50% of diesel energy density, so for obtaining the same energy output as diesel, double amount of methanol needs to be introduced into combustion chamber, i.e., fuel supply rate, refraction volume, and the effective area of nozzle and orifice need to increase by approximately two times as compared to diesel-fueled CI engine, and to maintain the flow timing of methanol, spring–mass, and stiffness will be increased by 2.26 times and high-pressure pipe length needs to be decreased by 0.7 times.

6.3.2 *Cold Start*

Methanol high heat of vaporization requires more amount of energy to convert methanol into liquid as compared to diesel, which leads cold start problem in the CI engine. The cold start problem can be dealt with by preheating the methanol. Glow plug technology in the CI engine is a favorable solution to deal with the cold start problem. The glow plug has a heating component which preheats the methanol before injection. The autoignition temperature of methanol is 843 K, i.e., difficult to achieve by increasing the compression ratio but it easily ignites on a hot surface, as a results glow plug technology is a promising option to ignite the methanol in the combustion chamber.

In 1990, Caterpillar developed a heavy-duty diesel engine, which was methanol-fueled. In this engine, they used glow plug technology for the smooth combustion of methanol in the engine (Richards 1990). Mueller and Musculus (2001) have done an investigation on glow plug assisted, 1720 cc, four-stroke, heavy-duty, DICI engine. They performed a comparative study on CN 45 and M100-fueled DICI engines and found M100-fueled engine operates quieter than the CN 45-fueled engine due to the smooth combustion of methanol. Low soot emission was found in M100-fueled engine as compared to the CN 45 fuel due to the less carbon content and smooth combustion of methanol in the DICI engine. Figure 6.3 described the use of glow plug technology in the CI engine. It shows the heating element just adjacent to the tip of the fuel injector, which preheats the methanol for smooth combustion.

Fig. 6.3 Glow plug installed in the CI engine (Richards 1990)

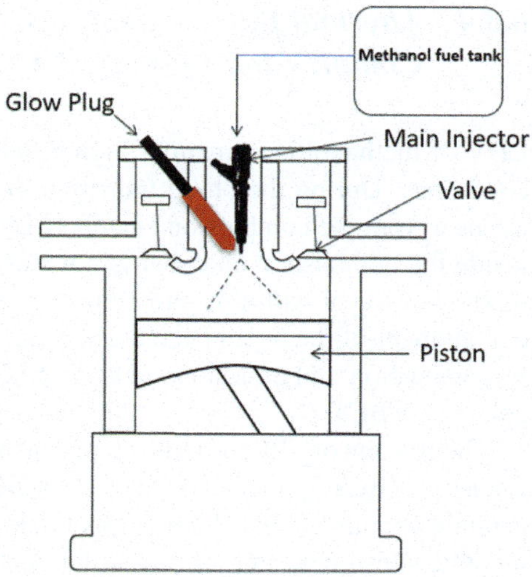

Suresh et al. (2010) conducted an experiment on a single-cylinder, 552.62 cc, four-stroke, DICI engine with and without glow plug. Without glow plug, they found less BTE and more unburnt hydrocarbon (UHC) compared to pure diesel. Again, investigation was done with glow plug and found BTE increased by 3%, and UHC, smoke emissions, and carbon monoxide (CO) reduced by 69%, 9%, and 10%, respectively.

6.3.3 Material Compatibility

Methanol's polarity leads to material compatibility issues in IC engine parts. Almost all alcohol is corrosive in nature which affects both ferrous and non-ferrous material. Light alcohol like methanol is highly reactive against metals as well as elastomers, used for sealing purposes in CI engines. Materials such as copper, magnesium, and aluminum are highly affected by methanol. But other ferrous metals and steels are normally less affected. Although M85 trial in California (Dutzi et al. 1983) did not find any material compatibility issues, but, at the end of the trials, some issues were reported in the fuel line supply system at the service station. Hence, material compatibility issues must be noted for the engine as well as the line supply system. Methanol Institute provides information regarding methanol-compatible material on its website (Methanol Institute 2016). McCoy et al. (1993) suggested that austenite stainless steel with nickel and zinc alloy coating can be used for mitigating corrosion. Instead of conventional elastomers, fluorocarbon elastomers or nitrile butadiene rubbers can be in methanol engines.

6.3.4 Thermal Barrier Coatings (TBCs) for Methanol-Fueled Compression Ignition (CI) Engines

LHV of methanol is four times higher than diesel which creates problems in cold conditions. During the phase transition of methanol from liquid to gas, it absorbs the heat from the combustion chamber. But in cold conditions, the heat availability inside the combustion chamber is not sufficient for the combustion of the fuel–air mixture. To deal with this situation, the combustion chamber can be insulated for containment of the heat generated within the engine that helps in the improvement of combustion and emissions reduction. This strategy is known as thermal barrier coatings (TBCs).

The concept of TBCs was developed in the early fifties by the American Aerospace Industry. TBCs applications lead to developments and technologies, comprising ceramic coatings (150–250 microns thick, typically 6–8% Y_2O_3 Stabilized ZrO_2), metallic substrates, and inter-metallic bond coats (typically NiAl, NiCr, NiCrAlY, etc.) and ceramic multilayers with a complex structure which are required to operate in a harsh and highly challenging environment of industrial and aircraft gas turbine engines (Yonushonis 1991).

Simultaneously, research in tapping the potential of TBCs in IC engines began (widespread by the early eighties) with the clear objective to thermally insulate the internal parts of the engine including the combustion chamber. These engines became widely popular and were known as Low Heat Rejection (LHR) engines. The concept involved the containment of the heat generated within the engine to be made available for improvement of combustion and emissions reduction. The operating temperatures of LHR engines were higher than conventional engines. This facilitated significant low emissions and fuel consumption leading to higher thermal efficiency. Kamo et al. (1989) developed an uncooled adiabatic engine in a 5 Ton army truck (thick TBCs, 0.76–6.0 mm) with 48% thermal efficiency. Some results of YSZ coated combustion chamber of IC engine included (a) a small reduction in fuel consumption with a maximum of ~6% at low engine power and (b) an increase of 8–10 bars in peak cylinder pressures in the LHR engine, especially at power outputs were observed. Nevertheless, reportedly the exhaust gas temperature was mostly low, unburned hydrocarbon matters were significantly high at low rpm and/or low power and the aluminum in the bond coat was found to be oxidized during the engine tests (Mendera 2000).

6.4 Methanol Engine Characteristics in CI Engine

This section represents the engine performance, combustion, and emissions when methanol is utilized as CI engine fuel instead of baseline diesel.

6.4.1 Methanol Performance Characteristics in CI Engines

This section describes the performance characteristics like BTE and SFC of methanol-fueled CI engines. It includes dual fuel; direct mixture as well as dedicated methanol-fueled CI engine. Different trends have been found in many reported studies. Some found improvements in thermal efficiency as methanol fraction increases in the methanol–diesel blend and others reported the opposite trends. The possible factors for high BTE are mentioned below:

- Higher oxygen content in methanol improves combustion efficiency, which leads to improved thermal efficiency.
- Low CN of methanol leads to longer ignition duration that allows more fuel to burn in the pre-combustion phase, that results in high in-cylinder temperature and higher BTE.
- The high LHV of methanol leads to absorbs more heat during compression stroke which reduces the wall heat loss. This absorbed heat may be transformed into useful work during the expansion stroke, which eventually improves efficiency (Saxena et al. 2021).

The possible reason for reducing thermal efficiency as methanol fraction increases in the diesel–methanol blend is the low calorific value of methanol. LCV of methanol is half of diesel and gasoline, thus for producing the same energy output as diesel, double the quantity of methanol needs to be injected. Since, the fuel-injected pressure is the same, which leads longer injection time. Thus, the combustion phasing of the M&D blend and baseline diesel will be different, which leads to low thermal efficiency (Saxena et al. 2021).To evaluate the engine performance of the methanol-fueled engine, Sayin (2010) have done an experimental study on a 770 cc, naturally aspirated, four-stroke, single-cylinder, DICI engine, with CR 17. The study was carried out by changing the engine speed from 1000 to 1800 rpm and keeping the load constant at 30 Nm. They have done a comparative study on M0, M5, and M10 fuels. For preventing methanol and diesel phase separation, dodecanol (by 1% volume) is added to the M&D mixture. Their study shows decreasing brake thermal efficiency and increases in BSFC as methanol % increases in the mixture.

Shamun et al. (2017) have investigated on 2124 cc, Scania D13 heavy-duty diesel engine, utilized at higher compression of 27 and fueled with pure methanol. The tests represent the effect on GIE by varying fuel injection pressure (FIP), EGR, λ, and CA50 at 6 bar IMEP and 1200 rpm crank speed. In Fig. 4a, increasing FIP has negative impact on GIE. When engine is running at higher CR, higher FIP results fuel accumulation in squish region which will not participate in the combustion process (Yanai et al. Apr. 2016). In Fig. 4b, increasing EGR % has a positive impact on GIE. EGR is a non-combustible mixture, which absorbs the heat from the surrounding charge and reduces the heat loss from the wall, and absorbed heat is converted into useful work during expansion stroke, which leads overall increment in GIE.

Najafia and Yusafb (2011) have done a comparative study on M10, M20, M30, and pure diesel by changing engine speeds from 800 to 2200 rpm with the step of

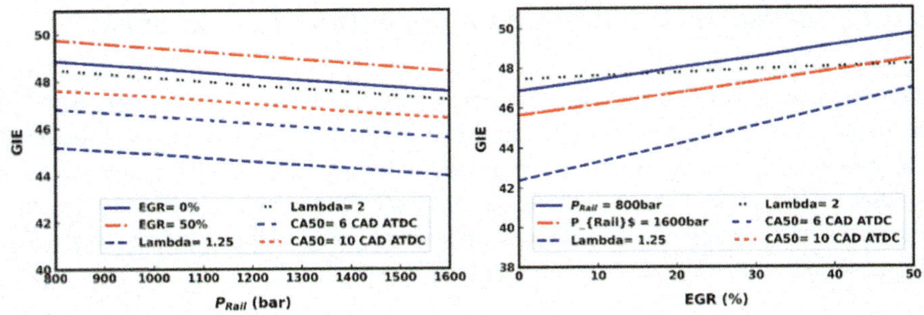

(a) (b)

Fig. 6.4 Effect of (**a**) variation of FIP and (**b**) variation of EGR on the gross thermal efficiency of methanol-fueled engine (Shamun et al. 2017)

200. They found higher torque (Fig. 5a) for the M&D mixture than in the baseline case. The input power (Fig. 5b) is high for the M&D mixture than in the baseline case because of complete combustion in the M&D case due to oxygen content in methanol. In Fig. 5c, the BSFC of the M&D blend is high due to the LCV of methanol. Due to high LHV, the exhaust gas temperature of the M&D blend is low, which results in low NO_x emissions (Fig. 5d).

Singh et al. (2020) have done an experimental study on a 512 cc, single-cylinder, four-stroke, research engine, which was converted to RCCI (reactive controlled compression ignition) mode and CI mode. In RCCI mode, diesel is injected through CRDI (common rail direct injector) system while methanol is introduced through the intake valve by PFI (port fuel injector) system. Diesel is a high reactive fuel, while methanol is low reactive fuel. In this experimental study, they varied FIP from 500 to 1000 bar and the premixed ratio (r_p) was taken at 0, 0.25, 0.50, and 0.70. They found that if r_p increases, the start of combustion retarded because of LRF fuel which reduces the global reactivity of the charge while increasing fuel injector pressure (FIP) resulted in the advanced start of combustion because, at high FIP, the droplet size of diesel and methanol is small. At high pressure of 1000 bar, the droplet size of fuel is very small which led to superior combustion. So, the in-cylinder pressure will be high and, as the $\mathrm{r_p}$ increases, LRF fuel quantity increases which leads to the reduction of in-cylinder pressure. In the performance characteristic for r_p 0 and 0.25, an increase in FIP leads to high knocking, which leads to reduced brake thermal efficiency but from 0.50 to 0.75 r_p, BTE increase because methanol is a highly anti-knocking agent.

Canakci et al. (2009) conducted an experimental investigation on a 395 cc, four-stroke, naturally aspirated, DICI diesel engine. Methanol blend in diesel varied from 0 to 15% with the step of 5%. It concludes that an increase of methanol in diesel reduces BTE and increases BSFC. The same trend was also concluded by Sayin et al. (2009) and Ciniviz et al. (2011).

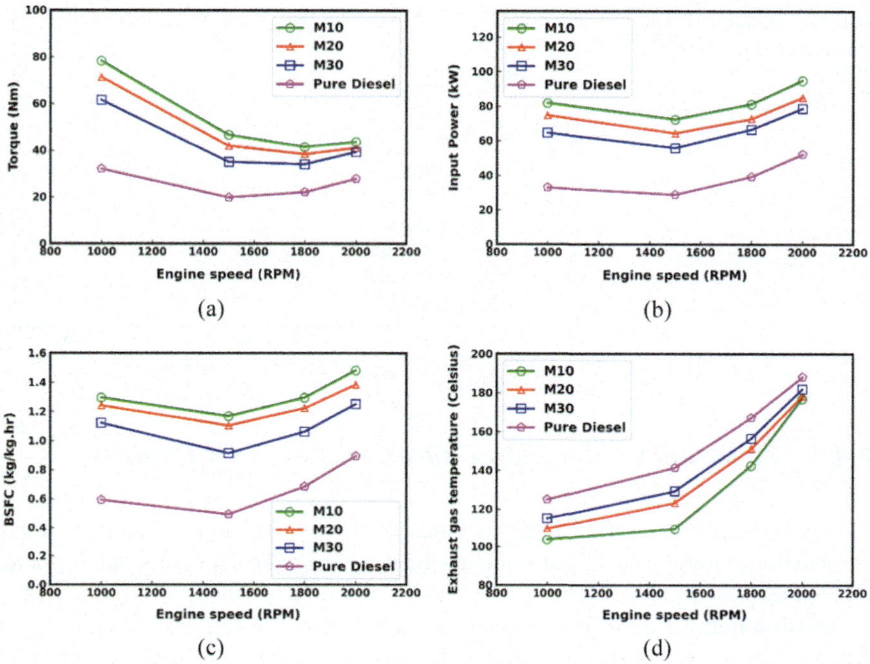

Fig. 6.5 Performance characteristics of methanol fueled engine. **a** Torque versus speed, **b** input power versus speed. **c** BSFC versus speed, and **d** EGT versus speed (Najafia and Yusafb 2011)

Methanol LCV is about 50% of diesel and gasoline, therefore for producing the same energy output as diesel, double quantity of methanol needs to be injected. Since, the fuel injection pressure is same, which results longer ignition delay in the M&D dual-fuel engine. Thus, the combustion phasing of the M&D blend and baseline diesel will be different, which leads to low thermal efficiency. Another possible reason for reducing thermal efficiency is the low cetane number of methanol, which leads longer ignition delay and a large amount of fuel burned in relief mode, which deteriorated the thermal efficiency of the engine (Saxena et al. 2021). Like thermal efficiency, a similar trend follows in BSFC. BSFC increased as methanol fraction in M&D mixture increased because of LCV of methanol and low cetane number of methanol, which results in combustion in relief mode, thus a small amount of useful work is waste.

Bayraktar (2008) have done an experimental investigation on the DICI engine. In their comparative investigation on pure diesel, M5, M10, and M15 at different speeds varying from 1000 to 1600 rpm found improved SFC up to 10% mixing of methanol in diesel. Figure 6.6 shows improved SFC, as methanol fraction increases at different speeds due to higher oxygen content in methanol, which results in proper combustion.

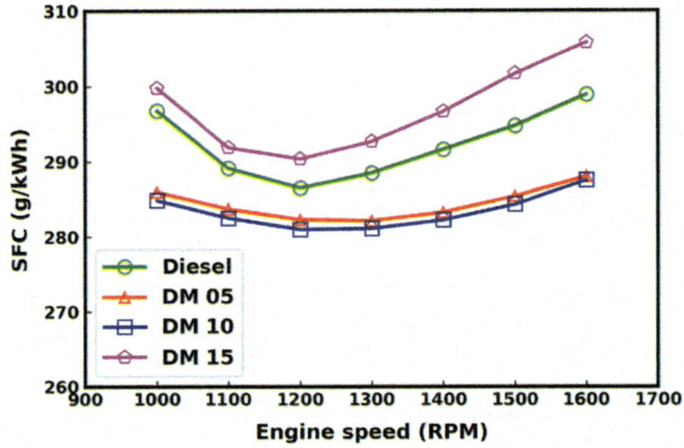

Fig. 6.6 Variation of SFC versus rpm (Singh et al. 2020)

6.4.2 Methanol Combustion Characteristics in CI Engines

In this portion, combustion characteristics of the engine such as ignition delay, combustion duration, and heat rejection rate are described for a methanol-fueled CI engine.

Ignition delay is described as the duration between SOI and start of combustion. Numerically, it can also be described as the crank angle duration between the crank angle at the SOI and the crank angle at which 10% of heat release takes place. After the injection, during the ignition delay, various physical and chemical phenomenon takes place like—droplet deformation and breakup, atomization of drops, evaporation, air and fuel turbulent mixing, and chemical reaction. As we have discussed earlier, methanol's LHV is four times higher than diesel and gasoline. So, during the vaporization, methanol takes heat from the combustion chamber, which results in a longer ignition delay as compared to the diesel-fueled CI engine. Another possible reason for longer ignition delay is the low CN of methanol. Cetane number shows reactivity of fuel. Hence, as methanol fraction in the D&M mixture increases, ignition delay increases. Ignition delay mainly depends on injection pressure and temperature, air intake pressure and temperature, speed, and load condition. Longer ignition delay is observed at low load and medium load but at high load, the in-cylinder temperature will be high, that results in the reduction in ignition delay (Mani and Šutinys 2016).

Combustion duration is described as the time or angle duration between the start of combustion and 90% combustion of fuel take place. Numerically, it can be explained as the crank duration between 10 and 90% of heat release from the fuel take place. In CI engines, combustion is completed in three different phases. These are—premixed combustion phase, diffusive combustion phase, and after-late combustion phase. In the premixed phase, due to longer ignition delay, proper mixing of air–fuel takes place, thus rapid combustion takes place. During the controlled combustion phase, fuel–air mixture and combustion take place simultaneously. The burning rate is low in the controlled combustion phase as compared to the premixed phase. In the late combustion phase, all the unburned charges present in the cylinder will burn during expansion stroke. The rate of burning in the late combustion phase is lower as compared to the premixed phase and controlled combustion phase. The combustion duration of methanol and methanol–diesel is shorter than the baseline diesel. It is because of the high flame velocity of methanol due to the high (H/C) ratio (Verhelst et al. 2019).

Heat release rate (HRR) is described as the rate of heat released from the fuel during combustion. It is generally measured in joule per second. Studies reported high peak of HRR in the MD blend as compared to baseline diesel. In methanol–diesel fueled IC engines, as discussed earlier ignition delay is longer, therefore, proper fuel–air mixing takes place, and most of the fuel burns in the premixed phase of combustion, which leads to high peak of HRR as compared to the pure diesel-fueled engine. The crank angle position of the peak of HRR is delayed in the case of the MD blend, because longer ignition delay (Yusaf et al. 2013; Saxena et al. 2021). Few of the studies also reported low peak of HRR in the methanol–diesel blend as compared to the pure diesel engine. The possible reasons are high LHV and low CV of methanol. The heat rejection rate (HRR) shows negative dip during the ignition delay because methanol absorbs heat from the combustion chamber to vaporize itself (Turkcan and Canakci 2011).

Zhen et al. (2019) have done a numerical study on a four-cylinder, intercooled, turbocharged, 1053.5 cc CI engine with a compression ratio of 17. They mainly focused on the injection system of the CI engine and used converge CFD-based simulation software. In their study, they took three modes D/M, M/D, and M/D/M mode. D/M mode means diesel will be injected first then methanol will be injected, same as the other mode. The injection pressure of diesel was 550 bar, whereas, for methanol, 180 bar in direct injection mode. In D/M mode, delay of diesel injection as well as increasing the dwell (crank angle between D_{SOI} and M_{SOI}) reduced the in-cylinder pressure. Early fuel injection resulted in low fuel economy and higher knocking intensity and too small dwell led to misfire and knocking. Too long dwell harmed the fuel economy. The optimal fuel strategy for D/M mode was $D_{SOI\text{-}10}/M_{SOI\text{-}8}$. In M/D mode delaying methanol injection after −25 aTDC caused intense knocking and delaying methanol injection resulted in an adverse effect on fuel economy and too small dwell led to misfire. The optimal injection strategy for M/D mode was $M_{SOI\text{-}25}$ and $D_{SOI\text{-}6}$. In M/D/M mode, methanol injection was set at −25 aTDC, and in this mode, the optimal mode was $M_{SOI\text{-}25}/D_{SOI\text{-}6}/M_{SOI\text{-}1}$. Out of all the modes M/D mode was the optimum mode then M/D/M followed by D/M mode (Zhen et al. 2019).

Datta and Mandal (2016) performed a numerical study on a 660 cc, single-cylinder, DICI diesel engine running at 1500 rpm. Injection timing was kept constant at −23 aTDC. Results indicated that delay in HRR and higher peak of HRR as methanol fraction in diesel–methanol increased (Fig. 7a). The possible reason for higher peak HRR is proper mixing of fuel–air during long ignition delay, which leads to more fuel burn in the premixed phase. In Fig. 7b, found that longer ignition delay as methanol fraction increases in the MD blend. The reasoning for the longer ignition delay is the high LHV and low calorific value of methanol.

Huang et al. (2004b) examined the combustion characteristics of a 903 cc, CI engine operating on an MD blend with a compression ratio of 18 and running at

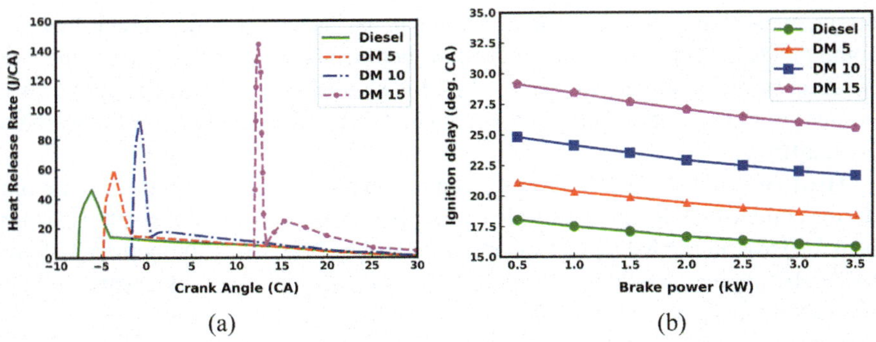

Fig. 6.7 **a** HRR versus CAD. **b** Ignition delay versus BP (Datta and Mandal 2016)

2300 rpm. Results indicated a reduction in combustion duration and an increase in ignition delay period as methanol fraction in the MD blend increases.

6.4.3 Methanol Emission Characteristics in CI Engine

In this section, methanol emissions characteristics such as NO_x, SO_x, CO, CO_2, unburned hydrocarbon (UHC), soot, and formaldehyde have been discussed. Both pure methanol and M/D blend fueled CI engine is discussed. The emissions in the CI engines depend on many parameters such as mass of fuel, cetane rating, latent heat of vaporization, injection timing of fuel, injected pressure and spray cone angle, etc.

Canakci et al. (2009) experimentally investigated about emission characteristics of diesel-fueled engine with methanol blending. Exhaust emissions like NO_x, smoke number (SN), CO, CO_2, and UHC are analyzed in this experimental work. NO_x emission is the main hazardous and challenging emission in CI engine. According to the European Environmental Agency (European Environment Agency 2019), around 40% of NO_x is emitted by road transport vehicles. NO_x emission is always in the form of oxide of nitrogen such as nitric oxide (NO) and nitrogen dioxide (NO_2).

NO_x formation mainly relies on engine in-cylinder temperature, oxygen availability, and sufficient time available for the reactions. In-cylinder temperature mainly depends on oxygen presence in the combustion chamber, CN of fuels, calorific value, and LHV of the fuel. Oxygen presence in the combustion chamber leads to higher combustion efficiency, which results in high in-cylinder temperature. Low CN leads to longer ignition delay which results in proper fuel–air mixing. Proper air–fuel mixing results in fast combustion of the fuel in the premixed phase of combustion that leads to increase in-cylinder temperature.

Ciniviz et al. (2011) also performed an experimental investigation and found that as the methanol fraction in the MD blend increase, NO_x emission increases. In this experimental work, NO_x emission was increased by 38%, 8%, and 2.5% for methanol blending of 15%, 10%, and 5%, respectively, as compared to baseline diesel engine.

Canakci et al. (2009) concluded in their experimental work that as the methanol ratio increases in dedicated diesel fuel, NO_x emission also increases. NO_x emissions were compared with M0 (Fig. 6.8), and found that NO_x emission increased by 7.59%, 16.79%, and 52.9% for M5, M10, and M15, respectively, for the load of 5 N m. Canakci et al. (2009) reported.

NO_x emission also enhances with an increase in engine load. As the FIP increases, atomized fuel particle diameter decreases, and this causes the vaporization of fuel particles very quickly. As the vaporization rate is very high, this results in a faster spreading of fuel particles throughout the combustion chamber. Because of this, the combustion rate increases and this leads to an increase in cylinder temperature. Therefore, as Injection pressure increases, NO_x emission also increases. It is investigated that for M5, as injection pressure decreases by 20 bar at 15 N m, NO_x emission decreases by 1.2%, and as injection pressure increases by 20 bar, NO_x emission increases by 4.5%.

Yao et al. (2008) also reported that as the engine load increases, NO_x emission also enhances. With the use of DMCC, there is a 50% decrement in NO_x emission as compared to baseline diesel engines. Huang et al. (2005) found that premixed phase of combustion increases with load and this leads to a lack of oxygen availability and therefore NO_x emission increase.

Smoke number (SN) Due to lack of air, smoke formation occurs especially in diesel engines. As the air–fuel ratio reduces, smoke formation increases. Due to the lack of oxygen, thermal breaking of longer chain molecules resulted in soot formation. Particulate matters contain soot, hydrocarbons, and SOF. Canakci et al. (2009) reported that enhancing methanol in the M&D blend improves oxygen quantity in fuel, which leads to proper combustion of the mixture therefore less smoke formation takes place. As FIP increases, fuel particle dia decreases that results in proper mixing of fuel–air therefore smoke formation decreases.

Yao et al. (2008) reported that as engine load increases, smoke opacity also increases in the diesel-fueled engine but with DMCC, the rate of increment in smoke opacity is lesser. Therefore, DMCC can reduce smoke emissions. Yao et al. concluded that in the case of DMCC there is a 50% decrement in smoke emission and a 80% decrement with the use of an oxidation catalyst as compared to a diesel-fueled engine.

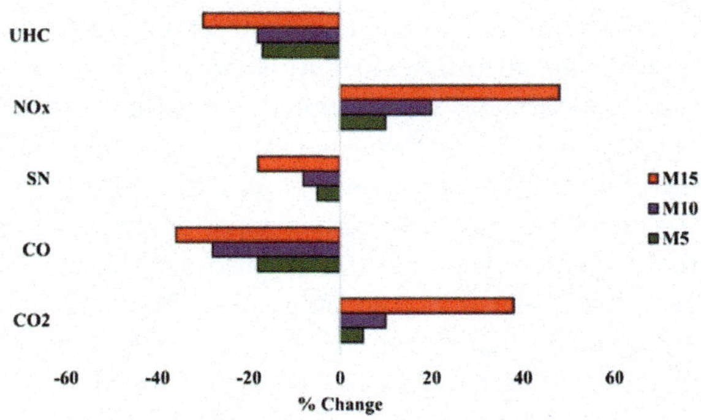

Fig. 6.8 NO_x emission at different blending of methanol (Canakci et al. 2009)

In methanol burning, there is no smoke, and the burning speed is very rapid which helps in smoke reduction. Huang et al. (2005) explained further smoke emissions can be reduced by more than 80% when DMCC is coupled with an oxidation catalytic converter.

Carbon monoxide (CO) emission CO comes out from the tailpipes of the vehicle due to improper fuel combustion in combustion chamber. Improper combustion occurs because of insufficient oxygen present in engine cylinder that leads in CO emission, and insufficient oxygen leads to incomplete conversion of CO into CO_2. As methanol fraction in methanol–diesel increases, CO emission reduces because of the oxygen present in methanol. Oxygen presence in methanol leads to improved combustion efficiency which eventually leads to low CO emission as compared to dedicated CI engine (Verhelst et al. 2019).

Canakci et al. (2009) reported that as methanol blend enhances in diesel fuel, CO emission decreases. In the comparison of dedicated diesel engines (M0) with M5, M10, and M15, reduction in CO emission was found by 14.87%, 24.34%, and 35.12% for M5, M10, and M15, respectively, at 5N-m load. It was observed that as injection pressure enhances, CO emission decreases. This is because, as FIP increases, air–fuel mixing is better, i.e., homogeneous mixture, and this leads smooth and proper combustion. Yao et al. (2008) investigated that, in the case of DMCC, CO emission is more compared to dedicated diesel engines because of lower charge temperature due to the high LHV of methanol.

Unburned hydrocarbon (UHC) emissions Unburned hydrocarbon emission shows the improper combustion of fuel inside the cylinder. Incomplete combustion occurs due to lack of oxygen presence inside the engine cylinder which results in UHC. UHC mainly results because of the unburned air–fuel mixture. There can be other UHC emission sources such as incomplete combustion and engine lubricant. UHC is always in gaseous state. Solid state UHC is known as particulate matter.

Due to oxygen present in methanol, it leads better combustion quality. Because of polar nature of methanol molecules, it cannot be absorbed by non-polar lubricating oils so this results in less UHC formation. Therefore, as methanol content increases in Diesel fuel, UHC decreases. Canakci et al. (2009) found that for M5, M10, and M15, UHC decreases by 18%, 33%, and 46%, respectively. Örs et al. (2020) also found in their experimental work that as the methanol blend in diesel increases, total hydrocarbon oxidation enhances. Due to higher in-cylinder temperatures, it is easier for fuel to react with oxygen. It is also observed that for methanol, laminar flame speed is higher compare to diesel fuel. As laminar flame speed is higher in the case of methanol fuel, this decreases combustion duration and simultaneously increases combustion temperature. This higher combustion temperature leads to proper combustion and therefore total hydrocarbon emissions are less. Sayin et al. (2009) also found in their research work that hydrocarbon emission decreases as the methanol blend in diesel fuel increases. Yao et al. (2008) also investigated that in the case of DMCC, HC emission is higher as compare to pure diesel engines. But by the use of a catalytic converter, this increase in HC emission can be reduced. Chao et al. (2001) also investigated, both at low and high loads and found that 90% HC emission increases in both cases as compared to baseline diesel engines.

6.5 Summary and Future Scope

Detailed review of literature on methanol fueled engines revealed that methanol is a well-suited fuel for CI engines for replacing fossil-based fuels. Properties like low carbon and high oxygen content in methanol lead to proper combustion and less carbon-based emissions. Also, methanol doesn't have sulfur content therefore it produces zero sulfur-based emissions. But, some properties of methanol like high LHV and lower heating value creates problem in the CI engine and can be rectified by modification in some parts of the CI engine. The polar nature of methanol leads to material compatibility issues in the CI engines, which can be mitigated by using methanol-compatible materials in the manufacturing of these engines. Various injection strategies of methanol fueling such as direct mixture, fumigation, and dual injection have been reviewed. Summary of major findings is as follows:

- In MD blend, as the methanol fraction increases, thermal efficiency improves due to the higher oxygen content, that results proper combustion of the fuel–air mixture. But some researchers found reduced thermal efficiency as methanol fraction increases, the possible reason for that is low CN and low calorific value of methanol.
- The CN of any fuel shows the reactivity of the fuel. CN of methanol is very low which makes methanol a low reactivity fuel. So, the properties like high LHV and low CN results in longer ignition delay. During a longer ignition delay, proper mixing of fuel–air mixture takes place, which results in fast combustion of a large quantity of fuel in premixed phase of combustion therefore combustion duration increases, as the methanol quantity in the M-D blend increases.
- In case of methanol fueled CI engines, NO_x emission is a major issue. In this research, found that as FIP increases, NO_x emission also increases. Increase in FIP results in a decrement in injected fuel particle size that ultimately increases the vaporization rate. Because of the higher vaporization rate, fuel particles spread at faster rate throughout the cylinder and this phenomenon increases rate of combustion. Therefore, in-cylinder temperature increases which results in higher NOx emission.

Being an oxygenated fuel with less carbon content, methanol produces less carbon-based emissions. The methanol production rate is high as compared to other renewable fuels which makes methanol best option to replace fossil-based fuels. Glow plug assisted methanol fueled CI engine produces less noise as compared to diesel-fueled CI engine due to the smooth combustion of methanol. But the properties like low energy density and high LHV create many challenges when using methanol as CI engine fuel. Due to the polar nature of methanol, it is highly reactive against the elastomers and metals, which need to be resolved by using methanol-compatible material. The low viscosity and lubricity of methanol cause leakage problems from gland seals, gaskets, O-rings, and packings that is also a challenge to be rectified before large scale usage of methanol.

References

Aakko-Saksa PT et al (2020) Renewable methanol with ignition improver additive for diesel engines. Energy Fuels 34(1):379–388. https://doi.org/10.1021/ACS.ENERGYFUELS.9B02654/ASSET/IMAGES/LARGE/EF9B02654_0008.JPEG

Agarwal AK, Sharma N, Singh AP, Kumar V, Satsangi DP, Patel C (2019) Adaptation of methanol-dodecanol-diesel blend in diesel genset engine. J Energy Resour Technol Trans ASME 141(10):1–9. https://doi.org/10.1115/1.4043390

Aigal AK, Pundir BP, Khatchian AS (1986) High pressure injection and atomization characteristics of methanol. SAE Trans 95:691–708. http://www.jstor.org/stable/44469082

Baranescu RA (1986) Fumigation of alcohols in a multicylinder diesel engine-evaluation of potential. SAE Technical Paper.https://doi.org/10.4271/860308

Bayraktar H (2008) An experimental study on the performance parameters of an experimental CI engine fueled with diesel-methanol-dodecanol blends. Fuel 87(2):158–164. https://doi.org/10.1016/j.fuel.2007.04.021

Canakci M, Sayin C, Ozsezen AN, Turkcan A (2009) Effect of injection pressure on the combustion, performance, and emission characteristics of a diesel engine fueled with methanol-blended diesel fuel. Energy Fuels 23(6):2908–2920. https://doi.org/10.1021/ef900060s

Chao MR, Lin TC, Chao HR, Chang FH, Chen CB (2001) Effects of methanol-containing additive on emission characteristics from a heavy-duty diesel engine. Sci Total Environ 279(1–3):167–179. https://doi.org/10.1016/S0048-9697(01)00764-1

Ciniviz M, Köse H, Canli E, Solmaz Ö (2011) An experimental investigation on effects of methanol blended diesel fuels to engine performance and emissions of a diesel engine. Sci Res Essays 6(15):3189–3199. https://doi.org/10.5897/SRE11.230

Datta A, Mandal BK (2016) Impact of alcohol addition to diesel on the performance combustion and emissions of a compression ignition engine. Appl Therm Eng 98:670–682. https://doi.org/10.1016/j.applthermaleng.2015.12.047

Dutzi E, Gershman R, Maynard D (1983) California methanol assessment. Altern Fuel II

Effect of ethers and ether/ethanol additives on the physicochemical properties of diesel fuel and on engine tests | Health & Environmental Research Online (HERO) | US EPA. https://hero.epa.gov/hero/index.cfm/reference/details/reference_id/1249169. Accessed 25 June 2022

European Environment Agency (2019) Nitrogen oxides (NOx) emissions european environment agency. https://www.eea.europa.eu/data-and-maps/indicators/eea-32-nitrogen-oxides-nox-emissions-1/assessment.2010-08-19.0140149032-3. Accessed 04 June 2022

Heisey JB, Lestz SS (1981) Aqueous alcohol fumigation of a single-cylinder di diesel engine. SAE Technical Paper. https://doi.org/10.4271/811208

Huang Z et al (2004a) Combustion behaviors of a compression-ignition engine fuelled with diesel/methanol blends under various fuel delivery advance angles. Bioresour Technol 95(3):331–341. https://doi.org/10.1016/j.biortech.2004.02.018

Huang ZH et al (2004b) Engine performance and emissions of a compression ignition engine operating on the diesel-methanol blends. Proc Inst Mech Eng Part D: J Automob Eng 218(4):435–447. https://doi.org/10.1243/095440704773599944

Huang ZH et al (2005) Engine performance and emissions of a compression ignition engine operating on the diesel-methanol blends. 218(4):435–447. https://doi.org/10.1243/0954407047733599944

Imran A, Varman M, Masjuki HH, Kalam MA (2013) Review on alcohol fumigation on diesel engine: a viable alternative dual fuel technology for satisfactory engine performance and reduction of environment concerning emission. Renew Sustain Energy Rev (Pergamon) 26:739–751. https://doi.org/10.1016/j.rser.2013.05.070

Kamo R, Woods ME, Bryzik W (1989) Thin thermal barrier coating for engines. https://www.osti.gov/biblio/5312271

Li Z et al (2020) Investigation of injection strategy for a diesel engine with directly injected methanol and pilot diesel at medium load. Fuel 266(January):116958. https://doi.org/10.1016/j.fuel.2019.116958

Mani K, Dr. Šutinys E (2016) Effect on the ignition delay and the characteristics of engine by using biodiesel fuel. Int J Eng Res V5(12):296–297. https://doi.org/10.17577/ijertv5is120288

Methanol Institute (2016) Compatibility of metals & alloys in neat methanol service, no C, pp 1–18. http://www.methanol.org/wp-content/uploads/2016/06/Compatibility-of-Metals-Alloys-in-Neat-Methanol-Service.pdf

Methanol Economy | NITI Aayog. https://www.niti.gov.in/methanol-economy. Accessed 27 June 2022

McCoy GA, Kerstetter J, Lyons JK (1993) Alcohol-fueled vehicles: an alternative fuels vehicle, emissions, and refueling infrastructure technology assessment. https://doi.org/10.2172/258222

Mendera Z (2000) Effects of plasma sprayed zirconia coatings on diesel engine heat release. J KONES. Internal Combust Engines 7(1–2):382–389

Mueller CJ, Musculus MP (2001) Glow plug assisted ignition and combustion of methanol in an optical di diesel engine, no 724. SAE Technical Paper.https://doi.org/10.4271/2001-01-2004

Najafia G, Yusafb TF (2011) Experimental investigation of using methanol-diesel blended fuels in. Int J Technol Eng Syst 2(March):81–84

Odaka M, Koike N, Tsukamoto Y, Narusawa K (1992) Optimizing control of NO_x and smoke emissions from di engine with EGR and methanol fumigation, no 412. SAE Technical Paper. https://doi.org/10.4271/920468

Örs İ, Sayin B, Ciniviz M (2020) An experimental study on the comparison of the methanol addition into gasoline with the addition of ethanol. Int J Autom Sci Technol 4(2):59–69. https://doi.org/10.30939/ijastech..713682

Pang X et al (2006) Characteristics of carbonyl compounds emission from a diesel-engine using biodiesel–ethanol–diesel as fuel. Atmos Environ 40(36):7057–7065. https://doi.org/10.1016/J.ATMOSENV.2006.06.010

Reitz RD et al (2020) IJER editorial: the future of the internal combustion engine. Int J Engine Res 21(1):3–10. https://doi.org/10.1177/1468087419877990

Richards BG (tran) (1990) Methanol-fueled Caterpillar 3406 engine experience in on-highway trucks. JSTOR. https://www.jstor.org/stable/44580435. Accessed 13 Oct 2021

Saxena MR, Maurya RK, Mishra P (2021) Assessment of performance, combustion and emissions characteristics of methanol-diesel dual-fuel compression ignition engine: a review. J Traffic Transp Eng (English Ed.) 8(5):638–680. https://doi.org/10.1016/j.jtte.2021.02.003

Sayin C (2010) Engine performance and exhaust gas emissions of methanol and ethanol-diesel blends. Fuel 89(11):3410–3415. https://doi.org/10.1016/j.fuel.2010.02.017

Sayin C, Ilhan M, Canakci M, Gumus M (2009) Effect of injection timing on the exhaust emissions of a diesel engine using diesel-methanol blends. Renew Energy 34(5):1261–1269. https://doi.org/10.1016/j.renene.2008.10.010

Shamsul NS, Kamarudin SK, Rahman NA, Kofli NT (2014) An overview on the production of bio-methanol as potential renewable energy. Renew Sustain Energy Rev 33:578–588. https://doi.org/10.1016/J.RSER.2014.02.024

Shamun S, Haşimoğlu C, Murcak A, Andersson Ö, Tunér M, Tunestål P (2017) Experimental investigation of methanol compression ignition in a high compression ratio HD engine using a Box-Behnken design. Fuel 209(May):624–633. https://doi.org/10.1016/j.fuel.2017.08.039

Shi X, Yu Y, He H, Shuai S, Wang J, Li R (2005) Emission characteristics using methyl soyate–ethanol–diesel fuel blends on a diesel engine. Fuel 84(12–13):1543–1549. https://doi.org/10.1016/J.FUEL.2005.03.001

Singh AP, Sharma N, Satsangi DP, Agarwal AK (2020) Effect of fuel injection pressure and premixed ratio on mineral diesel-methanol fueled reactivity controlled compression ignition mode combustion engine. J Energy Resour Technol Trans ASME 142(12):1–11. https://doi.org/10.1115/1.4047320

Suresh R, Durgaprsad B, Senthil Kumar P, Albert M (2010) Performance analysis for emission and combustion of methanol blend diesel fuel in CI engine using glow plug. In: Proceedings of the international conference on frontiers in automobile and mechanical engineering 2010, pp. 164–169. https://doi.org/10.1109/FAME.2010.5714825

State-of-the-art report on the use of alcohols in diesel engines on JSTOR. https://www.jstor.org/stable/44468053?seq=1. Accessed 25 June 2022

Turkcan A, Canakci M (2011) Combustion characteristics of an indirect injection (IDI) diesel engine fueled with ethanol/diesel and methanol/diesel blends at different injection timings. In: Proceedings of the world renewable energy congress—Sweden, vol 57, Linköping, Sweden, pp 3565–3572. https://doi.org/10.3384/ecp110573565

Verhelst S, Turner JW, Sileghem L, Vancoillie J (2019) Methanol as a fuel for internal combustion engines. Prog Energy Combust Sci 70:43–88. https://doi.org/10.1016/J.PECS.2018.10.001

Yanai T, Aversa C, Dev S, Reader G, Zheng M (2016) Investigation of fuel injection strategies for direct injection of neat n-butanol in a compression ignition engine. SAE Int J Engines 9(3):1512–1525. https://doi.org/10.4271/2016-01-0724

Yao C, Cheung CS, Cheng C, Wang Y, Chan TL, Lee SC (2008) Effect of diesel/methanol compound combustion on diesel engine combustion and emissions. Energy Convers Manag 49(6):1696–1704. https://doi.org/10.1016/J.ENCONMAN.2007.11.007

Yonushonis TM (1991) Barrier For diesel components coatings energy conservation and renewable office of propulsion systems, p 131. http://ntrs.nasa.gov/archive/nasa/casi.ntrs.nasa.gov/19940024847.pdf

Yusaf T, Hamawand I, Baker P, Najafi G (2013) The effect of methanol-diesel blended ratio on ci engine performance. Int J Automot Mech Eng 8(1):1385–1395. https://doi.org/10.15282/ijame.8.2013.26.0114

Zhen X, Wang Y (2015) An overview of methanol as an internal combustion engine fuel. Renew Sustain Energy Rev 52:477–493. https://doi.org/10.1016/J.RSER.2015.07.083

Zhen X, Li X, Wang Y, Liu D, Tian Z (2020) Comparative study on combustion and emission characteristics of methanol/hydrogen, ethanol/hydrogen and methane/hydrogen blends in high compression ratio SI engine. Fuel 267(December 2019):117193. https://doi.org/10.1016/j.fuel.2020.117193

Chapter 7
The Impact of Renewable Fuels and Fuel Additives (Dodecanol) on Particulate Mass Emission for Sustainable Mobility

Jyothi Jayakumar, Priyanka Gupta, Nisha Yadav, Jitendra Dixit, and Nikhil Sharma

Abstract Among IC engines, diesel vehicles are the major contributors to particulate mass emissions. Renewable fuels play a major role in reducing both gas phase and particulate mass emissions. In addition to renewable fuels, fuel additives are also investigated by scientists to compensate for the limitations of using renewable fuels. In this book chapter, a review of renewable fuels to be used in compression ignition (CI) engine was carried out. In addition to this, an experimental investigation was carried out in a single cylinder, water cooled variable compression ratio (VCR) diesel engine to make a comparative analysis among baseline diesel fuel and diesel + dodecanol (1% v/v) without renewable fuels. Dodecanol was chosen due to its advantages over other fuel additives. Results pertaining to in-cylinder pressure, hydrocarbon (HC) emission, carbon mono-oxide (CO) emission and particulate mass emission are presented. Soot was collected on a 47 mm quartz filter paper using a dilution tunnel and later analysed for soot morphology using scanning electron microscope (SEM). Energy dispersive spectroscope (EDS) analysis was also performed to find out the presence of metals in soot. Gravimetric analysis shows that Diesel + dodecanol (1% v/v) resulted in 2.62 mg of soot deposited on filter paper whereas diesel fuel showed 1.46 mg of soot deposited on filter paper for a duration of 30 min.

Keywords Renewable fuels · Sustainable mobility · Additive · Particulate mass emission · Filter paper

Abbreviations

BSFC	Brake Specific Fuel Consumption
BTE	Break Thermal Efficiency
CLD	Chemiluminescence Detector
CNG	Compressed Natural Gas

J. Jayakumar · P. Gupta · N. Yadav · J. Dixit · N. Sharma (✉)
Department of Mechanical Engineering, Malaviya National Institute of Technology Jaipur, Jaipur, Rajasthan 302017, India
e-mail: nikhil.mech@mnit.ac.in

P. C. Shukla et al. (eds.), *Renewable Fuels for Sustainable Mobility*, Energy, Environment, and Sustainability, https://doi.org/10.1007/978-981-99-1392-3_7

CO	Carbon Monoxide
CO_2	Carbon Dioxide
DMC	Dimethyl Carbonate
EDS	Energy Dispersive Spectroscope
EGR	Exhaust Gas Recirculation
EGT	Exhaust Gas Temperature
EHN	2-Ethylhexyl Nitrate
FID	Flame Ionisation Detector
H_2	Hydrogen
HC	Hydrocarbon
HHO	Hydrogen-Hydrogen-Oxygen
IC	Internal Combustion
IMEP	Indicated Mean Effective Pressure
LPG	Liquefied Petroleum Gas
NDIR	Non-Dispersive Infrared
NO_x	Oxides Of Nitrogen
PM	Particulate Matter
PODE	Polyoxymethylene Dimethyl Ether
SEM	Scanning Electron Microscope
SI	Spark Ignition
UHC	Un-Burnt Hydrocarbon
VCR	Variable Compression Ratio
WHO	World Health Organisation

7.1 Introduction

World is facing twin crises of fossil fuel shortage and environmental degradation. Globally, diesel engines are a significant source of carbon monoxide (CO), nitrous oxide (NO_x), and hydrocarbon (HC) emissions, contributing to global warming. Diesel engine plays a vital role in current economic development. Diesel engine is still leading a role mainly in long distance cargo, construction, agriculture and transportation sector (Reitz, et al. 2020). The majority of the world population is inhaling unhealthy air containing fine particulates, having a size below the limit suggested by World Health Organisation (WHO). This is now identified as the fourth most probable reason for early deaths and health issues. With the growth of the human population, the use of this conventional energy source is leading to higher environmental degradation.

Diesel engine contributes many harmful emissions such as CO, oxides of nitrogen (NO_x), carbon dioxide (CO_2) and particulate matter (PM). Among all these emissions, PM and NO_x emissions is critical to human health. In developing countries transportation sector is one of the fastest growing sectors and a major contributor of the greenhouse gas emissions. PM is the mixture of all solid and liquid particles suspended in the air. Fine particles travel to the lungs and can also act as carriers for many potentially toxic substances (Eastwood 2008). Alternative fuels are one of the probable reasons to control emission from diesel engines. As compared to the traditional fuels methanol is recommended as a new generation of substitute fuel for internal combustion (IC) engine in the upcoming time and the government of India is promoting the methanol as a fuel. Methanol also known as methyl alcohol is a saturated, single carbon straight chained compound. Methanol is an alcohol and is a polar, neutral, colourless and highly flammable liquid. Methanol mix's with H_2O, alcohols, esters and furthermost other organic solvents. It is only slightly soluble in oils and fats. Methanol contains oxygen which helps in the complete combustion of fuel resulting in eliminating the un-burnt hydrocarbon. It has excellent thermo-physical properties like high latent heat of vaporisation which helps in the reduction of the in-cylinder temperature.

Methanol is one of the friendliest fuels for IC engines, for illustration, (1) methanol can be used for different/high compression ratio internal combustion (IC) engine that could replace diesel in compression ignition (CI) engine; (2) methanol can be used in spark ignition (SI) engine due high good octane number (Zhen et al. 2015). Han et al. (Lu et al. 2019) examined NO_x emission influencing factor in methanol/diesel dual-fuel CI engine. Author conducted a numerical simulation followed by an experiment in which the author observed the effect of methanol concentration and exhaust gas recirculation (EGR). Author found that NO_x emission increases when methanol is present in the premixed reason and the temperature is high in the cylinder. But in the experiment, the author found that the NO_x emission first reduces and then increases with methanol addition. Jamrozik (2017) performed investigation on performance and emission characteristics of a direct injection single cylinder four stroke CI engine that works on the methanol/diesel and ethanol/diesel blend. The concentration of the alcohols varies from 0 to 40% by volume. The increase in the methanol concentration up to 30% in methanol/diesel blend had an optimistic effect on the BTE of the engine. No significant change was observed in mean effective pressure. Furthermore, a reduction was observed in CO, HC and CO_2. Similarly, when ethanol and diesel blend was used in the same engine it was found that the break thermal efficiency (BTE) was increased with constant mean effective pressure inside the cylinder. Also, a reduction was observed in CO and CO_2 and THC emission remained constant.

Elfasakhany and Mahrous (2016) conducted an investigation on the performance and emission characteristics of n-butanol/methanol/gasoline mixtures in spark ignition engines. In this experiment, author prepared the blend in which the concentration of butanol/methanol was 0, 3, 7, and 10% v/v in gasoline. These blends were examined in an engine operating at a speed range of 2600–3400 rpm. Furthermore, the performance and emissions of the methanol and butanol/gasoline mixtures were compared with the pure gasoline fuel. With the increase in the concentration of the butanol/methanol to the pure gasoline, the author found the performance of the engine was degraded. Furthermore, the authors observed that CO_2 emissions were reduced where CO and UHC were increased. Li et al. (2017) investigated emission, performance and combustion characteristics of a single cylinder four stroke

SI engine which worked with different concentrations of methanol, ethanol, and n-butanol blended with pure gasoline. The author found that with alcohols addition in the pure gasoline, combustion phasing get advanced; butanol and gasoline blends exhibited the lesser brake specific fuel consumption (BSFC), ethanol and gasoline blends formed the lowermost UHC emission, and methanol and gasoline blends exhibited the lowermost NO_x emission.

Sharudin et al. (2017) examined the effects of using butanol additives on performance, combustion, and emission characteristics in conventional SI engine. In this experiment, author prepared a blend in which the concentration of butanol was 5, 10, and 15% v/v in methanol/gasoline. These blends were examined in an engine operating at a speed range from 1000 to 2500 rpm fuelled with butanol, methanol and gasoline blends. The author found that with iso-butanol addition to methanol/gasoline blends, the break thermal efficiency (BTE) was enhanced, break specific fuel consumption (BSFC) was reduced and EGT increased with increase in load. Furthermore, for iso-butanol, methanol and gasoline blends, CO and un-burnt hydrocarbon (UHC) emissions reduced, and CO_2 and NO_x emissions were found to increase. Duraisamy et al. (2020) investigated the effect of diesel/methanol and Polyoxymethylene dimethyl ether (PODE)/methanol blend on the performance and combustion characteristics in CI engine. This experiment was conducted on the three cylinders, four strokes and turbocharged direct injection CI engine. This engine was operated at 1500 rpm and a constant break mean effective pressure of 3.4 bar. It was found that the increase in the methanol concentration mass fraction extended the ignition delay and diminished the in-cylinder pressure for both diesel and methanol blends as well as methanol and PODE blend. At the 80% of premixed mass fraction of the blend, the maximum BTE of both diesel and methanol blend as well as methanol and PODE blend was 3.5% higher as compared to the conventional diesel combustion.

Renewable fuels, such as alcohols, compressed natural gas (CNG), Hydrogen (H_2), vegetable oils, biogas, and liquefied natural gas (LNG), have the prospect of substituting petroleum-based fuels in modern IC engines with little to no modification. Compared to fuels made from crude oil refining process, these fuels burn clean, have a lower reliance on conventional fuels and have less production cost (Nouni et al. 2021). Alcohols include methanol, ethanol, propanol and butanol. The specific heating value of gasoline and methanol is less than ethanol (26.8 MJ/kg) and methanol (19.7 MJ/kg). Methanol is also known as wood alcohol derived from natural gas and is considered an IC engine alternative fuel. This fuel is usually produced by steam-reformed natural gas to form synthesis gas. Methanol has chemical properties similar to ethanol as an engine fuel. The properties of methanol, such as lower flammability, make it safer than gasoline. Ethanol is a renewable fuel produced from biomass. It is clear colourless alcohol and is derived from various biomass materials. Ethanol has a boiling point of 78.5 °C and a specific density of 0.789 g/ml at 20 °C. LPG, commonly referred to as "Autogas" in some countries, is a mixture of propane and butane that can also be used as a fuel for vehicles. Auto LPG's development has only been moderately successful in India despite the economic advantages for the user and being more ecologically friendly than petrol and diesel because there isn't a supportive governmental system in place (Nouni et al. 2021). Biogas primarily comprises methane and carbon dioxide produced from organic materials such as agricultural waste, manure, plants, green waste, and food waste. Biogas is used as a fuel in stationary engines. However, the main problem is storing the biogas. Another problem with biogas plants is their size, as they require large areas and can only be used for stationary engines (Nouni et al. 2021). Biodiesel is a renewable fuel produced domestically from recycled grease or vegetable oils. Among the numerous bio-fuels discussed above, alcohols are now considered as possible future IC engine fuel when blended with conventional fuels. Krzemiński et al. (2017) reported that ethanol and methanol can be blended with diesel oil to propel the engines. However, at low-temperature diesel blended ethanol have low solubility index.

Sharma et al. (2020) reported a reduction in CO, HC, and smoke emissions by 58%, 60%, and 49%, respectively. However, due to the addition of HHO, an increase in O_2 concentration resulted in a higher reaction temperature, resulting in a minor NO_x emission increase. Bose and Maji (2009) used EGR technology to lower NO_x levels. The NO_x value for hydrogen enrichment without EGR was 1211 ppm at 80% load, but the NO_x value for hydrogen enrichment with 20% EGR was 710 ppm. Yadav et al. investigated the emissions from a direct injection C.I. engine operating in dual-fuel mode (hydrogen-diesel) with an exhaust gas recirculation system. In the engine, hydrogen-enriched air was used as the intake charge, with 10% and 20% exhaust gas recirculation (EGR), respectively. The outcomes were as follows: At 80% load, the NO_x value for hydrogen enrichment without EGR was 470 ppm, whilst it was 447 ppm for hydrogen enrichment with EGR. The decrease in peak combustion temperature caused by the inert gas in the EGR was attributed to the reduction in NO_x. At 80% load, neat diesel operation had an O_2 concentration of 15.1% by volume. Without EGR, O_2 had a volume concentration of 15.2, 10.4 with 10% EGR, and 9% with 20% EGR. No hydrogen carbon atoms result in low CO and CO_2 emissions (SinghYadav et al. 2012; Baltacioglu et al. 2016). Baltacioglu et al. (2016) conducted an experimental investigation to compare pure Hydrogen and HHO (hydroxy) enriched biodiesel (B10) fuel. This study aims at pilot injection of diesel engine performed when alternative fuels such as pure Hydrogen, HHO, and biodiesel were added. Some of the findings that could be drawn from the diesel engine experiment results: when compared to diesel fuel, H_2 and B10 and HHO and B10 achieved CO emission reductions of 29% and 22%, respectively. H_2 and B10 fuels emit 8.72 and 22.3% less CO_2 than HHO and B10 and ordinary diesel fuels. Saravanan et al. (2008) employs an EGR system based on the principle of recirculating exhaust gases back into the inlet manifold, which combines with new air and dilutes by the intake charge (a diluent). It lowered the combustion chamber's peak combustion temperature, reducing NO_x production. The minimum NO_x concentration was 464 ppm with a 25% EGR (Saravanan et al. 2008).

In another experimental investigation, nano-fuel additives were used in a single cylinder direct injection diesel engine to explore performance and emission characteristics of jatropha biodiesel (Ganesh and Gowrishankar 2011). Interestingly, Dimethyl carbonate (DMC) is well-thought-out as a choice for meeting the oxygenate specifications on gasoline (Pacheco and Marshall 1997). Fayyazbakhsh et al. (2017) investigated the performance of the engine and the properties of the fuel in IC engine with additives. Authors found that the pollutants reduced with increase in engine speed. Le et al. (2020) investigated the effect of cetane number booster mainly 2-ethylhexyl nitrate (EHN) on the performance and emission in IC engine. Authors found that the using EHN CN booster additives improves the overall cetane number of the blend resulting in improved performance of the engine and reduction in the detonation tendency. Cetane number is the measurement of the combustion speed of diesel fuel and diesel fuel components. As a matter of fact, lower cetane number fuels tend to have longer ignition delay periods relative to higher cetane number fuels. Additives such as alkyl nitrate have capability to enhance cetane number significantly. Researchers are working on additives such as 1-tetradecane, nitro ethane, 1-octadecanol, and isobutyl stearate. Yanowitz et al. (2017) have investigated different types of methods to calculate the cetane number of different additives. This author has also suggested some additives have high CN and cost effective which can be used to improves the performance of the IC engine along with able to suppress detonation. At low temperature, solubility of alcohol like fuels changes in diesel like fuels. Additives can enhance the solubility characteristics of diesel with alcohol fuels such as diesel. Dodecanol is known to increase the solubility characteristics of alcohol such as ethanol and methanol with diesel fuel. Table 7.1 shows a comparative analysis for ethanol port fuel injection + gasoline direct injection vs gasoline direct injection engine for engine out emission. There are advanced engine research centres working on the advanced concept and engine emission control using various renewable fuels and advanced engine technologies (Shamun et al. 2020; Belgiorno et al. 2019; Ianniello et al. 2021; Blasio et al. 2022; Dimitrakopoulos et al. 2017; Wang et al. 2015).

From the above discussion, it can be said that both "renewable fuels" and "additives" play a vital role in improving the characteristics of fuel. In this book chapter, a review of renewable fuels to be used for CI engine was done. There are many research articles that discuss the comparative analysis of diesel/renewable fuel/diesel

Table 7.1 Comparison of emission characteristics for ethanol port fuel injection + gasoline direct injection vsersu gasoline direct injection engine

Authors (ethanol port fuel injection + gasoline direct injection vs. gasoline direct injection engine)	Emissions characteristics				
	CO	HC	NO_x	PM	PN
Kim et al. (2015)	Increase	Mixed trend	Mixed trend	Decrease	Decrease
Liu et al. (2015)	N.A	N.A	N.A	N.A	Decrease
Qian et al. (2019)	Increase	Decrease	Decrease	N.A	N.A
Sun et al. (2019)		Decrease	N.A	N.A	Decrease
Kalwar et al. (2020)	Decrease	Increase	Decrease	Decrease	Decrease

+ renewable fuels/diesel + renewable fuels + fuel additives. There is limited literature available on diesel + additives fuel (without the addition of bio-fuels) to be used in IC engine. The present experimental investigation explores the use of additives in diesel fuel and its effect on particulate mass emissions. A single cylinder 661.5 cc diesel engine was selected to perform experimental investigation. A comparative analysis among baseline diesel fuel and diesel + dodecanol (1% v/v) without renewable fuels was carried out. Many of the industry issues (purity, blending, transport and storage of these fuels) with renewable can be addressed with the use of appropriate fuel additives. In this experimental investigation, dodecanol was chosen due to its advantages over other additives. Results pertaining to in-cylinder pressure, HC emission, CO emission and particulate mass emission are presented. Soot was collected on a 47 mm quartz filter paper and later analysed for soot morphology using scanning electron microscope (SEM). Energy dispersive spectroscope (EDS) analysis was also performed to find out the presence of metals in soot. This study provides valuable insights into particulate mass emission from I.C engine.

7.2 Experimental Setup and Procedure

This section is divided into two parts. (a) Engine experimental setup, (b) Experimental procedure.

Engine Experimental Setup

A single cylinder 661.5 cc diesel engine was selected for this experimental investigation. The specifications of the engine to be used for the experiments are given in Table 7.2. Schematic of the experimental setup is shown in Fig. 7.1. An eddy current dynamometer was coupled to the engine. A standalone panel box is used, which contains an airbox, two fuel tanks, a fuel measurements unit, a manometer, transmitters for measuring air and fuel flow rates, an engine indicator and rotameters for measuring water flow in a calorimeter and cooling water. Brake power, indicated power, frictional power, brake mean effective pressure (BMEP), indicated mean effective pressure (IMEP), thermal efficiency, mechanical efficiency, volumetric efficiency, specific fuel consumption, and air/fuel (A/F) ratio could be used to measure the performance of VCR engines using the setup. Additionally, the AVL digital analyzer measures the concentration of smoke emissions, nitric oxide, carbon monoxide and hydrocarbons.

Experimental procedure

Engine was made to operate for 20 min prior to taking any reading. At a given test condition, exhaust gas temperature (EGT) and coefficient of variance (CoV) in IMEP were monitored to achieve steady state condition. Two vacuum pumps each of 15 LPM were used to suck exhaust gas from the engine tail pipe and made to pass through a dilution tunnel. The filter paper was kept in between the two flanges of the dilution tunnel. The exhaust was diluted with 2 LPM flow rate to avoid any condensation in the

Table 7.2 Engine specifications used in this experiment

Parameters	Specifications
MFG and model of the engine	Kirloskar Engine oil Ltd., Model TV1
Number of cylinders	Single
Fuels used	Diesel, diesel doped with dodecanol
Bore × Length	8.75 cm × 11 cm
Connecting rod's length	23.4 cm
Maximum rating of power	3.5 kW
Number of strokes	Four
Revolution per minute	1500
Injection point	0°–25° BTDC
Compression ratio	12:1 to 22:1

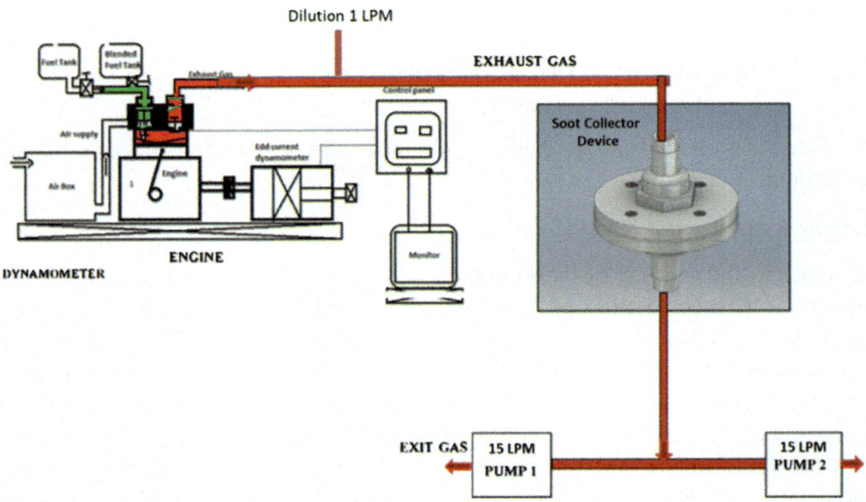

Fig. 7.1 Schematic of the experimental set for soot loading

dilution tunnel. Stick protocol was followed and it was ensured that no dust particles entre into the filter paper or sampling device. Upper half of the tunnel was attached with the exhaust pipe of the engine using a metal pipe and lower half is connected with inlets of two parallel vacuum pumps using high temperature resistance tubes. Table 7.3 shows the experimental points and test procedure. A comparative analysis among baseline diesel fuel and diesel + dodecanol (1% v/v) without renewable fuels was carried out at 16 Nm engine load@1500 rpm. A solenoid injector was used with a single fuel injection per cycle.

To check the outlet velocity of exhaust gases passed through filter paper, an anemometer was used. Air velocity at the exhaust must be between 0.35 and 0.80 m/s.

Table 7.3 Experimental parameters

RPM	1500	1500
Load (Nm)	16	16
Running time (min)	30	30
Fuel used	Diesel	Diesel + Dodecanol (1% v/v)
Flow rate of pump (lpm)	30	30

Anemometer showed 0.75 m/s air velocity at the end of exhaust pipe. This difference in air velocity is because of pressure drop across the filter paper. The engine was operated for 30 min for each sample (diesel fuel and diesel fuel with dodecanal). The PM collector tunnel is connected with the exhaust pipe of the engine using a metal pipe having half inch diameter and is fastened to the top part of the tunnel. The weight of the soot loaded in 30 min was noted through a high precision weighing machine. The weight of the unloaded filter paper was also noted to know the weight difference for the gravimetric test. To know the morphology and chemical composition of each PM-loaded filter paper, SEM and EDS tests were performed. Soot is essentially non-conductive in nature. To make soot samples conductive, a sputtering machine was used for gold coating on the samples. Once the soot sample becomes conductive, emission of secondary electrons takes place and soot sample conducts evenly and creates a homogeneous surface for analysis and imaging. Thereafter, SEM and EDS were performed.

7.3 Results

Physical chemical properties of dodecanol and renewable fuel play a vital role on emissions. To improve the solubility of ethanol in diesel fuel, additive is required. One such additive widely used for this purpose is dodecanol ($CH_3(CH_2)_{11}OH$). Chemical structure of dodecanol is shown in Fig. 7.2. 1-dodecanol molecule contains a total of 38 bonds. There are 12 non-H bond, 10 rotatable bonds, 1 hydroxyl group and 1 primary alcohol.

Using additive in diesel increases the solubility index of primary alcohol in diesel fuel at low temperatures. Dodecanol has higher octane number than the diesel so it is also used as an octane number booster by mixing it into low cetane fuel

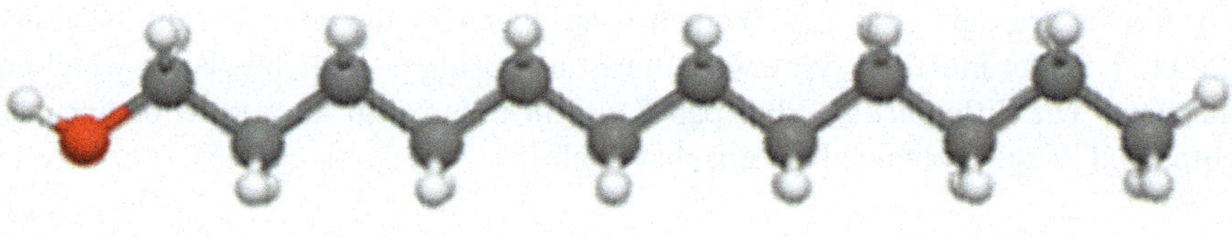

Fig. 7.2 3D chemical structure of 1-dodecane

Table 7.4 Comparison of properties between dodecanol and diesel (Gangwar et al. 2017)

Property	Diesel	1-Dodecanol
Molecular formula	$C_{12}H_{24}$	$C_{12}H_{26}O$
Molecular weight (kg/kmol)	170	186
Cetane number	48	63.6
Stoichiometric fuel/air ratio	0.6924	0.07462
Flash point (°C)	52	127
Ignition temperature (°C)	235	527
Lower heating value (MJ/kg)	42.74	39.86
Density (g/cm^{-3})	0.84	0.831
Oxygen (% wt)	0	8.6

Table 7.5 Measurement methods for different emissions (Hu et al. 2012)

Gas Component	Measurement principle
CO	Non-dispersive infrared (NDIR)
HC	Flame ionisation detector (FID)
NO_x	Chemiluminescence detector (CLD)
PM	Partial/Full flow Dilution Tunnel

(diesel) (Duraisamy et al. 2020). Table 7.4 shows a comparison of properties between dodecanol and diesel.

In order to control different types of emission, first emission should be measured accurately. These emissions increase environmental and health risk. A wide range of measurement techniques is available for regulated emissions. A few of the important methods are given in Table 7.5.

Combustion data was acquired for both the test fuels as explained in the experimental section. Figure 7.3 shows in-cylinder combustion pressure for diesel and diesel + dodecanol (1% v/v) along with 47 mm soot loaded filter paper. In-cylinder pressure for both the test fuel was found to be nearly the same. At TDC, diesel + dodecanol was found to have slightly higher in-cylinder pressure. On combustion of diesel fuel, the engine produces a mixture of gases containing elemental carbon, sulphur compounds and hydrocarbons. The collected samples using diesel and diesel + dodecanol (1% v/v) as a fuel are shown in Fig. 7.3. As seen in Fig. 7.3, uniform deposition of PM particle was observed and soot loading looks nearly the same.

Gravimetric Test Results

For gravimetric test, samples obtained from operating the engine with diesel and diesel + dodecanol (1% v/v) were measured using a high precision weighing machine. The difference in filter paper weight before and after soot loading was compared. Results obtained are given in Table 7.6.

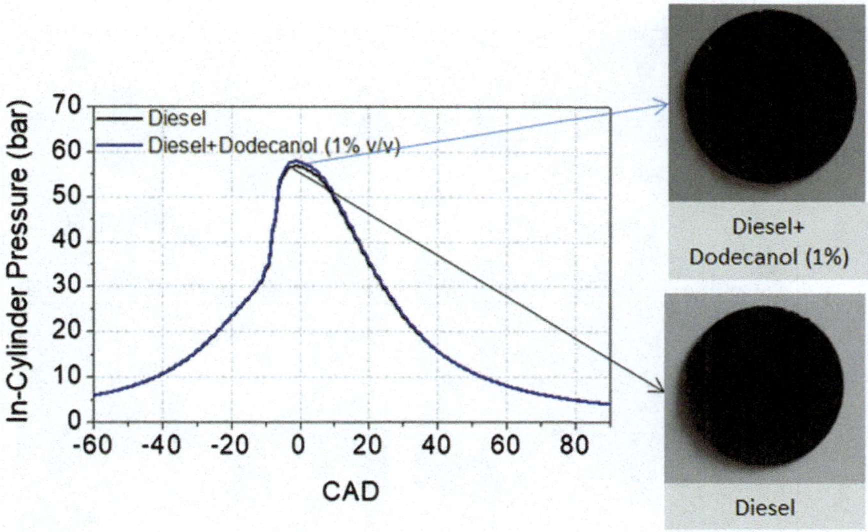

Fig. 7.3 In-cylinder combustion pressure for diesel and diesel + dodecanol (1% v/v) along with 47 mm soot loaded filter paper

Table 7.6 Gravimetric result for the samples collected using diesel and DD as the fuel

Fuel Used	Sample-1 (Diesel)	Sample-2 (Diesel + dodecanol (1% v/v))
Weight before loading the PM (mg)	142.84	142.66
Weight after loading the PM (mg)	144.93	145.28
Total weight of the PM collected (mg)	2.09	2.62
% change in weight	1.46	1.84

Morphological Test Results

To know the morphology and chemical composition of each PM-loaded filter paper, SEM and EDS were performed. The results of SEM and EDS testing for sample 1, i.e. diesel as a fuel and sample-2, Diesel + dodecanol (1% v/v) as the fuel are given in Fig. 7.4.

SEM Result for Sample Number-1

For the diesel sample, SEM imaging was done for different magnifications for observing the morphological properties of the PM. PM particles tend to take a spherical shape where the particle size was observed to be in the range of 50–80 nm. But the individual particles bind with the adjacent particles and form a chain-like

Fig. 7.4 SEM images of soot samples for two different test fuels

structure. The individual size of particles is in nanometre. Due to the aggregation of PM particles their size and shape both change. The shape and size of PM particles depend on the chemical composition of the test fuel.

SEM Result for Sample Number-2

When additive was mixed in diesel engine, chemical composition of the fuel changed. Because of this change in chemical composition, PM morphology also got changed. The particle size in this case varies from 60 to 90 nm evidently greater compared to the diesel fuel.

Comparison of the SEM Results

The comparison morphology of both the samples collected is given in Fig. 7.4. Change in PM morphology can be seen clearly. As diameter of PM particles are larger than the PM particles of diesel engine. Aggregation of PM particles in case of diesel + dodecanol (1% v/v) fuel is more than diesel fuel, whereas PM particles of diesel fuel are forming a cylindrical shape after aggregation, PM particles of diesel + dodecanol (1% v/v) fuel are forming a more complex structure which indicates that the aspect ratio of diesel PM particles is greater than that of DDPM particles.

The particulate emissions from diesel engines consist of carbon, condensed hydrocarbons and sulphate. The PM generation is due to various factors like improper (or) incomplete combustion, fuel injection pressure, etc. *SEM* images of particulates showed that at higher magnification clear particle boundaries were visible, which was neither fused nor fully matured.

EDS Test

Metals are adsorbed on the soot emitted from the engine. Metals are essential for the survival of humans, animals and plants. However, metals inhalation with soot in large quantity may result in dangerous health effects. Cd, Hg and As are poisonous. Fe and Zn are known to induce cancer. Therefore, it is important to investigate the presence of metals in particulate emitted by diesel engines, especially when they are being fuelled with additives (dodecanol). Sources of metals in soot are metals in fuel itself, additives present in lubricating oil and wear debris emanating from engine components. EDS test was performed to know the chemical composition of soot particles. Other than unburned carbon, PM contains silicon, various hydrocarbon, lead, sulphur compounds, etc. EDS test was performed on the same machine where the previous test was performed.

EDS FOR SAMLE Number-1

For filter paper loaded with particles of diesel fuel, EDS result is shown in Fig. 7.5. It represents the peak composition of materials observed during the test.

The sample of diesel fuel PM contains carbon; silicon and oxygen. The weight percentage of various elements of PM is represented in Table 7.7.

Apart from carbon, oxygen and silicon, there are many other elements present in the soot sample which are not shown in the EDS testing graph as these are very less in amount as compared to those mentioned above. The sources of these elements are lubricant used in the engine, leakages, metal elements because of wear and tear.

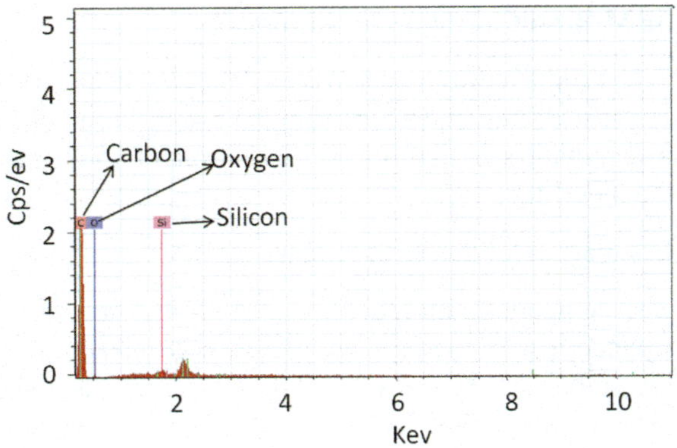

Fig. 7.5 Chemical composition of PM collected from diesel fuel

Table 7.7 Composition of PM sample of diesel fuel

Element	Element (% weight)	Atom (%)
Carbon	78.6	96.22
Oxygen	1.77	1.63
Silicon	1.52	0.8

EDS for Sample Number-2

EDS result obtained with the sample-2 using Diesel + dodecanol (1% v/v) as the fuel is represented in Fig. 7.6.

The sample of Diesel + dodecanol (1% v/v) fuel PM contains carbon and oxygen. The weight percentage of various elements of soot is given in Table 7.8. The amount of increase in the elemental percentage is evidently seen in the results confirming the conclusion that the addition of additives increases the soot emissions. The silicon peak was not observed in EDS using sample-2.

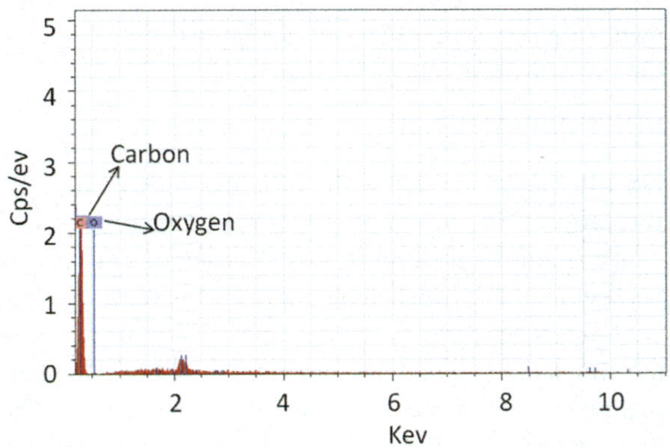

Fig. 7.6 Chemical composition of PM collected from DD fuel

Table 7.8 Composition of PM sample of DD fuel

Element	Element (% weight)	Atom (%)
Carbon	81.11	97.68
Oxygen	1.12	1.02

There is an observed difference of carbon content by 2.51% and oxygen content is reduced implying more hydrocarbon compounds (PM is majorly un-burnt hydrocarbon) being released when the additive is mixed even if it increases the combustion rate.

Figure 7.7 shows CO and HC emission for both the test fuels. It was found that, for both the test fuel, CO emission increases with an increase in engine load. Also, Diesel + dodecanol (1% v/v) showed relatively less CO emission compared to diesel fuel. At high engine load, a trend was found to be the opposite, i.e. Diesel + dodecanol (1% v/v) showed higher CO emission compared to diesel fuel. Throughout the engine operating condition, HC emission was found to be lower for Diesel + dodecanol (1% v/v) relative to diesel fuel. Also, HC emissions were range bound independent of the engine load. The possible reason for the decrease in HC and CO emission for Diesel + dodecanol (1% v/v) is the presence of oxygen in dodecanol which helps in relatively complete combustion.

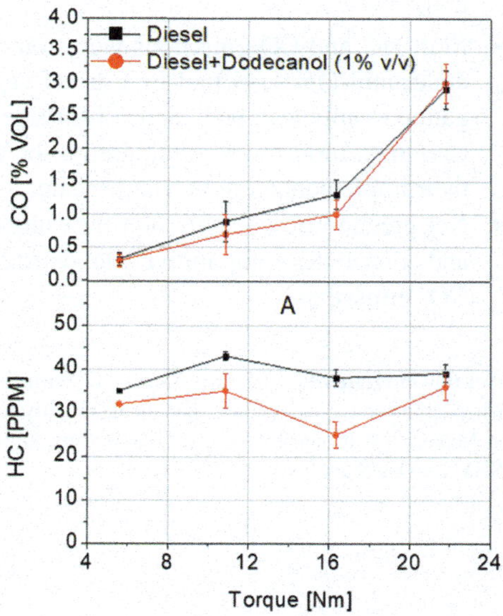

Fig. 7.7 CO and HC emission from test fuels

7.4 Conclusions

Experimental investigation was carried out in a single cylinder, water cooled variable compression ratio (VCR) diesel engine to make a comparative analysis among baseline diesel fuel and diesel + dodecanol (1% v/v) without renewable fuels.

- In the gravimetric test, a significant weight increase is seen in PM-loaded filter of Diesel + dodecanol (1% v/v) fuel compared to that of diesel fuel. This shows that altering the composition of fuel results change in particulate emission. When additives are mixed in the fuel, certainly it improves the combustion but it also increases the emission.
- In morphological analysis, it is found that the morphology of PM particles collected from diesel fuel is different from Diesel + dodecanol (1% v/v) fuel. This indicates that the composition of fuel affects the morphology of the particles. Here in this case, when the additive is mixed in diesel fuel, the tendency of PM particles to get aggregated increases and as a result of its morphology of PM particles of Diesel + dodecanol (1% v/v) fuel is more complex than that of diesel fuel.
- In EDS test analysis it is found that the amount of carbon in PM collected from Diesel + dodecanol (1% v/v) fuel is more than that of diesel fuel which shows the alteration of the chemical composition of exhaust gas on adding additives in fuel.
- Both HC and CO emissions were found to be lower with the use of Diesel + dodecanol (1% v/v) fuel in a diesel engine. Overall, this experimental investigation results may be beneficial for scientists working in fuel-related industries. Certainly, more work is required to be done to explore the use of additives and their implementation on a large scale.
- Net greenhouse gas emission by using renewable fuels is comparatively lower and contributes comparatively lesser to climate change and emits relatively lower CO_2 emissions.

Acknowledgements The assistance received from research staff Mr. Ramesh Chand Meena and Mr. Mahaveer to conduct the experiment is highly acknowledged. Materials Research Centre (MRC) at Malaviya National Institute of Technology Jaipur, Jaipur is also acknowledged for performing SEM and EDS analysis.

References

Baltacioglu MK et al (2016) Experimental comparison of pure hydrogen and HHO (hydroxy) enriched biodiesel (B10) fuel in a commercial diesel engine. Int J Hydrog Energy 41(19):8347–8353

Belgiorno G, Blasio GD, Beatrice C (2019) Advances of the natural gas/diesel RCCI concept application for light-duty engines: comprehensive analysis of the influence of the design and calibration parameters on performance and emissions. Natural gas engines. Springer, pp 251–266

Di Blasio G et al (2022) Application of clean fuels in combustion engines. Springer

Bose PK, Maji D (2009) An experimental investigation on engine performance and emissions of a single cylinder diesel engine using hydrogen as inducted fuel and diesel as injected fuel with exhaust gas recirculation. Int J Hydrog Energy 34(11):4847–4854

Dimitrakopoulos N et al (2017) PPC operation with low ron gasoline fuel. A study on load range on a euro 6 light duty diesel engine. In: The proceedings of the international symposium on diagnostics and modeling of combustion in internal combustion engines. The Japan Society of Mechanical Engineers

Duraisamy G, Rangasamy M, Govindan NJRE (2020) A comparative study on methanol/diesel and methanol/PODE dual fuel RCCI combustion in an automotive diesel engine 145:542–556

Eastwood P (2008) Particulate emissions from vehicles. Wiley

Elfasakhany A, Mahrous AF (2016) Performance and emissions assessment of n-butanol–methanol–gasoline blends as a fuel in spark-ignition engines. Alex Eng J 55(3):3015–3024

Fayyazbakhsh A, Pirouzfar V (2017) Comprehensive overview on diesel additives to reduce emissions, enhance fuel properties and improve engine performance. Renew Sustain Energy Rev 74:891–901

Ganesh D, Gowrishankar G (2011) Effect of nano-fuel additive on emission reduction in a biodiesel fuelled CI engine. In: 2011 International conference on electrical and control engineering. IEEE

Gangwar JN, Saraswati S, Agarwal S (2017) Performance and emission improvement analysis of CI engine using various additive based diesel fuel. In: 2017 international conference on advances in mechanical, industrial, automation and management systems (AMIAMS). IEEE

Hu J et al (2012) Real-world fuel efficiency and exhaust emissions of light-duty diesel vehicles and their correlation with road conditions. J Environ Sci 24(5):865–874

Ianniello R et al (2021) Ethanol in dual-fuel and blend fueling modes for advanced combustion in compression ignition engines. Alcohol as an alternative fuel for internal combustion engines. Springer, pp 5–27

Jamrozik A (2017) The effect of the alcohol content in the fuel mixture on the performance and emissions of a direct injection diesel engine fueled with diesel-methanol and diesel-ethanol blends. Energy Convers Manag 148:461–476

Kalwar A, Singh AP, Agarwal AK (2020) Utilization of primary alcohols in dual-fuel injection mode in a gasoline direct injection engine. Fuel 276:118068

Kim N, Cho S, Min K (2015) A study on the combustion and emission characteristics of an SI engine under full load conditions with ethanol port injection and gasoline direct injection. Fuel 158:725–732

Krzemiński A et al (2017) Effect of dodecanol additive on auto-ignition properties of diesel oil and ethanol blends. Combust Engines 56

Le MD et al (2020) Experimental and numerical investigation of the promoting effect of a cetane booster in a low-octane gasoline fuel in a rapid compression machine: a study of 2-ethylhexyl nitrate 222:36–47

Li Y et al (2017) Experimental comparative study on combustion, performance and emissions characteristics of methanol, ethanol and butanol in a spark ignition engine 115:53–63

Liu H et al (2015) Comparative study on alcohol–gasoline and gasoline–alcohol Dual-Fuel Spark Ignition (DFSI) combustion for engine particle number (PN) reduction. Fuel 159:250–258

Lu H et al (2019) An investigation on the characteristics of and influence factors for NO_2 formation in diesel/methanol dual fuel engine. Fuel 235:617–626

Nouni M et al (2021) Alternative fuels for decarbonisation of road transport sector in India: options, present status, opportunities, and challenges. Fuel 305:121583
Pacheco MA, Marshall CL (1997) Review of dimethyl carbonate (DMC) manufacture and its characteristics as a fuel additive. Energy Fuels 11(1):2–29
Qian Y et al (2019) Engine performance and octane on demand studies of a dual fuel spark ignition engine with ethanol/gasoline surrogates as fuel. Energy Convers Manag 183:296–306
Reitz RD et al (2020) IJER editorial: the future of the internal combustion engine. SAGE Publications Sage UK, London, England, pp 3–10
Saravanan N et al (2008) An experimental investigation on hydrogen as a dual fuel for diesel engine system with exhaust gas recirculation technique. Renew Energy 33(3):422–427
Shamun S, Belgiorno G, Blasio GD (2020) Engine parameters assessment for alcohols fuels application in compression ignition engines. Alternative fuels and their utilization strategies in internal combustion engines. Springer, pp 125–139
Sharma PK et al (2020) Characterization of the hydroxy fueled compression ignition engine under dual fuel mode: experimental and numerical simulation. Int J Hydrog Energy 45(15):8067–8081
Sharudin H et al (2017) Investigation of the effects of iso-butanol additives on spark ignition engine fuelled with methanol-gasoline blends 114:593–600
SinghYadav V, Soni S, Sharma D (2012) Performance and emission studies of direct injection CI engine in duel fuel mode (hydrogen-diesel) with EGR. Int J Hydrog Energy 37(4):3807–3817
Sun P et al (2019) Comparative study on the effects of ethanol proportion on the particle numbers emissions in a combined injection engine. Energies 12(9):1788
Wang Z et al (2015) Comparative study on alcohols–gasoline and gasoline–alcohols dual-fuel spark ignition (DFSI) combustion for high load extension and high fuel efficiency. Energy 82:395–405
Yanowitz J et al (2017) Compendium of experimental cetane numbers. National Renewable Energy Lab. (NREL), Golden, CO (United States)
Zhen X, Wang YJR (2015) An overview of methanol as an internal combustion engine fuel. Reviews 52:477–493

Chapter 8
A Bibliometric Review of Alcohol–Diesel Blend in CI Engines

Mukesh Kumar, Chandan Kumar, Umesh Kumar Das, Praveen Saraswat, and K. B. Rana

Abstract A large number of diesel engine applications in the transportation and agriculture sector has become a major concern for environmental pollution. It is quite difficult to improve diesel engine performance and at the same time controlling its emissions effectively. There are various methods to achieve the same e.g., changes in engine design, fuel blending with additives, engine exhaust treatment, etc. The most practical technique for controlling high emissions without degrading engine performance is to modify fuel with additives. Alcohols are the most promising additives which have been used by various researchers worldwide. In the current study, a bibliometric-based study and the role of various alcohol additives in improving combustion, performance, and fumes outflow qualities of CI engines were comprehensively reviewed. Due to higher oxygen levels in the combustion zone, it was found that combining diesel with alcohol reduces the emissions of CO_2, CO, HC, soot, and particle matter.

Keywords Diesel-fuel engine · Compression ignition engine · Alcohol · Emission · Performance

Nomenclature

ACS	American Chemical Society
°C	Degree centigrade
BP	Brake power
EGR	Exhaust gas recirculation

M. Kumar · U. K. Das
University of Engineering & Management, Jaipur, India

C. Kumar (✉) · P. Saraswat
Swami Keshvanand Institute of Technology, Management & Gramothan, Jaipur, India
e-mail: chandanpink1988@gmail.com

K. B. Rana
Rajasthan Technical University, Kota 324010, India

P. C. Shukla et al. (eds.), *Renewable Fuels for Sustainable Mobility*, Energy, Environment, and Sustainability, https://doi.org/10.1007/978-981-99-1392-3_8

BTE	Brake thermal efficiency
HC	Hydrocarbon
BSFC	Brake-specific fuel consumption
IP	Injection pressure
CI	Compression ignition
IT	Injection time
CN	Cetane number
NO_x	Oxides of nitrogen
CO_2	Carbon dioxide
PM	Particulate matter
CR	Compression ratio
RPM	Revolution per minutes
cSt	Centistokes
T&F	Taylor and Francis

8.1 Introduction

To maintain a clean environment and explore clean fuel for engines, the government is continuously promoting the use of alternative fuels and emphasizing reducing CO_2 emissions. Diesel engines are increasingly in demand in the transportation and agricultural industries because of their beneficial qualities, including dependability, durability, fuel efficiency, and specific power production (Shamun and Belgiorno 2020). The continuous use of diesel engines also leads to an increase in fuel consumption and emissions (smoke, CO, CO_2, NO_x, HC, and PM) (Moghaddam et al. 2012). It is very essential to increase the ignition characteristics of diesel engines for the development of economic growth and environmental management (Maricq 2007). The ignition characteristics of an engine can be improved by changing the injection timing (IT), compression ratio (CR), and injection pressure (IP) (Kumar et al. 2018a). However, it is both difficult and costly due to the complicated design characteristics of the CI engine (Huang et al. 2016). As a result, using blended diesel is a better alternative for improving performance while simultaneously lowering emissions from a CI engine. Alternative fuels/additives, such as biodiesel/vegetable oil, can enhance the engine's combustion characteristics, albeit it may result in a modest increase in NO_x emissions (Ameer and Gopal 2012; Behcet 2011).

Additives are a type of chemical that can be mixed into a tiny amount of diesel to improve its combustion qualities (Giakoumis et al. 2013). Many additives for blending with diesel are easily accessible and can be utilized in C.I. engines. Methanol, n-butanol, ethanol, DEE, EEA, NM, NE, n-heptane, etc. are some examples of additives (Hellier et al. 2013; Karabektas and Hosoz 2009). Additives include various advantages such as without using any reagents, additives can be blended directly into diesel since additives have a high amount of oxygen in their molecular composition (Sukjit et al. 2012). The use of additives is feasible without altering the engine (Nabi and Hustad 2010a). The preparation of diesel-additive blends poses no risks to human health or safety (Kumar et al. 2014). The components of a diesel engine are not damaged by the use of additives (Goyal et al. 2015). Additives can improve the combustion properties of blended fuels since they are oxygenated fuels (Kumar et al. 2019). It is possible to increase the diesel-additives blend's CN (Kumar et al. 2018b). Up to 50% or more of the exhaust emissions may be decreased. Without compromising other factors, additives could increase engine performance by up to 18% (Kumar et al. 2020).

According to certain reports, alcohols were added to diesel fuel at a concentration of less than 10% by volume to improve the results for brake thermal efficiency (BTE) and brake-specific fuel consumption (BSFC) (Padwa et al. 2016). Alcohol also improves combustion efficiency when mixed with diesel–biodiesel blends (Shamun et al. 2018). The addition of alcohols to pure diesel fuel increased carbon monoxide (CO) and oxide of nitrogen (NO_x) emissions but decreased particulate matter (PM) and hydrocarbon (HC) emissions (Kumar et al. 2015; Khalife et al. 2017).

The motivation behind this article is to give a complete audit of the physicochemical properties, a system of activity, and impacts of different alcohol-added additives (methanol, ethanol, and n-butanol) that are utilized related to diesel fuel to further develop engine performance and fumes outflow qualities.

8.2 Methodology for Bibliometric Analysis

This study determined to review the alcoholic diesel fuel mixtures used in CI engines. The first stage in doing a bibliometric analysis is to select a suitable database from which to retrieve the necessary documents. The Scopus database is used to define conceptual boundaries using the keywords "Diesel-fuel blend" AND "CI engine" AND "Alcohol". The Scopus data retrieved from 814 publications encompasses all articles published between 2010 and 2021 (until 14 September). All matching review articles and technical papers were enhanced and reviewed in accordance with the objectives, and the papers' adherence to the rules was assessed. On the Scopus database, a search for associated terms was conducted. The search string was ("Diesel Fuel Blends" AND "Engine") AND ((alcohol)) AND (ci AND engine).

The most complete database is Scopus, which includes articles from respected publishers including Elsevier, Springer, Taylor & Francis, Inderscience, IGI Global, Wiley, IEEE, IOP Science, Emerald, and more (Jamwal et al. 2021). Only papers related to this subject were included in the search. Reviews of conferences, books, and undefined and editorial research pieces were excluded to assure the study's excellence. The research insertion and elimination standards are listed in the Table 8.1.

Table 8.1 Insertion and elimination criteria

Insertion criteria	Document type: journal article only Peer review papers Strongly related to the alcohol-based diesel blends for CI engine The period from 2010 to September 2021
Elimination criteria	Non-english (NE) Published articles before 2010 No full text (NF) Other than journal papers, i.e., conference abstracts, posters, others

8.3 Bibliometric Analysis

The bibliometric analysis enables the systematic study of bibliographic data derived from the Scopus database. In previously conducted bibliometric analyses using a variety of software packages, each with its own set of advantages and disadvantages (Baker et al. 2020). The three main categories of the bibliometric analysis used in this study were: (1) the number of papers published annually; (2) the top authors, institutions, and countries active in this subject; and (3) the top journals and keywords in this field. We analyze the authors' keywords using the Vosviewer software.

Year-by-year publications: Fig. 8.1 shows year-by-year publication data for diesel fuel blends, CI engines, and alcohol. Within this specified period of Scopus coverage years from 2010 to September 2021, a total of 12 years (approximately) including 814 documents have been published. It has been found that in the year 2020, 189 articles were published at the highest rate of 23.21%. The lowest number of articles were published in 2010, with 14 articles at a rate of 1.71%. The average publication per year is 67 articles.

Fig. 8.1 Year-wise publications

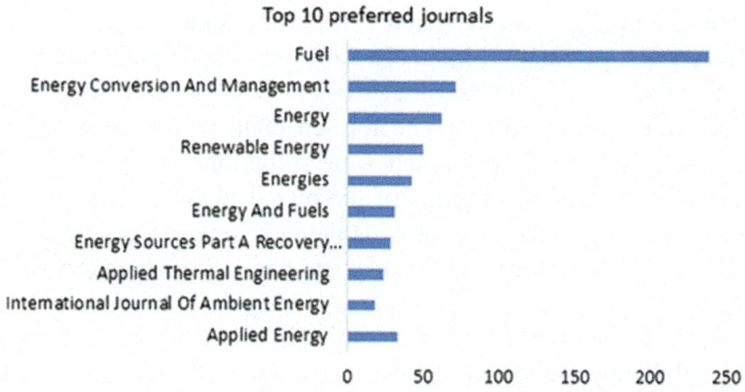

Fig. 8.2 Top 10 preferred journals

The study revealed that four different publishers were among the top ten most productive journals, as illustrated in Fig. 8.2. Elsevier published the top four journals, totaling six. Taylor & Francis (T&F), the American Chemical Society (ACS), and MDPI published the remaining four journals.

The most productive journal was Fuel (Elsevier) with 238 articles covering 29.23% of the total publications, followed by Energy conversion and management (Elsevier) (72, 8.84%), Energy (Elsevier) (62, 7.61%), Renewable energy (Elsevier) (50, 6.14%), Energies (MDPI) (43, 5.28%), Energy & Fuels (ACS) (32, 3.93%), Energy Sources, Part A: Recovery, Utilization, and Environmental Effects (Taylor & Francis) (29, 3.56%), Applied thermal energy (Elsevier) (24, 2.70%), International Journal of Ambient Energy (T&F) (19, 2.33%), and Applied Energy (Elsevier) (33, 4.05%).

Top 10 leading authors: Fig. 8.3 shows the 10 most prolific authors in our results, affiliated to six countries as follows; Malaysia (4 authors), USA (1 author), Turkey (1 author), India (2 authors), Iran (1 author), and China (1 author). Masjuki, H.H. from the International Islamic University, Malaysia had 19,508 times citations, 79 h-index, and 255 i10-index. Kalam, M.A. Faculty of Engineering, University of Malaya

Fig. 8.3 Top 10 leading authors

(20,926 times citations, 77 h-index, and 308 i10-index) and Agarwal, A.K. Indian Institute of Technology Kanpur, India (18,980 times citations, 58 h-index, and 202 i10-index) also had a good number of citations.

Top 10 Countries Publications: The top ten contributing countries worldwide between 2010 and 2021 were determined using affiliations received from each author in the contribution database. In terms of the overall number of papers published, Fig. 8.4 shows the 10 countries that provide the most research articles. India has the most documents (354), followed by Turkey and China, which each have 39 and 39 articles.

As illustrated in Fig. 8.5, research activities in this field are growing at a rapid pace. Among all scientific topics, engineering had the highest proportion of literature (65.9%). The research topic encompassed all new sectors, with energy emerging as the second most popular area of study.

Affiliation-wise publications: Fig. 8.6 shows affiliation-wise publication of articles in various journals. Out of 814 contributions, the University of Malaya has

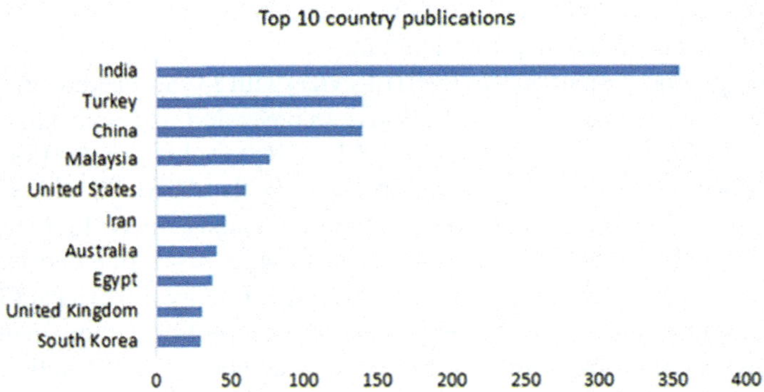

Fig. 8.4 Top 10 country publications

Fig. 8.5 Area-wise distribution

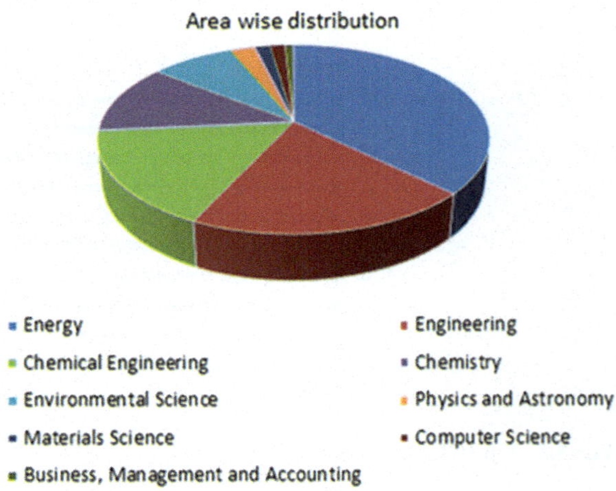

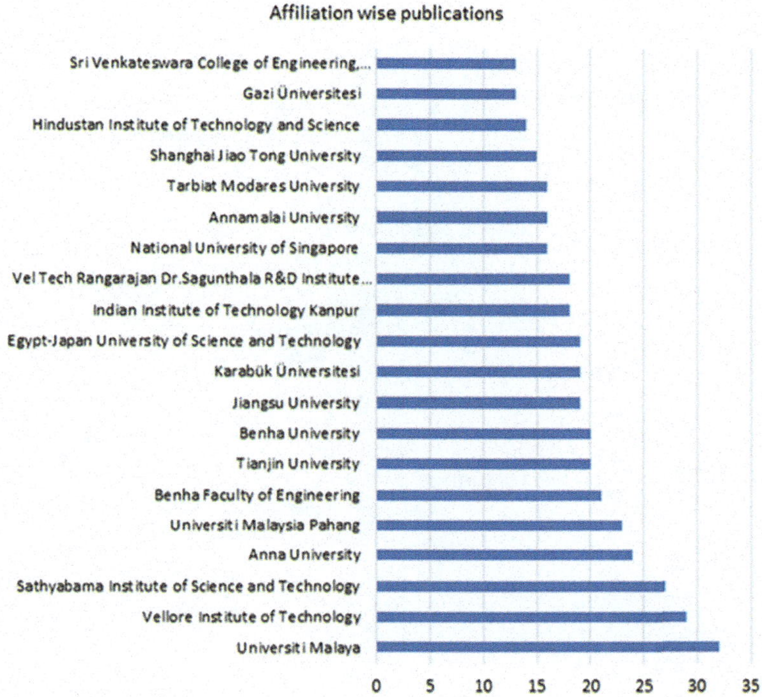

Fig. 8.6 Affiliation-wise publications

contributed the highest article 32 (3.93%), Vellore Institute of technology comes to the second position contributing 29 articles (3.56%), and Sathyabama Institute of science and technology in the third position 27 articles (3.31%), followed by Anna university 24 articles (2.94%), University Malaysia Pahang 29 (2.82%), Benha Faculty of engineering 21 articles (2.57%), Tianjin University and Benha University every 20 articles (2.45%).

Using the co-occurrence of author keywords in the overlay display mode, Fig. 8.7 shows a snapshot of the bibliometric grid that was produced. A keyword must appear 10 times minimum to be considered used. In comparison to other keywords, the term 'biodiesel' appeared the most frequently with 262 occurrences.

The co-occurrences of keywords, links, and the overall link strength for each cluster are listed in Table 8.2. As indicated in the preceding table, biodiesel and diesel engines have the highest frequency and strength of linkage, followed by combustion and performance. This shows that biofuels have a substantial impact on engine performance and fumes outflow.

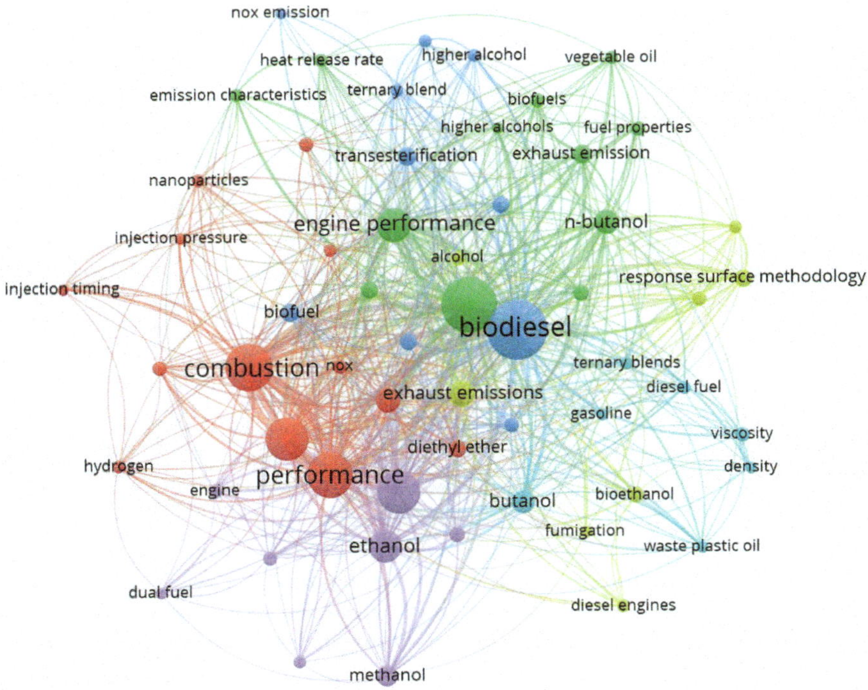

Fig. 8.7 Authors' keywords occurrences network

8.4 Various Useful Properties of Alcohol

The properties of alcohol are crucial in forecasting the combustion characteristics of CI engines. Understanding the useful properties of alcohol additives listed in Table 8.3 is essential for determining the best combination of additives with pure diesel that will improve engine performance while lowering emissions. The physicochemical and combustion parameters of several alcohol additions are compared in Figs. 8.8 and 8.9.

It has been revealed that n-butanol has the highest boiling point and density, which may help to improve engine combustion processes, fuel efficiency, and power. It also has the greatest cetane number, which can help in improving combustion properties. Methanol has a high oxygen concentration, which may enhance the engine's ignition quality.

Table 8.2 The top 20 keywords and the overall link strength

S. No	Keyword	Occurrences	Total link strength
1	Biodiesel	262	530
2	Diesel engine	227	491
3	Combustion	169	401
4	Performance	157	424
5	Emissions	139	331
6	Emission	137	326
7	Engine performance	90	201
8	Ethanol	82	206
9	Exhaust emissions	49	109
10	Diesel	46	100
11	n-butanol	45	95
12	Butanol	40	92
13	Methanol	33	79
14	Exhaust emission	27	62
15	Biofuel	26	48
16	Transesterification	26	51
17	Combustion characteristics	24	47
18	Waste cooking oil	24	52
19	Compression ignition engine	23	35
20	Diethyl ether	22	52

8.5 Alcohols as Additives for CI Engines

Alcohol is a fuel additive with many advantages like better combustion, energy saving, and emission reduction. It has a high oxygen concentration, is inexpensive to produce, and burns cleanly. It may be used as a diesel engine's alternative fuel (Desjardins et al. 2008; Kumar et al. 2018c; Nabi and Hustad 2010b). Alcohols like methanol, ethanol, n-butanol, and others (Alcohol compound additives) are improved combustion characteristics fuels that can be used to raise the performance and reduce emissions when the fuel is burned (Datta and Mandal 2016; Fayyazbakhsh and Pirouzfar 2017). As per a review conducted by Khalife et al. (2017), BTE and BSFC values have reportedly improved after alcohols were added to diesel fuel at a concentration of less than 10% by volume. Pure diesel fuel with alcohol additives added to it produced a substantial amount of oxygen in the combustion chamber, which led to a drop in hydrocarbon and particulate matter emissions but an increase in carbon monoxide and oxide of nitrogen emissions. According to Adelman (1979)

Table 8.3 Physico-chemical and combustion properties of alcohols

Properties	Methanol (Kumar et al. 2018a, b, 2019)	Ethanol (Kumar et al. 2018a; Padwa et al. 2016)	n-butanol (Kumar et al. 2018a; Goyal et al. 2015)
Chemical formula	CH_3OH	C_2H_5OH	C_4H_9OH
Flash point (°C)	11	08	35
Latent heat (kJ/kg)	1178	840	585
Density (kg/m^3)	729	792	810
Viscosity (cSt)	0.54	1.04	2.20
Boiling point (°C)	64	78	118
Octane number	112	92	96
Cetane number	05	07	25
Lower heating value (MJ/kg)	20	27	33
Oxygen content (%)	50	34	22
Auto-ignition Temp. (°C)	470	434	385

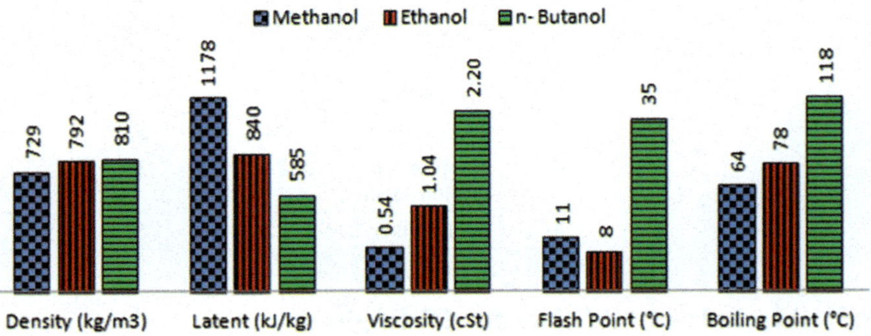

Fig. 8.8 Comparison of physico-chemical properties of alcohol additives

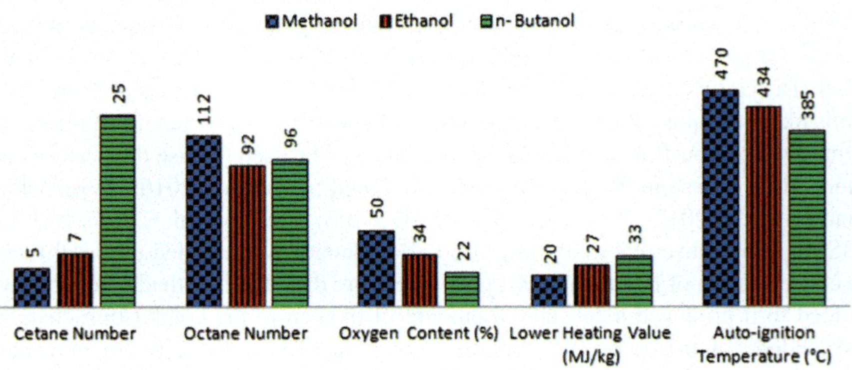

Fig. 8.9 Comparison of combustion properties of alcohol additives

and Wagner et al. (1979), adding alcohol to the fuel can improve engine performance because of:

- It is easily injected, atomized, and mixed with air because of its high oxygen concentration, high hydrogen/carbon ratio, and low sulfur content.
- Fewer pollution than diesel fuel due to its lower viscosity.
- Rapid laminar flame propagation can speed up the ignition process and increase the engine's performance.
- Effective evaporative cooling results in a compression stroke and cooler intake process. This increases the engine's volumetric efficiency while decreasing the required work input during the compression stroke.

8.5.1 Methanol

Methanol is a renewable, cost-effective, ecologically beneficial, and high-demand alternative to fossil-based fuels. In the current research, some scholars have looked into using methanol as a substitute for traditional CI engine fuels to address environmental and economic problems (Zhen and Wang 2015).

The methanol effects to diesel ratio on performance and fumes outflow were examined by Wei et al. (2016) who observed that adding a little amount of methanol to diesel did not affect BTE, but adding a large amount of methanol to diesel reduced BTE somewhat. Table 8.4 summarizes the mixing effects of methanol additions on CI engine performance and exhaust emission parameters.

BSFC increased, as the volume of ethanol or methanol in the fuel grew, although the increment in BSFC for diesel–methanol mixture was established to be greater as compared to ethanol–diesel blends (Nur et al. 2015; Fang et al. 2013; Pidol et al. 2012). The impact of suction air temperature on the combustion and ignition characteristics of a diesel-methanol combination was studied by Pan et al. (2015). The researchers noticed that as the fraction of fumigated methanol increased, CO, HC, formaldehyde emissions and unburned methanol increased, but these uncontrolled emissions reduced when intake air temperature increased.

Engine performance is significantly impacted by adding methanol to diesel. It has been noted that a rise in the mass fraction of methanol would result in a rise in the rate of heat emission. Methanol pilot injection is another effective method for improving an engine's combustion characteristics (Kumar et al. 2021). Even though there are several techniques for overcoming the difficulties of directly applying methanol to a diesel engine, methanol fumigation appears to be a viable approach for converting from diesel to a great methanol substitution mode (Wang et al. 2015).

Table 8.4 Summary of diesel-methanol additives blending performance

Literature	Test condition	Engine performance and Emissions
Sayin et al. (2009)	• Diesel, Methanol (5–15% vol.) • Different injection time (15–25° BTDC) • 5–20 Nm loads	• BTE decreased with an increase in methanol % • BSFC increased with an increase in methanol % • For all test scenarios, advanced injection timings yielded negative results • Smoke, CO, and HC were reduced with the rise in % of methanol • NO_x and CO_2 raised with the rise in % of methanol
Sayin (2010)	• M5, M10 Diesel-methanol • Various rpm (1000, 1800 rpm) and fixed torque at 30 Nm	• BTE concentration reduced as the methanol percentage increased, and it was minimum at M10 • BSFC is raised in all fuel blends • NO_x raised at all blends • Smoke, CO, and HC were reduced
Tutak et al. (2015)	• 20–90% volume Diesel-methanol • Various crank angles (350–400) • Various loads (34, 67, 100%)	• BTE is enhanced with blends as compared to diesel • CO, and HC raised • NO_x and soot decreased with methanol–diesel blend fueled at small loads
Chen et al. (2016)	• Diesel, methanol (10–30% by volume) • Various loads (25–100%) • Various engine rpm (1200–1950 rpm)	Increased engine efficiency was accomplished by increasing the methanol content of diesel at great engine speeds
Datta and Mandal (2016)	• Diesel, Methanol (DM5, DM10, DM15) • Different BP (25–100%)	• BTE raised as braking power and methanol concentration increased • BSFC decreased with increases in the load, and BSFC increased with increases in the methanol quantity • NO_x, CO_2, and smoke declined

(continued)

Table 8.4 (continued)

Literature	Test condition	Engine performance and Emissions
Chen et al. (2017)	• 10–30% vol. Diesel-ethanol • Different injection conditions • Different loads (50–100%)	• IMEP was reduced with an increase in engine load • Soot emission decreased slowly with all three fuel blends for 50 and 75% engine loads

8.5.2 *Ethanol*

Using biomass, such as low-quality cereal grains (wheat, barley, rice, and maize), potatoes, beans, cassava roots, field peas, and other plant materials, ethanol is produced as an alternative fuel (Prasad et al. 2007). Clean fuel is may be produced by combining different alcohols such as ethanol and diesel fuel (Putrasari et al. 2013). Methanol and ethanol both decrease the pollution in CI engines, but ethanol has the advantage of being a non-conventional fuel with strong mixing properties (Can et al. 2004).

Mohammadi et al. (2005) claim that blending ethanol with diesel fuel lowers particulate matter and oxide of nitrogen emissions. However, mixing causes the amount of HC and CO released to increase. According to Xu et al. (2007), increasing the percentage of ethanol (10–30% volume) in blended fuels increased BTE and fuel consumption when compared to diesel fuel. When ethanol mixed fuels were compared to pure diesel fuel, they produced less soot and NO_x emissions; nevertheless, NO_x emissions rose as the load increased.

Table 8.5 summarizes the impact of ethanol additions on diesel engine performance and emissions when blended.

Alptekin (2017) mixed 2–12% ethanol into pure diesel fuel and tested it on a six-cylinder CI engine at five rotational speeds (i.e., 1600–2000 rpm) in full load mode. They found an increase in ignition delay and irregularity in engine performance when the ethanol content in diesel fuel was increased by more than 8%. Gnanamoorthi and Devaradjane and (Taghizadeh-Alisaraei and Rezaei-Asl 2016) used an ethanol–diesel blend at higher compression ratio values and found a growth in the value of BTE.

Ethanol is a potential replacement for fossil fuels because of its low emissions in modern engine technologies. As compared to dual fuel and diesel combustion, ethanol developed high gross-indicated efficiency. The high quantity of ethanol in blends and exhaust gas recirculation systems employing a multiple-injection approach has been shown to have a positive effect on emissions reduction while maintaining higher efficiency than dual fuel and diesel combustion (Ianniello et al. 2021).

Table 8.5 Summary of diesel–ethanol additives blending performance

Literature	Test condition	Engine performance and emission
Ajav et al. (1999)	• 5–20% vol. Diesel–ethanol • Different loads (25–100%) • Constant speed (1550 rpm)	• Constant brake power (BP) was obtained for all blends • BSFC was raised by up to 9% with diesel and 20% with ethanol • Oxides of nitrogen and carbon monoxide emissions are lowered with ethanol–diesel
Cole et al. (2001)	• Diesel, Ethanol (10, 15% vol.) • Different torques (50–250 Nm)	• As compared to diesel, PM reduced up to 75% with a 10% ethanol mixture • Oxides of nitrogen declined up to 84%
Xingcai et al. (2004a)	• DE5, DE10, DE15, DE20 Diesel-ethanol 25–100% loads	• BTE improved with 15% ethanol • BSFC dropped as the engine load increased • Smoke declined up to 50% • CO increased • NO_x decreased with 50–60% and 32–40% loads
Xingcai et al. (2004b)	• Diesel–ethanol • Different speeds (1500–3500 rpm) • Different BMEP (0.2–0.7 MPa)	• BTE increased • BSFC reduced • Oxides of nitrogen and smoke are reduced • CO increased remarkably at all load conditions and reduced with CN improver
Rakopoulos et al. (2007)	• 5–15% by vol. Diesel-ethanol • –20 to 80 CA	• BTE and BSFC rates improved moderately • NO_x and CO decreased
Park et al. (2010)	• 10, 20% by vol. Diesel–ethanol • –60° to 60° ATDC crank angle • –30° to –10° ATDC pilot injection time	• NO_x emissions are lowered when an ethanol–diesel combination is used • EGR results in less soot emission • The amount of CO and HC increased when the pilot injection time was raised

(continued)

8.5.3 N-Butanol

In the second generation of biofuels, n-butanol (four-carbon alcohol) may be a better alternative for blending with diesel since the possibility of incorporating bio extenders into diesel engines is growing every day.

Table 8.5 (continued)

Literature	Test condition	Engine performance and emission
Datta and Mandal (2016)	• DE5, DE10, DE15 Diesel-ethanol • Different BP (25–100%)	• BTE increased as braking power was increased • BSFC reduced as the load increased • NO_x emissions were lowered more significantly with diesel–ethanol than with diesel–methanol
Pedrozo et al. (2017)	• Diesel, Ethanol (Energy fraction 0–80%) • 0 and 25% EGR rates	• Great efficiency is obtained through dual fuel combustion of ethanol and diesel • At 25% EGR, NO_x is successfully decreased
Nour et al. (2017)	• Diesel, Ethanol (11 mg/cycle) • Various crank angle • Various EGR rates	• The IMEP rose as the amount of ethanol injected into the exhaust manifold increased • The NO_x and soot were lowered by 88% and 30%, respectively
Alptekin (2017)	• 15% vol. Diesel–ethanol and isopropanol • Three engine speeds (1500–2500 rpm) • Four engine loads (BMEP: 3.3–8.3 bar)	• BSFC raised by 6.2% and 5.7%, respectively, when ethanol (15% by volume) and isopropanol (15% by volume) were added • CO and NO_x emissions increased as a result of DE15 use

Diesel–butanol mixes (10, 15, 20, and 25% butanol vol.) were also used in our printing studies due to their outstanding properties (Nayyar et al. 2017a, b). These experimental/analytical investigations looked at additional input parameters including injection pressure, engine loads, injection duration, and compression ratio, in addition to the blending ratio. The optimal n-butanol content in diesel was determined to be 19.82% vol. at half load and 18.84% vol. at full load, which is quite comparable to the experimental outcomes (20% by vol.). The effects of butanol additives on the performance and emissions of diesel engines are summarized in Table 8.6.

Zheng et al. (2015) compared the combustion and fumes outflow qualities of neat butanol to those of pure diesel fuel and found that engine performance was similar; however, soot and NO_x fumes were reduced. In the majority of cases, He et al. (2013) found that increasing the quantity of n-butanol in gasoline resulted in a modest reduction in NO_x emissions.

Diesel–butanol blends were found good option for the replacement of pure diesel. At full load conditions with the use of 20% butanol, the smoke and NO_x were reduced

Table 8.6 Summary of diesel–butanol additives blending performance

Literature	Test condition	Engine performance and emission
Rakopoulos et al. (2013)	• Diesel, Butanol (8–24% vol.) • Different loads (low, medium, and high) at a constant speed (2200 rpm)	• BTE and BSFC slightly increased with increasing butanol concentration in diesel • Oxides of nitrogen, Smoke, and CO are reduced • The concentration of HC increased as the butanol concentration increased
Dogan (2011)	• 5- 20% by vol. Diesel–butanol • Different loads (3.3–13.1 Nm)	• BTE and BSFC concentrations increased as butanol content increased • The opacity of NO_x, CO, and smoke reduced as the butanol level increased • HC increased with increasing butanol content
Choi et al. (2015)	• 10–20% by vol. Diesel–butanol • Different speeds (1000-–4000 rpm)	• NO_x emissions were reduced at all operating conditions using n–butanol–diesel fuel mixtures • Carbon monoxide and hydrocarbon concentrations raised

by 62.5% and 8%, respectively as compared to pure diesel (Sain et al. 2020). The review reveals that there is a lot of disagreement in the literature, and there is a lot of room for determining the best performance and emissions utilizing diesel-additives mixes in diesel engines.

8.6 Conclusions and Future Scope

The main goal of this study is to look at the physicochemical properties and combustion characteristics of various alcohol additives and their impact on diesel engine performance and emissions, from the above-discussed studies on the performance and fumes outflow qualities of CI engines running on alcohol–diesel blends versus diesel fuel, the following conclusions may be drawn:

- The addition of alcohols (methanol, ethanol, and butanol) to pure diesel fuel has resulted in increased BTE and BSFC values.
- Addition of alcohols to diesel fuel reduces emissions of CO, HC, soot, and particulates, owing to greater oxygen levels in the combustion zone. However, in few studies, it has been observed that NO_x and CO_2 emissions can increase.
- Numerous additives may be investigated with varying blend ratios in future studies to increase performance and minimize emissions. Engine settings, i.e., injection

pressure, compression ratio, injection timing, etc. may also be changed to improve combustion and decrease emissions.

Acknowledgements The authors are grateful for the facilities provided by Swami Keshvanand Institute of Technology, Management, and Gramothan (SKIT), Jaipur to conduct this study.

References

Adelman H (1979) Alcohols in diesel engines-A review. SAE Technical Paper

Ajav EA, Singh B, Bhattacharya TK (1999) Experimental study of some performance parameters of a constant speed stationary diesel engine using ethanol-diesel blends as fuel 17:357

Alptekin E (2017) Evaluation of ethanol and isopropanol as additives with diesel fuel in a CRDI diesel engine. Fuel 205:161

Ameer S, Gopal KR (2012) A review of the effects of catalyst and additive on biodiesel production, performance, combustion, and emission characteristics. Renew Sustain Energy Rev 16:711

Baker HK, Pandey N, Kumar S, Haldar A (2020) A bibliometric analysis of board diversity: current status, development, and future research directions. J Bus Res 108:232–246

Behcet R (2011) Performance and emission study of waste anchovy fish biodiesel in a diesel engine. Fuel Process Technol 92:1187

Can O, Celikten I, Usta N (2004) Effects of ethanol addition on performance and emissions of a turbocharged indirect injection diesel engine running at different injection pressures. Energy Convers Manag 45:2429

Chen Z, Yao C, Wang Q, Han G, Dou Z, Wei H (2016) Study of cylinder-to-cylinder variation in a diesel engine fueled with diesel/methanol dual fuel. Fuel 170:67

Chen Z, Yao C, Yao A, Dou Z, Wang B, Wei H (2017) The impact of methanol injecting position on cylinder-to-cylinder variation in a diesel methanol dual fuel engine. Fuel 191:150

Cole RL, Poola RB, Sekar R, Schaus JE, McPartlin P (2001) Effects of ethanol additives on diesel particulate and NOx emissions. SAE Technical Papers

Datta A, Mandal BK (2016) Impact of alcohol addition to diesel on the performance combustion and emissions of a compression ignition engine. Appl Therm Eng 98:670

Desjardins PP, Pitsch H, Malhotra R, Kirby SR, Boehman AL (2008) Structural group analysis for soot reduction tendency of oxygenated fuels. Combust Flame 154:191

Dogan O (2011) The influence of n-butanol/diesel fuel blends utilization on a small diesel engine performance and emissions. Fuel 90:2467

Fang Q, Fang J, Zhuang J, Huang Z (2013) Effects of ethanol-diesel-biodiesel blends on combustion and emissions in premixed low temperature combustion. Appl Therm Eng 54:541

Fayyazbakhsh A, Pirouzfar V (2017) Comprehensive overview on diesel additives to reduce emissions, enhance fuel properties and improve engine performance. Renew Sustain Energy Rev 74:891

Giakoumis EG, Rakopoulos CD, Dimaratos AM, Rakopoulos DC (2013) Exhaust emissions with ethanol or n-butanol diesel fuel blends during transient operation: a review. Renew Sustain Energy Rev 17:170

Goyal N, Nayyar A, Kumar C (2015) Experimental investigation of the performance of VCR diesel engine fuelled by n–butanol diesel blend. Int J Res Eng Technol 4:444

He BQ, Liu MB, Yuan J, Zhao H (2013) Combustion and emission characteristics of a HCCI engine fuelled with n-butanol-gasoline blends. Fuel 108:668

Hellier P, Ladommatos N, Allan R, Rogerson J (2013) Combustion and emissions characteristics of toluene/n-heptane and 1–octene/n-octane binary mixtures in a direct injection compression ignition engine. Combust Flame 160:2141

Huang H, Zhou C, Liu Q, Wang Q, Wang X (2016) An experimental study on the combustion and emission characteristics of a diesel engine under low-temperature combustion of diesel/gasoline/n–butanol blends. Appl Energy 170:219

Ianniello R, Belgiorno G, Di Luca G, Beatrice C, Di Blasio G (2021) Ethanol in dual-fuel and blend fueling modes for advanced combustion in compression ignition engines. In: Energy, environment, and sustainability. Springer, Singapore. ISNB 978-981-16–0931-2

Jamwal A, Agrawal R, Sharma M, Kumar A, Kumar V, Garza-Reyes JAA (2021) Machine learning applications for sustainable manufacturing: a bibliometric-based review for future research. J Enterp Inf Manag. https://doi.org/10.1108/JEIM-09-2020-0361

Karabektas M, Hosoz M (2009) Performance and emission characteristics of a diesel engine using isobutanol–diesel fuel blends. Renew Energy 34:1554

Khalife E, Tabatabaei B, Demirbas A, Aghbashlo M (2017) Impacts of additives on performance and emission characteristics of diesel engines during steady state operation. Prog Energy Combust Sci 59:32

Kumar C, Bafna M, Nayyar A, Parkash V, Goyal N (2014) Experimental investigation of the performance of VCR diesel engine fuelled by nm-diesel blend. IJETAE 4:122

Kumar C, Rana KB, Tripathi B, Nayyar A (2018a) Properties and effects of organic additives on performance and emission characteristics of diesel engine: a comprehensive review. Environ Sci Pollut Res 25(23):22475–22498

Kumar C, Rana KB, Tripathi B, Gupta P (2018b) Combustion characteristics of methanol blended diesel fuel in CI engine. Int J Pharm Sci Rev Res 50:101–104

Kumar C, Rana KB, Tripathi B, Nayyar A (2018c) A comparative study of oxygenated additives for diesel in compression ignition engine. Int J Renew Energy Technol 9:16

Kumar C, Rana KB, Tripathi B (2019) Effect of diesel-methanol-nitromethane blends combustion on VCR stationary CI engine performance and exhaust emissions. Environ Sci Pollut Res 26(7):6517–6531

Kumar C, Nayyar A, Bafna M, Agarwal A, Parkash V (2015) Analysis of emission characteristic of NM-diesel blend on VCR diesel engine. Int J Recent Adv Mech Eng 4:115

Kumar C, Rana KB, Tripathi B (2020) Effect of ternary fuel blends on performance and emission characteristics of stationary VCR diesel engine. Energy Sour Part A: Recovery Utiliz Environ Effects 1–20

Kumar C, Rana KB, Tripathi B (2021) Combustion, performance and emission analysis of diesel-methanol fuel blend in CI engine. In: Energy, environment, and sustainability. Springer, Singapore, pp 229–246. https://doi.org/10.1007/978-981-16-1280-0_9. ISBN: 978-981-16-1279-4

Maricq MM (2007) Chemical characterization of particulate emissions from diesel engines: a review. Journal of Aerosol Sci 38:1079

Moghaddam MS, Moghaddam MM, Aghili S, Absalon A, Najafi A (2012) Performance and exhaust emission characteristics of a ci engine fueled with diesel–nitrogenated additives. Int J Chem Eng Appl 3:363

Mohammadi A, Kee SS, Ishiyama T, Kakuta T, Matsumoto T (2005) Implementation of ethanol diesel blend fuels in PCCI combustion

Nabi MN, Hustad JE (2010a) Experimental investigation of engine emissions with marine gas oil-oxygenate blends. Sci Total Environ 408:3231

Nabi MN, Hustad JE (2010b) Experimental investigation of engine emissions with marine gas oil-oxygenate blends. Sci Total Environ 408:3231

Nayyar A, Sharma D, Soni SL, Mathur A (2017a) Experimental investigation of performance and emissions of a VCR diesel engine fuelled with n-butanol diesel blends under varying engine parameters

Nayyar A, Sharma D, Soni SL, Mathur A (2017b) Characterization of n-butanol diesel blends on a small size variable compression ratio diesel engine: modeling and experimental investigation. Energy Convers Manag 150:242

Nur A, Putrasari Y, Santoso WB, Kosasih T, Reksowardojo IK (2015) Performance characteristic of indirect diesel engine fuelled with diesel-bioethanol using Uniplot software. Energy Procedia 68:167

Nour M, Kosaka H, Bady M, Sato S, Abdel-rahman AK (2017) Combustion and emission characteristics of DI diesel engine fuelled by ethanol injected into the exhaust manifold. Fuel Process Technol 164:33

Padwa RS, Gupta S, Singh V, Kumar C (2016) Experimental investigation on the performance of VCR diesel engine fuelled by E-NM2-diesel blend. Int J Innov Res Sci Eng Technol 5:6794

Pan W, Yao C, Han G, Wei H, Wang Q (2015) The impact of intake air temperature on performance and exhaust emissions of a diesel methanol dual fuel engine. Fuel

Park SH, Youn IM, Lee CS (2010) Influence of two-stage injection and exhaust gas recirculation on the emissions reduction in an ethanol-blended diesel-fueled four-cylinder diesel engine. Fuel Process Technol 91:753

Pedrozo VB, May I, Zhao H (2017) Exploring the mid-load potential of ethanol-diesel dual-fuel combustion with and without EGR. Appl Energy 193:263

Pidol L, Lecointe B, Starck L, Jeuland N (2012) Ethanol-biodiesel-diesel fuel blends: performances and emissions in conventional diesel and advanced low temperature combustions. Fuel 93:329

Prasad S, Singh A, Joshi HC (2007) Ethanol as an alternative fuel from agricultural, industrial and urban residues. Resour Conserv Recycl 50:1

Putrasari Y, Nur A, Muharam A (2013) Performance and emission characteristic on a two cylinder DI diesel engine fuelled with ethanol-diesel blends. Phys Procedia 32:21

Rakopoulos CD, Antonopoulos KA, Rakopoulos DC (2007) Experimental heat release analysis and emissions of a HSDI diesel engine fueled with ethanol-diesel fuel blends. Energy 32:1791

Rakopoulos DC, Rakopoulos CD, Giakoumis EG, Dimaratos AM, Kyritsis DC (2013) Effects of butanol–diesel fuel blends on the performance and emissions of a high-speed DI diesel engine. Energy Convers Manag 51:1989

Sain NK, Nayyar A, Kumar C, Rana KB, Tripathi B (2020) Effect of nitromethane–*n*-butanol–diesel blends on diesel engine emissions. In: Advances in energy research (Springer Proceedings in Energy), vol 2, pp 457–466. ISBN: 978-981-15–2661-9. https://doi.org/10.1007/978-981-15-2662-6_42

Sayin C (2010) Engine performance and exhaust gas emissions of methanol and ethanol–diesel blends. Fuel 89:3410

Sayin C, Ilhan M, Canakci M, Gumus M (2009) Effect of injection timing on the exhaust emissions of a diesel engine using diesel–methanol blends. Renew Energy 34:1261

Shamun S, Belgiorno G, Di Blasio G, Beatrice C, Tuner M, Tunestal P (2018) Performance and emissions of diesel-biodiesel-ethanol blends in a light duty compression ignition engine. Appl Therm Eng 145:444–452

Shamun S, Belgiorno G, Di Blasio G (2020) Engine parameters assessment for alcohols fuels application in compression ignition engines. Alternative fuels and their utilization strategies in internal combustion engines. In: Energy, environment, and sustainability. Springer, Singapore, pp 125–129. https://doi.org/10.1007/978-981-15-0418-1_8

Sukjit E, Herreros JM, Dearn KD, Garcia-Contreras R, Tsolakis A (2012) The effect of the addition of individual methyl esters on the combustion and emissions of ethanol and butanol-diesel blends. Energy 42:364

Taghizadeh-Alisaraei A, Rezaei-Asl A (2016) The effect of added ethanol to diesel fuel on performance, vibration, combustion and knocking of a CI engine. Fuel 185:718

Tutak W, Lukacs K, Szwaja S, Bereczky A (2015) Alcohol–diesel fuel combustion in the compression ignition engine. Fuel

Wagner TO, Gray DS, Zarah BY, Kozinski AA (1979) Practicality of alcohols as motor fuel. SAE Technical Paper.

Wang Q, Wei L, Pan W, Yao C (2015) Investigation of operating range in a methanol fumigated diesel engine. Fuel 140:164

Wei L, Yao C, Han G, Pan W (2016) Effects of methanol to diesel ratio and diesel injection timing on combustion, performance and emissions of a methanol port premixed diesel engine. Energy 95:223

Xingcai L, Jian-Guang Y, Wu-Gao Z, Zhen H (2004a) Effect of cetane number improver on heat release rate and emissions of high speed diesel engine fueled with ethanol-diesel blend fuel. Fuel 83:2013

Xingcai L, Zhen H, Wugao Z, Li D (2004b) The influence of ethanol additives on the performance and combustion characteristics of diesel engines. Combust Sci Technol (Taylor Fr) 176:1309

Xu B, Qi Y, Zhang W, Cai S (2007) Fuel properties and emission characteristics of ethanol–diesel blend on small diesel engine. Int J Automot Technol 8:9

Zhen X, Wang Y (2015) An overview of methanol as an internal combustion engine fuel. Renew Sustain Energy Rev 52:477

Zheng M, Han X, Asad U, Wang J (2015) Investigation of butanol-fuelled HCCI combustion on a high efficiency diesel engine. Energy Convers Manag 98:215

Part III
Renewable Fuel Production

Chapter 9
Biomass and CO_2-Derived Fuels Through Carbon-Based Catalysis. Recent Advances and Future Challenges

Andreia F. Peixoto, Diana M. Fernandes, Ana B. Dongil, Elodie Blanco, and Cristina Freire

Abstract Liquid transportation fuels from biomass and CO_2 are considered a promising strategic alternative to simultaneously reduce greenhouse gas emissions and fulfill the massive energy demands. Pyrolysis allows the transformation of lignocellulosic biomass into bio-oil (liquid fraction) which can be further upgraded to hydrocarbon fuels by well-known methodologies including catalytic hydrodeoxygenation (HDO) and steam or aqueous phase reforming to transform bio-oil into hydrocarbons and H_2 as promising alternatives to fossil fuels in forthcoming future. The efficient conversion of CO_2 to fuels and useful chemicals is an essential step toward reducing the atmospheric concentration of CO_2. (Electro)chemical catalytic CO_2 reduction is a promising route to convert CO_2 back into valuable chemicals and fuels. The use of carbon materials for biomass and CO_2 transformation has gained impact as an alternative to conventional oxides due to their excellent features including high surface area, electroconductivity, and low cost. This chapter describes the recent works reported in the literature on the use of carbon-based catalysts to convert biomass and CO_2 into sustainable biofuels. An overview of the most promising carbon-based catalysts and processes will be presented, including the main

A. F. Peixoto (✉) · D. M. Fernandes · C. Freire
LAQV-REQUIMTE, Departamento de Química e Bioquímica, Faculdade de Ciências, Universidade do Porto, 4169-007 Porto, Portugal
e-mail: andreia.peixoto@fc.up.pt

D. M. Fernandes
e-mail: diana.fernandes@fc.up.pt

A. B. Dongil
Instituto de Catálisis y Petroleoquímica, CSIC, Calle de Marie Curie 2, Cantoblanco, 28049 Madrid, Spain
e-mail: a.dongil@csic.es

E. Blanco
Departamento de Ingeniería y Gestión de la Construcción, Pontificia Universidad Católica de Chile, Vicuña Mackenna 4860, Santiago, Chile

Departamento de Ingeniería Química y Bioprocesos, Pontificia Universidad Católica de Chile, Vicuña Mackenna 4860, Santiago, Chile

Millenium Nuclei on Catalytic Processes Towards Sustainable Chemistry (CSC), Santiago, Chile

P. C. Shukla et al. (eds.), *Renewable Fuels for Sustainable Mobility*, Energy, Environment, and Sustainability, https://doi.org/10.1007/978-981-99-1392-3_9

challenges to improve carbon-based catalysts' performance and to reduce costs to make biomass and CO_2-derived fuels an effective alternative for future mobility.

Keywords Biomass valorization · CO_2 conversion · Carbon-based catalysts · Renewable fuels

9.1 Introduction

Alternative fuels are extremely important to the future of energy and transportation sectors which play an important role in today's economy and society with a huge impact on growth and employment. Modern industry depends on fossil resources for producing fuels and chemicals and the forecasts for 2050 rely on emissions of about 200 million tons of CO_2 produced from the massive used of these fossil feedstocks contributing to cumulative greenhouse gas emissions and environmental issues (De Luna et al. 2019). Concerning this topic, the production of renewable fuels from alternative resources is an extremely important request and, the growing interest in new technologies for the valorization of renewable biomass and CO_2, in the last 10 years is well documented in literature reviews (Sudarsanam et al. 2018; Kondratenko et al. 2013; Lu et al. 2021). Biomass is not only the major carbon-containing sustainable resource for supporting the exponentially growing energy demand but also prompts less greenhouse gas emissions. The best policy to compete with fossil-based fuel refineries is the integration and upgrading of biomass-derived feedstocks in a similar (bio)refinery approach. Biorefineries use several technologies to convert different biomass feedstock into biofuels and bio-based chemicals, able to replace a large fraction of industrial chemicals and materials currently derived from fossil resources (Ruiz et al. 2013).

There is a large number of technologies to convert biomass into chemicals and fuels including biological, thermal, and chemical processes (Sudarsanam et al. 2018). Pyrolysis is a widely used process to sustainably convert biomass into liquid fuels. The search for advanced fast pyrolysis processes has grown in the last years in order to convert low-value biomass materials into bio-oil and numerous useful products (Amenaghawon et al. 2021). However, bio-oil is a complex mixture of numerous oxygenated compounds with a significant content of water, depending on the biomass source.

Compared with traditional fossil energy sources, bio-oil presents unstable physical–chemical characteristics and low calorific value. In order to make bio-oil a vital part of energy infrastructures, as a future transportation fuel, their upgrading through oxygen removal and molecular weight reduction is imperative in order to simultaneously increase thermal stability and reduce volatility and viscosity (Jacobson et al. 2013). There are several examples of bio-oil upgrading methods, Fig. 9.1, although they present limitations to their large-scale application (Qu et al. 2021). The hydrodeoxygenation (HDO) process is considered a fundamental strategy of biomass upgrading to produce transportation fuels. The main challenge of HDO technology

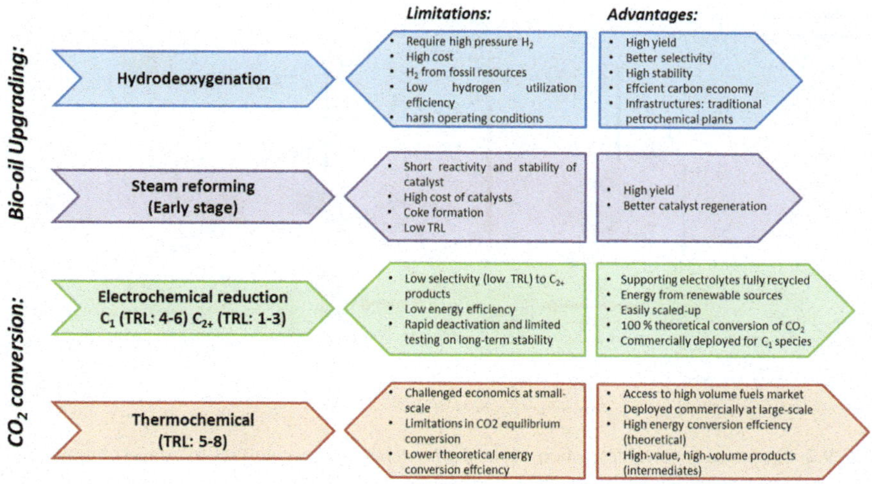

Fig. 9.1 Limitations and advantages of promising routes for (biomass) bio-oil upgrading and CO_2 conversion. Adapted from (Grim et al. 2020; Panwar and Paul 2021)

implementation at the commercial scale is its dependence on high-pressure hydrogen, expensive and mainly obtained from fossil resources and, the safety issues of the high H_2 pressure. The use of hydrogen-free alternatives such as in-situ production using bio-oil derivatives is an attractive option (Jin et al. 2019a).

Indeed, steam reforming allows converting the oxygenated and non-oxygenated hydrocarbons from bio-oil by reaction with steam at high temperatures, to obtain syngas (a mixture of hydrogen and carbon monoxide) and CO_2. This route to bio-oil upgrading has been extensively studied as illustrated in Fig. 9.1 (Tan et al. 2020), where the main advantages and limitations are highlighted.

The quest for a more sustainable society and a cleaner environment grounded based on green chemistry and engineering principles has prompted the development of new studies combining CO_2 capture and its transformation into high-added-value chemicals through catalytic or electrocatalytic reactions (Rafiee et al. 2018; Roy et al. 2018; Wu et al. 2017), Fig. 9.1. Currently, the typical thermochemical routes of CO_2 conversion produce methanol, CO, dimethyl ether (DME), etc., (Rafiee et al. 2018) and, even more advantageous, the electrochemical processes can provide valuable or energy-dense chemicals such as formic acid and ethanol (Kumar et al. 2017). CO_2 electrochemical reduction (CO_2RR) has recently attracted huge attention as a promising and clean technology to convert CO_2 into useful chemicals and low-carbon fuels using electrons as reductants at mild conditions.

Lu et al. (2021) have published recently a review about the use of efficient single-atom catalysts to convert biomass and CO_2 into renewable fuels and chemicals and presented a scheme illustrating the integrated carbon cycle of both biomass and CO_2 simultaneous conversion, Fig. 9.2 has a strategy to minimize CO_2 emissions and to find alternative renewable feedstocks for the intensive future demands of renewable energy and chemicals.

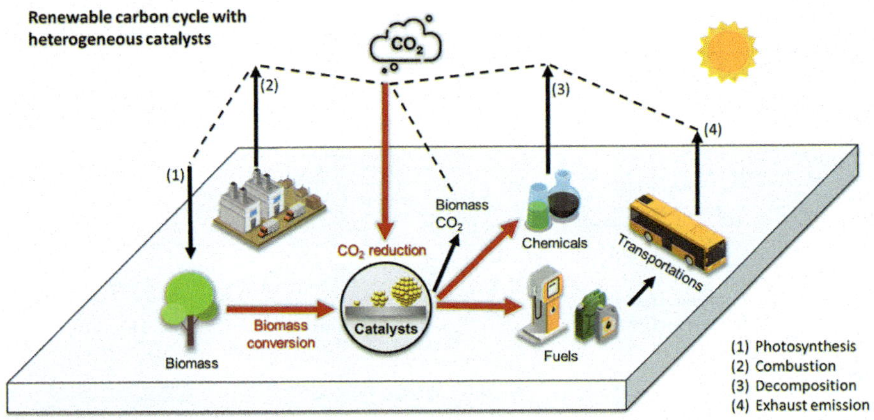

Fig. 9.2 Proposed renewable carbon cycle with heterogeneous catalysts to convert biomass and CO_2 into renewable fuels and chemicals simultaneously. Reproduced with copyright from (Lu et al. 2021)

The choice of the catalyst is a fundamental role in the upscaling and industrialization of biomass and CO_2 process conversion. Efficient catalysts have been designed and used for both biomass and CO_2 valorization, although a technology at commercial maturation is still a challenge due to the high costs of currently developed catalysts; the low selectivity control toward the target products; and most importantly, the low stability of the catalysts. The development of cost-efficient, highly active, and selective catalytic systems is essential to a significant step forward in this research topic, Table 9.1.

Carbon-based (nano)materials are important building blocks for a multitude of advanced applications including catalysis and energy conversion. Carbon materials have been widely used as catalysts or catalyst support in several liquid and gas phase reactions. Generally, carbon materials possess distinctive physicochemical properties including high specific surface area, variable porosity, excellent stability and superior electronic conductivity, and interesting tunable surface functional groups for incorporation of additional moieties. Also, the combination of these features with their structure, interconnected pores, and heat stability make them key materials to be used as catalysts or catalyst support for the conversion of biomass and CO_2 into fuels (Gawande et al. 2020; Fernandes et al. 2019).

Carbon-based materials exist in different allotropic forms such as carbon black (CB) and activated carbon (AC, including the biomass-derived materials), carbon aerogel, diamond, fullerene, carbon nanofibers (CNF), carbon nanotubes (CNT), graphite and graphene (Seelam et al. 2010; Lam and Luong 2014). The high surface area, low cost, and availability of AC and CB make them the materials of choice to support catalysts. Advanced engineering synthesis of various nanostructured carbon materials including nanoparticles, nanotubes (one-dimensional materials), graphene sheets (2D), and mesoporous carbons (3D) provide new opportunities for the development of advanced carbon-based catalysts not only for fuel production but also

Table 9.1 An overview of different technologies to convert bio-oil and CO_2 into derived fuels. The typical catalysts and supports and the future challenges for commercialization

Method	Catalysts/supports	Future challenges for commercialization
Hydrodeoxygenation • *Increases the heating value, reduces the viscosity of bio-oil*	• Sulphide transition metals (W, Co, Ni, Mo); • Reduced metals: Noble (Pd, Pt, Rh, Ru); and transition metals (Ni, Co, Mo, W); bimetallic (PtSn, Pt-Pd, RhPd, NiW, NiMo, Co-Mo; • Supports: SiO_2, SiO_2–Al_2O_3, TiO_2, ZrO_2, zeolites and carbon-based materials	• studies use model compounds; bio-oil stability is an issue • Catalysts are prone to deactivation in bio-oils • Economic viability is still a problem: the cost of noble metals prevents the large-scale application • Alternatives to current harsh conditions
Steam reforming • *Produces hydrogen by steam reforming of bio-oil*	• Metal catalysts (base and noble)-Ni, Rh, Pt, Pd, Co, Cu, Ir, Fe • Supports: MgO, MgO–Al_2O_3, CeO_2, ZrO_2, Calcite, Dolomite, Al_2O_3 Zeolites-Y, ZnO and carbon-based materials	• Use low-cost metals • Improve water tolerance of the low-cost active phase (Ni, Co, Fe) • Reduce reaction temperature to reduce thermal requirement by for example operating in the liquid phase • Improve TRL (currently at stage 1–2)
Thermochemical reduction • *Produces low-carbon fuels and methanol by* CO_2 hydrogenation	• Metal and metal oxide-based catalysts: Cu/ZnO/Al_2O_3, Fe_3O_4, Co/Al_2O_3; • Metal promoter molecules such as K, Mn, Na, and Cu	• Develop multifunctional water and CO_2 tolerant catalysts • Improve product selectivity • Improve catalysts performance
Electrochemical reduction • *Produces chemicals and low-carbon fuels using electrons as reductants under mild conditions*	• Noble metals: Pd, Pt, Ru • Transition metal elements and related compounds (metal oxides/metal complexes): Cu, Fe, Co, Ni, Cr, Mn	• Economic viability is still a problem: the cost of noble metals prevents the large-scale application • Electrocatalysts improve performance: high activity and stability • Reduce energy consumption and increase the energy efficiency of the process

for fine chemistry applications (Pérez-Mayoral et al. 2020). Graphene, one of the most promising carbon-based materials, offers high stability in water media, in both acid and basic conditions, which are typical conditions found in biomass feedstocks. Moreover, their high surface area also allows better dispersion of the active phases improving the efficiency of the catalysts in biomass and CO_2 upgrading and other catalytic applications. For instance, surface functionalized of carbon support could dramatically increase selectivity in HDO and significantly boost Faradaic efficiency toward CO and other valuable chemicals, such as CH_3OH, during electrochemical CO_2 reduction (CO_2RR). Carbon-based materials are also excellent hosts for metal-nanoparticles, working as hybrid catalysts with multifunctionalities to be used in biomass and CO_2 conversion.

There are several reviews recently published reporting the advances in biomass conversion/bio-oil upgrading (Khosravanipour Mostafazadeh et al. 2018; Cordero-Lanzac et al. 2021; Pujro et al. 2021; Attia et al. 2020) and on thermochemical conversion/electrochemical reduction of CO_2 (Fernandes et al. 2019; Qiao et al. 2014; Feng et al. 2019; Wang et al. 2017; Xie et al. 2018; Zhu et al. 2016; Khezri et al. 2017; Abdelkader-Fernandez et al. 2020; Handoko et al. 2018; Roy et al. 2018; Esteve-Adell et al. 2017b); with a few examples focusing on the of using carbon-based materials as catalyst support (Cordero-Lanzac et al. 2021; Lam and Luong 2014) and several examples exploring the physico-chemical and textural properties and surface functionalization methodologies and this is why we will not focus this topic in the present chapter (Lam and Luong 2014; Benzigar et al. 2018; Pérez-Mayoral et al. 2020).

However in this chapter, we present the recent advances in the application of carbon-based materials (including the ones prepared from biomass) to covert biomass/bio-oil into fuels giving special emphasis to the most promising technologies developed regarding their potential of industrialization as an alternative to the current unsustainable fossil fuels energetic scenario. Considerable attention has focused on the main technical challenges and future opportunities of bio-oil upgrading using HDO and steam reforming (hydrogen production); and, electro-or thermo-catalytic reduction of CO_2, Fig. 9.3. The design of promising catalysts and their structure-performance relationship are important points for future sustainable processes with the main goal of lowering the production costs providing a viable alternative to future mobility. The combination of carbon materials features with more sustainable processes can be the key to revolutionize the production of fuels from biomass and CO_2 although the limitations presented in Fig. 9.1 are still significant and struggle the large-scale application. We will make an overview of the future challenges to overcome these problems in the next sections.

9.2 Catalytic Carbon-Based Processes to Biomass-Derived Fuels

Conversion of biomass into liquids fuels and chemicals has attracted considerable interest in the last years. Several routes have been developed including pyrolysis as a typical sustainable process to convert biomass into liquid fuels and one of the most promising attempts to develop alternatives to energy production (Jin et al. 2019b; Carrasco et al. 2017). Fast pyrolysis is by far the most interesting industrial approach (Amenaghawon et al. 2021). The resulting oil also called bio-oil is produced under typical heating of 450–600 °C in the absence of oxygen (Dhyani and Bhaskar

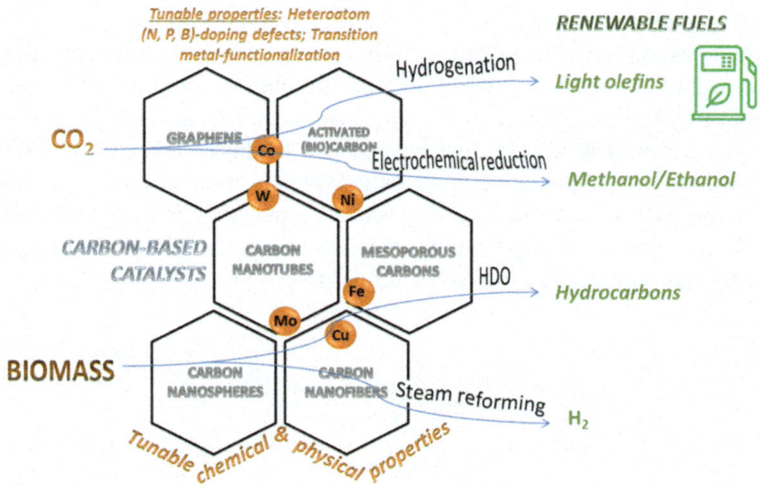

Fig. 9.3 Schematic representation of carbon-based catalysts' role in the future of biomass and CO_2-derived fuels

2018). Bio-oil is a complex mixture of compounds comprising sugars, carboxylic acids, hydroxyaldehydes, hydroxyketones, and phenolics. The presence of oxygen (typically 40–50%) is one of the key differences between bio-oils and other hydrocarbon fuels, causing lower heating value, thermal and chemical stability, higher viscosity and immiscibility, higher acidity, and coke formation during processing when compared with conventional fuels (Guo et al. 2018; Pang 2019). HDO and steam reforming are examples of downstream bio-oil upgrading, Fig. 9.1. In the next section special emphasis will be given to HDO reactions as the most promising and effective method for bio-oil upgrading to hydrocarbons. The focus will be given on the recent works employing carbon-based materials as a catalyst or catalyst support and alternatives to the conventional H_2 from fossil fuels, the "H_2-free" HDO reactions, as a sustainable strategy for HDO industrial application. An overview of the state of the art of HDO of real bio-oil upgrading will also be addressed. Furthermore, due to the importance of H_2 production from alternative sources the steam reforming of bio-oil to syngas (a mixture of hydrogen and carbon monoxide) will also discuss as other interesting routes to bio-oil upgrading.

9.2.1 *Hydrodeoxygenation of Bio-Oil*

Hydrodeoxygenation is an important catalytic reaction to transform biomass derivatives into potential fuels (De et al. 2015). During HDO process bio-oil C–O bonds are broken and by hydrogenation and/or dehydration eliminated decreasing the oxygen content. The catalysts should be carefully designed to meet the key challenge of

HDO processes and promote a high degree of oxygen removal under minimum hydrogen consumption. Several processes have been intensively studied for the selective removal of oxygen from bio-oil, lignin, and other substrate models under high pressure of hydrogen in the presence of a heterogeneous catalyst (Jacobson et al. 2013; Mortensen et al. 2011). In fact, and due to the complexity of the bio-oil molecules, the use of model compounds is typically used to discern mechanistic information, useful guidelines for the catalysts, and process design (Arora et al. 2020; Bjelic et al. 2019). The typical reaction routes for HDO of various model compounds are summarized in Fig. 9.4 (Qu et al. 2021).

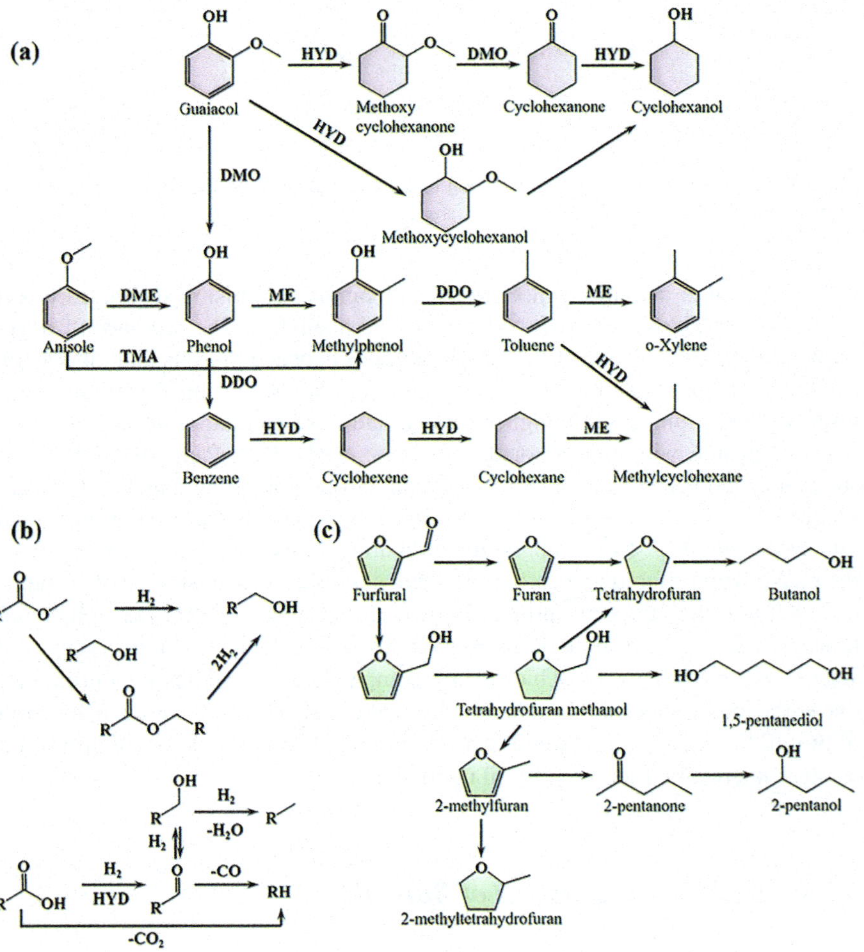

Fig. 9.4 Reaction routes for HDO of various model compounds: **a** containing phenol, guaiacol, and anisole; **b** acids and esters; and **c** furan and furfural. DDO: deoxygenation; HYD: hydrogenation; DME: demethylation; TMA: methyl transfer; ME: methylation. Reproduced from (Qu et al. 2021) with permission from the Royal Society of Chemistry

The HDO efficiency can be accessed by several reaction parameters including temperature, hydrogen pressure, type of reactor, solvent, time, catalyst selection, weight hourly space velocities (WHSV), etc. Among those, catalyst selection plays a central role and typically a good HDO catalyst must have appropriate surface oxophilicity to promote C–O bond cleavage, combined with moderate hydrogenation ability and to reach a selective hydrogenation of the bio-oil derivatives to the target products (Hita et al. 2020). The synergistic effect of support and the active phase (metal) can be crucial to the efficiency of HDO catalysts playing a role in the stabilization and dispersion of metallic particles. Moreover, the supports can also have activity, can mediate reactants activation, and can facilitate the active phase-reactant encounter enhancing metal-support interaction (Jin et al. 2019a). Carbon materials arise in several biomass conversion processes, including in HDO reactions, not only as promising catalyst supports, but also as a metal-free active phase catalyst, due to their distinctive physicochemical, low cost, and lower affinity for coke deposition compared to acidic supports (Mukundan et al. 2015).

The use of alternative HDO methodologies for bio-oil upgrade avoiding, for example, the use of external H_2, is attaining a significant interest as a potential economic approach to transform biomass into biofuels (Jin et al. 2019a). Due to the growing interest in more sustainable processes, this discussion will be centered on the performance of the use of carbon-based materials as catalyst support and the use of H_2 alternative sources in an attempt to serve as a helpful guide toward the development of efficient carbon-based catalysts and sustainable HDO processes and their potential of industrialization. To date, the HDO processes use mainly external H_2 and guaiacol is the mostly studied biomass derivative model. Noble metals (Pt, Pd, Rh, Ru) (Santos et al. 2018; Thompson and Lamb 2018), transition metals (Ni, Fe) (Guo et al. 2018) and metal sulfides, molybdenum carbides (Cai et al. 2017; Blanco et al. 2020c) and rhenium carbides (Blanco et al. 2020b, 2021) are usually used as metal active phases. Bimetallic multifunctional catalysts (Pd/Fe, Ru/Re) (Lee et al. 2020; Arora et al. 2020) and additive promotors can also act as an efficient catalyst for HDO catalysis. CB and AC have been used as carbon-based supports for the synthesis of efficient heterogeneous catalysts due to their low production cost and high specific surface area. Nevertheless, the use of CNT, CNF and GO provide new opportunities for the development of more efficient and stable catalysts. The use of commercial carbon-supported noble metal catalysts for HDO reactions have also been extensively explored (Bjelic et al. 2019; Thompson and Lamb 2018; Guo et al. 2018).

In a recent review, Sharma et al. (2020) presented a variety of carbon catalysts to convert lignin and its derivatives models by HDO reaction as a powerful tool for the generation of non-petroleum chemicals and fuels. The authors presented structurally different carbon-supported metal catalysts and discussed the effect of surface properties in mechanistic insights. Li and co-authors (2021) also focus their studies on the progress effects of the microenvironment of carbon-based catalysts on the HDO of biomass. The authors focus the review on the effects of the surface of carbon support microenvironment (structure and functionalization) in the catalytic performance and the mechanism of HDO reactions and concluded that carbon-based catalysts with mesoporous structure and high surface area favors the formation of small particle size and consequently higher dispersion and distribution of metallic active sites leading to high catalytic activity. They also found an effect of pore size distribution on the adsorption of reactants and transport of products which also affects the activity and selectivity of HDO reaction. These reviews represent a very good overview of the work done in the last years in HDO reactions using carbon-based materials and are important future guidance to the design and development of efficient carbon-based catalysts for this type of reaction. However, must of the published works use harsh reaction conditions, including (200–400 °C) and high H_2 pressure (4–20 MPa) in batch or fixed-bed reactors which means high energy consumption. Furthermore, the current commercialize H_2 gas is typically produced from fossil fuels which is an expensive and non-renewable source. Therefore, the use of milder HDO processes is invariably desirable to design an economic way to biomass conversion to fuels. Noble metal-supported carbon-based materials are the mostly used in the works presented in the literature, although significant efforts are being made to increase the economics of HDO technology motivating the transition to non-noble metal catalysts and mild reaction conditions (low H_2 pressure and temperature) to reach high selectivity for the desired products. The demonstrated evidence of comparable catalytic activities to noble metal-based catalysts obtained with some non-noble (Cu, Co, and Ni) metal-based catalysts is a promising signal to the economic scale-up of this process (Kordouli et al. 2017; Guo et al. 2018; Feitosa et al. 2019; Bjelic et al. 2019; Mendes et al. 2020; Lopez et al. 2020; Jin et al. 2021; Blanco et al. 2020a).

In general, the surface area and the pore size distribution of carbon materials play an important role in the particle size, dispersion, and distribution of metal active phases increasing the accessibility to the reactants and consequently the HDO conversion and selectivity.

9.2.1.1 "H_2-Free" Alternatives

Jin et al. (2019a) presented novel approaches to avoid the external H_2 supply, discussing the challenges and research trends of novel HDO technologies including catalytic transfer hydrogenation (CTH) and the combination of reforming and HDO. The development of hydrogen-free alternatives has been attracting the interest of the catalysis community and, although a challenge, a variety of alternatives are available at different degrees of maturity. Attempts to perform HDO at atmospheric H_2 pressure afforded excellent results for the production of aromatic compounds (Wu et al. 2021). Nevertheless, the design of highly active and stable catalysts remains a challenge and a significant drawback to the implementation of biorefinery schemes. Other "*H_2-free*" HDO strategies use as an alternative to hydrogen sources alcohols, formic acid (FA), and water-assisted in-situ HDO which is perhaps the most promising from an economic point-of-view, due to the possibility of using the water content of bio-oil as a source to generate the hydrogen in-situ. Seeking for economically viable biomass upgrading alternatives, Reina and co-authors (Jin et al. 2019b, 2021;

Parrilla-Lahoz et al. 2021) developed a novel in-situ HDO strategy using H_2O as a hydrogen source suppressing the external high-pressure hydrogen. The first approach (Jin et al. 2019b) used a series of Ni-based catalysts supported on CeO_2 with/without AC for the guaiacol HDO. The catalyst Ni/CeO_2–C showed to be the most promising reaching more than 20% conversion using only water as an H_2 source. Although the low conversion, the suppression of the high-pressure H_2 evidenced the importance to further optimize the design of the catalysts, reactor and operating conditions to improve the results and develop a new and economically advantageous route for HDO reactions.

More recently the same authors (Jin et al. 2021) used Ni supported nitrogen-doped AC catalysts (Ni/N-AC) and water as a hydrogen donor for the in-situ HDO of guaiacol. Nitrogen-doped carbons were more active than the undoped carbons in the "H_2-free" HDO process. The conversion of guaiacol increases by 8% when Ni/N-AC was used (prepared using polyaniline as a nitrogen source) comparing with non-doped Ni/AC catalysts. Therefore the excellent performance of Ni/N-AC was ascribed by the authors to the acid-base properties and to the modified electronic properties of the carbon support which favors the C–O cleavage, the water activation, and the dispersion of Ni particles on the catalysts' surface. In order to develop multifunctional catalysts able to activate water and the subsequent use of the in-situ generated hydrogen, the authors also presented a sequence of efficient Pt and Ni-based catalysts supported on N-doped graphene combined with ceria (Parrilla-Lahoz et al. 2021). NiCeO_2/GOr-N presented the best activity/selectivity balance and it was elected as the most promising catalyst to conduct the "H_2-free" HDO reaction. The as-prepared non-activated Pt-based catalysts showed superior performance than the Ni ones, although the reductive pre-treatment of the catalysts resulted in a superior performance for NiCeO_2/GOr-N with up to 30% conversion which was attributed to the optimal properties and better metal dispersion obtained with N-doped systems. Previously Yang et al. prepared a Co supported in an N-doped carbon catalyst (Co@NC) (Yang et al. 2017) by one-pot carbonization of biomass-derived glucose in the presence of melamine, as nitrogen source, and $CoCl_2$ as the catalyst precursor. The optimized Co@NC-700 (pyrolyzed at 700 °C) shows the best activity for vanillin HDO exhibiting 15.4 times higher activity than the non-doped Co/AC with >95% vanillin conversion and 100% selectivity to 2-methoxy-4-methylphenol (MMP) at 180 °C. The authors also compared the efficiency of Co@NC-700 catalyst under molecular hydrogen concluding that this catalyst is more active and selective for MMP under FA than under H_2. The superior performance of Co@NC-700 under H_2-free (external) conditions could be assigned to N-derived defective sites on carbon-based support, which potentially play a relevant role as base additive in FA dehydrogenation and metal like active center for the hydrodeoxygenation of vanillin.

Although the use of noble metals is not very attractive from an economic point of view, the fact of using milder conditions and other alternatives to molecular hydrogen can be advantageous. Emphasizing the work done by Nie et al. (2017) which prepared by one-pot carbonization of biomass-derived glucose, melamine as nitrogen source and $ZnCl_2$ as porogen nitrogen-enriched highly mesoporous carbons

(NMCs), with a specific surface area of 1017 $m^2 g^{-1}$ and mesopore volume proportion of 92.1%. This NMC allowed a well-dispersed Pd catalyst and consequently high activity (100% conversion), stability, and selectivity for HDO of vanillin to MMP (>99%) in water phase with formic acid (FA) as hydrogen donor, at 150 °C and 0.5 MPa N_2. The excellent results were ascribed to the high mesostructure, N-doped species on the surface of NMC, and their relationship with Pd particles. Pu et al. (2020) prepared for the first time N-doped hierarchically porous carbons (HPC) using MOF as a sacrificial template to control the material morphology and composition. The authors observed that by just varying the nitrogen precursors the HPC morphology, hydrophilicity, and nitrogen configuration can be fine-tuned. Particularly, a honeycomb-like morphology was obtained using dicyandiamide (DCD) as a nitrogen source. The resulted Pd@HPC-DCD catalyst showed excellent activity and stability to the HDO of vanillin in aqueous solution, using FA as hydrogen source under 0.5 MPa N_2 at 140 °C with >99% yield of unsaturated hydrocarbons. The significant catalytic activity of Pd@HPC-DCD in water was attributed not only to its hierarchical structure, but also to the well-dispersion of Pd species, even to the favorable hydrophilicity. De Luna et al. (2019) have previously presented a bimetallic catalyst Pd–Ag supported on Fe_3O_4/nitrogen-doped reduced graphene oxide (N-rGO) for HDO of lignin-derived compounds including vanillin with 100% conversion and 99% selectivity for MMP at 130 °C, using FA as a hydrogen source in water.

These published works showed that N-doped carbon materials are promising supports to enhance the catalytic performance of HDO due to: Basicity and hydrophilicity of the carbon-based supports which could improve the interaction between substrate and support under the aqueous reaction conditions; N-dopants on carbon matrix can be assumed as defects and anchoring sites favoring the particle formation and to the reduction of their size (Jin et al. 2021).

9.2.1.2 Toward Real Biofuels

The economic limitations are the main drawbacks to the industrialization of HDO and, in fact, HDO is only considered viable when renewable hydrogen is accessible and cheap enough to be used in large quantities or when alternative H_2 sources are employed. Recent publications have been using real bio-oil as feedstock and carbon-based supports taking advantage of their physicochemical features in HDO reactions, particularly on the conversion of bulkier and highly refractory components (Hita et al. 2020; Cordero-Lanzac et al. 2020). Noble metal catalysts, such as Ru, Rh, Pd, and Pt, showed high efficiency in the HDO of pyrolysis oil producing upgraded oil with higher yield and quality. Carbon surface functionalization can also improve the catalyst acidity contributing to their stability (Cordero-Lanzac et al. 2017, 2020). Cordero-Lanzac and co-authors promoted the HDO of real bio-oil using a phosphorous-functionalized activated carbon (ACP) bifunctional Pt–Pd catalyst. The authors observed that the presence of hydrothermally resistant acidic sites allowed for high selectivity to aromatics yields, maintaining the catalyst stability

(Cordero-Lanzac et al. 2020). However, the noble metal catalysts' high prices and their short-live when used for treating bio-oils due to sulfur poisoning (bio-oil generally contains sulfur) are unattractive for industrialization and clearly demonstrate the importance to develop inexpensive and highly active non-noble catalysts for HDO.

Inexpensive activated carbon-supported nickel and cobalt phosphide catalysts (varying the metal/P molar ratio) were prepared and applied in the HDO reaction of wood-derived pyrolysis oil (Guo et al. 2018). The effect of phosphorus content on HDO performance was investigated under 300 °C and 50 bar of H_2 for 3 h and it was possible to observe that the properties of the upgraded bio-oils were significantly affected. Both Ni and Co-based catalysts allow obtaining of upgraded bio-oils with lower O/C and H/C ratios increasing P content (up to $M/P = 3/2$) suggesting that the increasing of HDO activity is associated with the hydrogenation activity deterioration. High-quality bio-oil products (similar to the obtained with the commercial 5 wt% Ru/C) were achieved with the addition of the small amount of Ru as a co-catalyst. AC-supported NiP was selected as the most promising inexpensive catalyst allowing comparable activities to the expensive Ru/C catalyst. The same authors compared previously the activity of different catalysts under the same conditions presented above and concluded that AC-supported NiP was the most promising regarding the yields, the composition of upgraded oil, and the prices of the catalysts (Guo et al. 2016).

Mendes et al. (2020) prepared nickel phosphides supported in AC and mineral charcoal and compared the effect of using these two types of carbon with carbon-covered alumina (CCA) support (Mendes et al. 2020). HDO reactions were performed using a commercial bio-oil under 150 and 250 °C and H_2 pressures between 50 and 100 bar. The effect of temperature and H_2 pressure on the efficiency of deoxygenation and quality of bio-oil were evaluated. Ni_2P-AC showed better performance at 150 °C than Ni_2P on mineral charcoal and CCA, giving bio-oils with lower oxygen content and consequently higher heating value. On other hand, at higher temperatures (250 °C) Ni_2P/CCA showed the best results compared with the other carbon-based materials giving bio-oils with a lower O/C molar ratio and higher H/C molar ratio. These differences were justified by a pore blocking of the non-covered carbon supports. For the first time, Remon et al. (2021) evaluated the effect of the initial H_2 pressure and temperature, reaction time, and catalyst/bio-oil ratio in the HDO of bio-oil (lignocellulosic) using a Mo_2C/CNF catalyst to produce liquid biofuels and value-added chemicals. Process optimization revealed that using an initial H_2 pressure of 40 bar and temperature of 350 Remon et al., and 0.19 $g_{cat}\ g^{-1}{}_{bio\text{-}oil}$ for 1 h, allowed to convert 65% of the organic content of the bio-oil into a liquid bio-fuel which represents a deoxygenation degree of 70% and an energy efficiency of 62%. The authors proposed a schematic representation for the composition of the upgraded bio-oil and the potential reactions that contribute to the formation of solids (mainly char), aqueous, and gaseous species through the upgrading process, Fig. 9.5.

It is noteworthy to mention the great progress already made in the area of bio-oil upgrading contributing to enhance the knowledge about the potential use of biomass/waste biomass as an energy source, in future bio-refineries. The use of

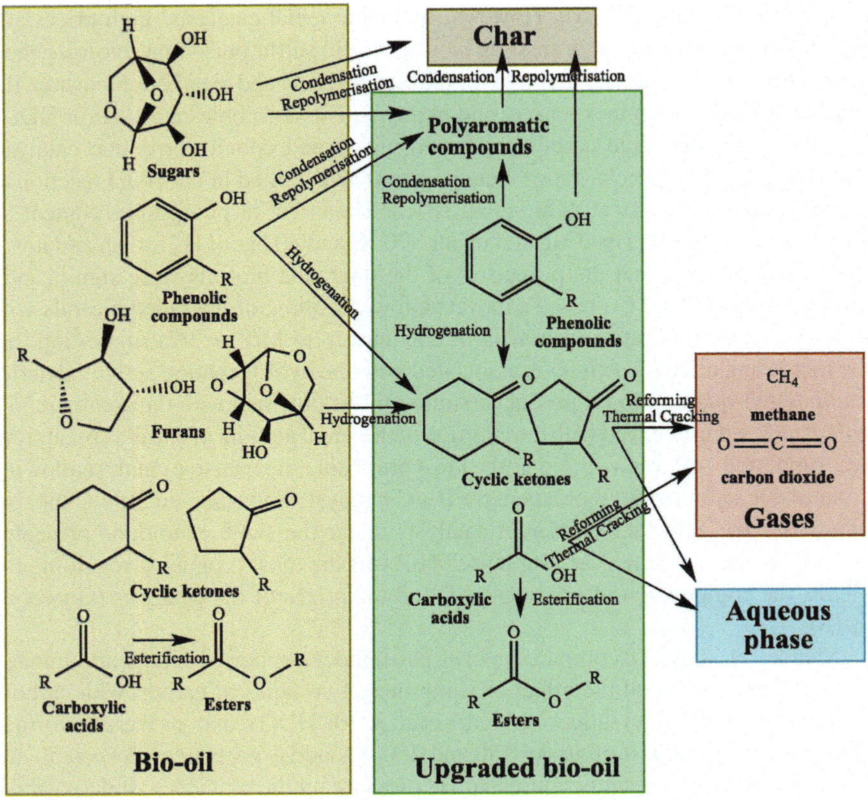

Fig. 9.5 Schematic pathway showing the chemical transformations of the bio-oil during the HDO process. Reproduced with copyright from (Remon et al. 2021)

"H_2-free" processes combined with high efficient (non-noble)-carbon-based catalysts with balanced acidity and basicity should be considered a key step to improve HDO catalytic performance and a strategy to achieve the economic viability of the industrial production of fuels from biomass. The use of low cost and renewable precursors to prepare the carbon-based catalysts is also highly recommended (Santos et al. 2020). Alcohols consider attractive alternatives to replace external H_2 supply since they can act as hydrogen donor by releasing H_2 via reforming reactions. The potential of H_2 production from steam reforming (SR) and aqueous phase reforming (APR) of bio-oil will be discussed in Sect. 2.2.

9.2.2 *Steam Reforming of Bio-Oil for Hydrogen Production*

Hydrogen (H_2) can be generated from biomass according to several processes that involve thermochemical or biological processes. Among the thermochemical processes, pyrolysis, liquefaction, and gasification are the main routes considered. All these processes transform biomass into a mixture of gas containing H_2, CO, CH_4, and other products. The gas phase can generally be enriched in H_2 using common processes like methane steam reforming or decomposition, water gas-shift reaction, or partial oxidation reaction. During pyrolysis, the bio-oil produced contains a fraction of heavy aromatics, namely tar, which must be removed as they are quite problematic for downstream equipment, causing clogging and blockage. Typically, tar can be removed by reforming over biochar-based materials and has been widely discussed in several reviews (Xu et al. 2020; Ren et al. 2021; Tan et al. 2020). On the other hand, in hydrothermal liquefaction, a bio-oil is also produced together with an aqueous phase that can be further upgraded by steam or aqueous phase reforming to H_2.

The general bottleneck of those processes is catalyst deactivation. The present part will give a sample of several recent studies where carbon materials can be employed to improve the H_2 production and/or catalyst stability from the bio-oil fraction depending on the process considered.

Steam reforming is mainly employed in small oxygenates easily vaporizable like methanol, ethanol, acetic acid, or glycerol. Methanol is not currently a bio-based product, and thus, it will be omitted in this part.

Steam reforming generally involves metal-supported catalysts, and there are only a few reports where carbon materials are employed as support (Özkan et al. 2011; Janas et al. 2015; Augusto et al. 2020; Seelam et al. 2010). The literature review displays that one of the critical factors for hydrogen production is the metallic particle size. Indeed, Silva et al. have reported that the turnover frequency (TOF) of H_2 production was improved by decreasing the particle size of Co in the case of ethanol steam reforming (ESR) (da Silva et al. 2014). Such behavior was explained by the presence of the lower amount of terraces responsible for the coke deposition. Hence reducing the particle size limited carbon deposition over the metal particle, enhancing the ESR activity. Based on this study, Augusto et al. have evaluated the effect of the carbon support for ESR in the case of Co-based catalysts (Augusto et al. 2020) and found that the carbon support affects the dispersion of Co. The better dispersion was obtained on carbon nanofibers (CNF) and AC, however, the latter suffer strong deactivation due to particle sintering, while over CNF the particle size was maintained with a slight deactivation due to partial oxidation of Co.

On the other hand, the use of carbon nanotube (CNT) as support was compared to more traditional support, Al_2O_3 by Seelam et al. (2010) and found that higher activity for ESR could be reached (Seelam et al. 2010).On the other hand, CNT was compared to graphite carbon as a support for Ni in the SRE, and results have shown that the CNT was the best support for the generation of H_2 (Rautio et al. 2015). However, the post-reaction analysis showed a significant carbon content increase after the reaction ascribed to CNF and CNT formation. Recently, a composite of CNT-silica fiber (SF) was also evaluated in ESR (Prasongthum et al. 2017). The combination of the fibrous structure of silica and the high surface area of CNT has been shown to significantly improve Ni's stability and metal dispersion, probably due to the presence of oxygen-functional groups (OFG). Another strategy employed to prevent Ni from sintering/oxidation was an encapsulation of the metal by a graphene shell (Chen et al. 2019). In this study, the thickness of the graphene shell was optimized and the resulting catalyst showed to enhance the H_2 yield product compared to a classical Ni/Al_2O_3 catalyst, with no deactivation reported during the experiment.

In the case of acetic acid steam reforming (ASR), similar fibrous composite platelet CNF-silica fiber (SF) was used as a support for Co and reported a higher ASR activity and better long-term stability compared to Co/SF and Co/Q10 that was ascribed to better metal dispersion and the unique structure (Natewong et al. 2018).

The effect of biochar activation was also evaluated in the case ASR over Ni catalyst (Chen et al. 2018b; Wang et al. 2020c; Stasi et al. 2020; Di Stasi et al. 2021). Indeed, it was reported that in the case of Ni supported over biochar, the activation using KOH-HNO_3 provided the highest specific surface area, and therefore, the Ni dispersion was enhanced (Chen et al. 2018b).

Interestingly authors also found that the activity could be improved by increasing the sulfur content. Recently, activation of biochar by HNO_3 using different concentrations has been carried out and showed to affect the hydrophilicity of the resulting support and the pore structure and the OFG, leading to a different particle size of Ni. Among these aspects, hydrophobicity is the most relevant parameter for ASR. Indeed, high hydrophobicity results in low activity and high coke formation while moderated hydrophobicity improved adsorption of the reaction intermediate on C–O–C-aliphatic OFG. In contrast, the gasification of reaction intermediates was proposed to occur over carbonyl OFG (Wang et al. 2020c). Di Stasi et al. have activated biochar using pure CO_2 at 700 °C and screened several metals in ASR and reported low deactivation of the metal phase during the reaction (Stasi et al. 2020). However, in a more recent study, they found that using a mixture of several pyrolysis bio-oil model compound deactivation was observed due to metal poisoning of heavy products, nevertheless such deactivation could be inhibited using bimetallic (Di Stasi et al. 2021). As far as glycerol steam reforming (GSR) is concerned, only a few reports are using carbon-based materials (Veiga and Bussi 2017; Liu et al. 2018c; Chen et al. 2020b). Again it appears that the use of carbon material improves the metal dispersion essentially. For instance, in the case of Ni supported over AC) oxidation of the support has been shown to enhance the Ni anchoring into smaller particle size (Veiga and Bussi 2017), while in the case of CNT support, the well-dispersed metallic Ni showed to promote the consecutive glycerol decomposition and the water gas-shift reactions to hydrogen production (Liu et al. 2018c). Recently, the same strategy to encapsulate Ni into a graphene shell and attached to a skeleton of SiO_2 was also investigated for GSR protecting Ni from oxidation and sintering with a high yield of H_2 (*ca.* 5.09 mol H_2/mol glycerol) and stability (Chen et al. 2020b). Glycerol reforming can take place in an aqueous phase and is generally a preferred process as it occurs at a lower temperature (Fasolini et al. 2019).

9.2.2.1 Aqueous Phase Reforming (APR)

APR was first introduced by Dumesic et al. as a promising process for producing H_2 in an aqueous phase under moderated conditions compared to the gas phase (Davda et al. 2005). In this respect, model compounds were often considered like glycerol, acetic acid, ethylene glycol, etc. Carbon materials were mainly considered support due to their high surface, variety of textural and structural properties, and good stability in the aqueous phase. Furthermore, their surface chemistry can be easily modified by introducing functional groups at the surface that can introduce hydrophilic properties to the hydrophobic carbon (Lam and Luong 2014). For instance, it has been reported that in the case of several ordered-mesoporous carbons (OMCs), the APR activity of ethylene glycol was directly correlated to the metal dispersion (Pt), which was affected by the different pore structures of the OMCs (Jeong et al. 2014). In this sense, nano-sized OMCs displayed higher Pt dispersion than micro-sized OMCs, and among the nano-sized OMCs, hollow-type framework configurations were better than rod-type. On the other hand, recently, Ru-supported over N-doped mesoporous carbon was investigated for glycerol APR (Gogoi et al. 2020). It was found that the incorporation of N into the carbon could significantly enhance the catalytic properties. Indeed N creates some basic sites over the support and improves anchoring over the support preventing it from sintering. Recently, the use of graphene as a support to Pt has shown to also allows narrow distribution of Pt particle size (1–5 nm) (Elias Bamaca Saquic et al. 2021).

Over CNT, superior yields toward H_2 in glycerol APR compared to oxide supports and associated with higher metal-support interactions (Rahman 2015). On the other hand, the introduction of OFG by acidic treatment can improve Pt dispersion; however, a lower TOF was reported but could be recovered by annealing the remaining OFG. Hence, it was proposed that OCG was detrimental when the reaction occurs in a binary mixture with different hydrophilicity due to competitive adsorption (Wang et al. 2010b). Similar conclusions could be obtained using bimetallic Pt–Co supported on single-wall CNT (Wang et al. 2010a).

Metal-free carbon materials have also been considered for the production of H_2 by APR (Esteve-Adell et al. 2017b, 2017a). Indeed a series of boron-doped graphene has been active for glycerol and glucose APR (Esteve-Adell et al. 2017b). The same group has also reported that ethylene glycol could be successfully decomposed into H_2 and CO_2 over metal-free graphene obtained from alginate pyrolysis and proposed that acid-base Lewis pairs act as the dehydrogenative center (Esteve-Adell et al. 2017a).

To summarize, the use of carbon-based material for improving the H_2 production has been shown to be effective especially in limiting the deactivation of the metallic site. In this respect, several strategies could be employed offering a myriad of possibilities to limit the deactivation. For instance, in the case of deactivation by particle sintering, the use of carbons with high surface area or specific morphology (e.g., CNF) enhances the metal dispersion thus, delaying the deactivation, while in some other cases, a graphitic shell prevents particles from sintering (Karapinar et al. 2019). On the other hand, the surface chemistry of the carbon is of great importance and can be easily tuned by for example acidic-treatment (Jeong et al. 2014), thermal treatment, or doping with heteroatoms like N (Kibria et al. 2019), or B (Kumar et al. 2017). In this sense, the hydrophobicity can be adjusted as well as the acid-base properties. All these parameters are essential in preventing deactivation.

9.2.3 Economic Analysis of Bio-Oil-Derived Fuels

The main challenge of bio-oil upgrading is to produce economical alternatives to light/heavy fuels for transport or heat generation. Recent tests on bio-oil combustion at an industrial scale demonstrated that bio-oil is a suitable alternative to conventional heavy fuel oil. The major issues affecting the bio-oil production/upgrading costs are the low product yield and the feedstock/equipment (upgrading) high costs, suggesting an expensive final upgraded bio-oil is still unable to be competitive as a commercial liquid fuel (Kumar and Strezov 2021). The application of alternative hydrogen sources is an effective strategy to decrease the high cost of HDO processes since generally this process occurs at less severe operation conditions (Kumar and Strezov 2021). H_2 produced from steam reforming of bio-oil is a clean and sustainable fuel as its combustion produces water and no harmful gases. In addition, it can be also used in HDO of bio-oil, as an alternative to H_2 from fossil derivatives in a biorefinery approach. Syngas can be further converted via Fischer-Tropsch process for the production of hydrocarbons.

9.3 Catalytic Carbon-Based Processes to CO_2-Derived Fuels

9.3.1 Thermochemical CO_2 Hydrogenation Using a Carbon-Based Catalyst

According to recent reports, around 50% of anthropogenic emissions can be absorbed by oceans, land, and forests and close to 85% of the remaining 50% could be used to obtain fuels and chemicals, the production of fuels being the main market (De et al. 2020). The following subchapter summarizes the research published in the last years, 2017–2021, on CO_2 hydrogenation using carbon-based catalysts through thermocatalysis. Although CO_2 can be transformed into several useful products such as CO, methane, formic acid, methanol, dimethyl ether, carbonates, lower olefins, etc., the majority of the literature is found for methanol and formic acid given their current interest as a hydrogen carrier. The information has been divided based on the target product, namely, CO_2 to methanol/formic acid, CO, and hydrocarbons (CH_4, CO_2^+, , and olefins). Transformations involving CO_2 hydrogenation are highly important especially for H_2 transportation and storage. However, the conversion of CO_2 has some thermodynamic challenges. On the one hand, C–O bonds are more stable than C–H bonds and, on the other hand, the hydrogenation products are normally liquids and, hence, entropically unfavored, since the transition from gas to liquid decreases the entropy.

9.3.1.1 Hydrogenation to Methanol and Formic Acid

Hydrogenation of CO to methanol is a well-known process for which commercial catalysts consist of Cu/ZnO. Based on this system, several research works have appeared regarding CO_2 hydrogenation to methanol using Cu/ZnO or Cu/ZnO/ZrO_2 in combination with different carbon structures. For example, Din et al. (2018) reported the use of Cu/ZrO_2 supported on carbon nanofibers with several Cu loadings (5–25 wt.%) prepared by deposition precipitation. The best catalytic results in the liquid phase hydrogenation using a CO_2/H_2 feed volume ratio of 1:3 was obtained with a Cu loading of 15 wt.% reaching a methanol yield of 20 g kg^{-1}cat h^{-1} at 180 °C and 30 bar. The catalytic performance was correlated with the concentration of basic sites and copper exposed area. The authors also studied CO_2 hydrogenation in the liquid phase at 30 bar and 180 °C using Cu–ZrO_2-carbon nanofibers-based catalysts prepared by deposition precipitation (Din et al. 2019). The catalysts were promoted with 1–4 wt.% ZnO which improved the activity and selectivity up to an optimal value of 3 wt.%, for which the measured yield to methanol was 45 kg^{-1}cat. h^{-1} with a methanol selectivity of 92% (vs. 78% without promoter). The increased yield to methanol was due to the better selectivity compared to the unpromoted catalysts, since the activity was indeed lower due to the worse dispersion of copper on the ZnO-promoted sample.

Also, Witoon et al. (2018) studied the effect of adding graphene oxide into CuO–ZnO–ZrO_2. Increasing GO loadings of 0.5–2.5 wt.% improved the yield of methanol. However, higher loadings hampered the co-precipitation of mixed metal oxides, this leading to the formation of isolated oxides nanoparticles and higher CuO particle size, and lower yield to methanol compared to the catalyst without GO. Nevertheless, the selectivity was still higher compared to the GO-free catalyst. The optimum GO loading was found to be 1 wt.% at 200 °C and 20 bars. The positive effect of GO was ascribed to the higher amount of active sites for CO_2 activation and to the hydrogen spillover promoted by GO from the copper nanoparticles to the metal oxide particles.

Deerattrakul et al. (2018) employed a different synthetic method to prepare Cu–Zn/reduced graphene oxide catalysts by impregnation of Cu and Zn on hydrothermal reduced graphene oxide. The researchers assessed the effect of the hydrothermal reduction temperature and found that 140 °C provided the highest surface area (458 m^2g^{-1}) among the tested temperatures. The catalysts with 15% weight loading of Cu–Zn afforded the highest yield to methanol (94.53 mgMeOHgcat–1 h^{-1}) at 250 °C and 15 bar.

The role of nitrogen doped on carbon supports has been also addressed by several authors on CO_2 hydrogenation to methanol and formic acid. For example, 10%Cu/Zn catalysts were prepared over an N-doped reduced graphene oxide (N-rGO) previously prepared by chemical exfoliation of graphite with hydrazine (Deerattrakul et al.

2017). Using in-situ X-ray absorption near-edge spectroscopy, the authors described the influence of the reduction time which was found to be optimal, among the materials tested, at 90 min and 350 °C. The best methanol yield at 250 °C and 15 bar was 591 mg MeOH/g cat/h.

More recently, Deerattrakul et al. (2019) reported that pyridinic nitrogen was relevant on the system CuZn/graphene to improve the metal dispersion, H_2 dissociation, as well as the activation of CO_2 to produce methanol. Also, the synthesis using urea as a nitrogen source provided the highest N-pyridinic among other N precursors tested such as ammonia and hydrazine hydrate.

Using a different experimental approach Ma et al. (2019b) also evaluated the effect of nitrogen-doped graphene. In this work, the authors synthetized several catalysts by co-precipitation of nitrogen-doped graphene (NG) and Cu/ZnO/Al_2O_3 with a different weight percentage of NG. The results showed that the addition of NG improved both CO_2 conversion and CH_3OH selectivity up to an optimal NG loading of 10 wt.%. The results were correlated with the better CO_2 and H_2 adsorption promoted by NG. Apparently, at low NG loading the system Cu/ZnO/Al_2O_3 is not well dispersed, the dispersion improving upon increasing the NG content. Also, the graphene is capable of transferring hydrogen to the copper particles forming copper hydride, this being relevant in the hydrogen spillover mechanism which may occur between the oxides and copper.

Additional theoretical calculations were performed by Esrafili et al. (2017) for the CO_2 hydrogenation with Pt- and Ni-doped graphene for which a termolecular mechanism was proposed. According to this mechanism, the H_2 molecule could be activated by two preadsorbed CO_2 molecules, one H atom is then transferred to one CO_2 molecule forming a −OCOH intermediate which would be then converted into HCOOH. The formation of formic acid would be the rate-determining step, which is more energetically favored over Pt than Ni. On the contrary, the bimolecular mechanism is described by two steps, which start with the co-adsorption of both reactants on the surface. Then, also a −OCOH intermediate is formed which finally is transformed into formic acid through the creation of a C–H bond.

More recently, Gao et al. also studied the CO_2 hydrogenation to formic acid by DFT, on Pt_4 cluster doped single-vacancy graphene (Xu et al. 2020). Among the reactions mechanism studied, Langmuir-Hinshelwood, Eley-Rideal, and termolecular Eley-Rideal, the activation energy results also suggested that the most favored reaction path was the termolecular Eley-Rideal.

Also by DFT, Yodsin et al. (2019) investigated the platinum supported on defective carbon nanocones (Pt/dCNC) during CO_2 hydrogenation to formic acid. The calculations suggested that formic acid yield depended on the concentration of H_2, and that dissociation of H_2 followed by H spillover is more energetically favorable, while the desorption of formic acid is the rate-determining step.

Some investigations evaluated the potential of Cu single metal atoms on C_2N monolayer for the CO_2 hydrogenation to formic acid using first-principles calculations (Ma et al. 2019a). The results indicated that two main mechanisms can take place. While the path I start with the co-adsorption of H_2 and CO_2, path II is initiated by the adsorption of H_2. The latter path is more energetically favored and the reaction could even take place at room temperature.

Sredojević et al. (2018) also studied CO_2 hydrogenation to formic acid by DFT using Ru and Cu single atoms trapped in graphene sheets. The calculations showed that a high activation energy is required to transfer hydrogen atoms to CO_2. Alternatively, the pre-dissociation of H_2 may occur on the metal surface before CO_2 adsorption, this being preferred energetically for Ru catalyst. For this system, the formic acid desorption is the rate-limiting step. However, for the Cu-based catalyst the strong affinity of H atoms for defective graphene hinders the transfer of the H atom to CO_2.

Molybdenum carbide supported on carbon materials has also been tested in CO_2 hydrogenation to formic acid and methanol as a low-cost catalyst with similar properties to noble metals. Examples in the liquid and gas phases can be found. Both theoretical and experimental results showed that the use of nitrogen-doped carbon creates an interface between molybdenum carbide and the support which enhances the adsorption and activation of CO_2 and H_2 (Wang et al. 2020a). The best system of those reported that MoC@N5.6C provided a turnover frequency of 8.20 molFAmolMoC^{-1} h^{-1} at 140 °C and 20 bar, in the liquid phase hydrogenation.

More recently, Dongil et al. (2021) evaluated Cu-molybdenum carbide systems supported on commercial high surface area graphite. The authors reported the synthesis of molybdenum carbide through the carbothermal method using the support as the hydrogen source. The formation of both molybdenum oxycarbide and carbide was confirmed by X-ray absorption near-edge spectroscopy. Deuterium desorption thermal experiments also showed that molybdenum hydride species were formed. The formed hydride species were suggested to be the reason for transformations involving $MoO_xC_yH_z \leftrightarrow \beta\text{-}Mo_2C$. Also, the formation of metallic Mo when the synthesis was performed at 70 °C under H_2, resulting in less active and selective catalysts. The addition of copper improved the catalytic performance, acting as a promoter since monometallic Cu was not active in the studied system.

9.3.1.2 RWGS and Methanation (or Hydrogenation to $C2^+$)

Ni and Ru single metal atoms supported on carbon nanotubes have also been tested by Rivera-Cárcamo et al. (2020) for CO_2 hydrogenation at atmospheric pressure and 340 °C. The results indicated that while nickel-based catalysts were selective to CO, ruthenium-based catalysts were more selective to CH_4. In addition, for ruthenium catalysts a relationship between selectivity and electronic properties was observed, so that electron-rich particles were more selective to CH_4, while electron-deficient were more selective to CO.

The effect of nitrogen has been studied by some authors. For example, Liu et al. (2018b) used zeolitic imidazolate frame-works (ZIFs) as precursors to obtain N-doped carbon supports. Two catalysts were prepared by pyrolyzing Fe and Fe/K modified ZIF and they were employed on the hydrogenation of CO_2 at 320 °C and 30 bar with an H_2/CO_2 ratio of 3. Results were also compared with carbon spheres

(CS) and AC. While conversion was just slightly higher on the Fe-ZIF catalysts (26.8–29.3%) and higher for the K-doped Fe(K) (34%), the selectivity obtained with the Fe/Zn-NC and Fe(K)Zn-NC was remarkable, reaching 22.1–27.4% to C5+ (compared to just 0.4–5.4% obtained with CS and AC) and 30–32% to C_2–C_4 olefins (compared to <3% reached with CS and AC).

Potassium-doped iron catalysts supported on SWCNT and MWCNT were also tested for CO_2 hydrogenation to hydrocarbons at 613 K, 20 bar, and H_2/CO_2 = 3 (Wang et al. 2020b). The different results with both types of CNT highlight the importance of the electronic properties of carbon materials. The authors reported high selectivity to olefins (C5+) being 39.8% and 21.8% for SWCNT and MWCNT respectively. However, the different conversions reported should also be considered when comparing the data since the value obtained with SWCNT was also higher (52.7%) compared to MWCNT (43.6%). The authors also reported that FeK/MWNTs catalyst favored the formation of light olefins (C2–C4 = selectivity, 30.7%) and that FeK/SWNTs catalyst led to low concentration of the unwanted products CO and CH_4 which resulted in an outstanding yield to of heavy olefins of 27.6 μmolCO_2 gFe^{-1} s^{-1}. The better selectivity of FeK/SWCNTs to heavy olefins (C5+ selectivity = 39.8%) was explained by the effect of the higher electron concentration on the outer surface of SWCNT, larger proportion of Hägg carbide and the ability to interact strongly with light olefins. Apparently, the electron donation of SWCNT due to their larger curvature, given by the lower diameter of SWCNT compared to MWNTs, i.e., 0.47 versus 2.2 nm, favors the scission of the C–O bond, this also promoting the formation of carbon monomers and the optimal carbon to hydrogen surface ratio to obtain olefins.

On the other hand, Wu et al. (2018) have investigated FeK supported over novel honeycomb-structure graphene and have observed that the support was able to confine the Fe particle preventing them from agglomerates, while the mesoporous-microporous architecture could reduce effectively mass transfer limitation achieving 73 μ mol CO_2 g^{-1} s^{-1} yield of light olefins (C2–C4 selectivity = 59%).

N-doped graphene synthetized by pyrolysis in argon of chitosan at 900 °C proved to be active and selective to methane at temperatures of about 500 °C with estimated turnover frequencies of 73.17 s^{-1} (Jurca et al. 2018). The better catalytic performance of this system compared to other doped graphenes catalysts seem to be related to the presence of N-pyridinic atoms which are able to adsorb CO_2 forming carbamate adsorbed species.

Also, Wang et al. (2018a) studied Ni supported on nitrogen-doped carbon nanotubes for CO_2 methanation at atmospheric pressure and several temperatures. The catalyst offered around 80% conversion and 100% selectivity to CH_4 at 360 °C using an H_2:CO_2 ratio of 4 and proved to be stable for more than 24 h.

Li et al. (2018) studied the promotion of Ni/CNT catalysts with manganese on CO_2 methanation and the effect of oxygen surface functionalization. The experimental and theoretical results indicated that, as calcination temperature increased from 350 to 550 °C, the amount of oxygen groups and, hence, the strength of interaction with metallic nanoparticles decreased.

The increase of calcination temperature from 350 to 550 °C had a negative impact on activity and was even more pronounced on the selectivity to CH_4. The presence of a larger proportion of defects (corner, edge, or terrace) on the catalyst prepared at 550 °C was given as an explanation for the less significant effect on the activity since these sites contribute to the activity.

Also, the addition of Mn as a promoter improved activity due to the better CO_2 adsorption as also confirmed by the CO_2-temperature programmed desorption and transient response experiment. The best catalysts of those tested were those promoted with Mn and calcined at the lowest temperature, 350 °C, which also proved to be stable during 140 h of operation.

Nitrogen doping carbon materials have also been tested on CO_2 hydrogenation to CO and CH_4. Unfortunately, it is difficult to find a comparison with analogous undoped systems or different carbon nanostructures.

Sikora et al. (2018) synthetized bamboo-like carbon nanotubes (BCNTs) by catalytic chemical vapor deposition which were oxidized by acid treatment to prepare calcium alginate-gelled BCNT spheres. After thermal treatment, Pd, Rh, and Ni nanoparticles supported on BCNT spheres were tested in CO_2 hydrogenation at 200–600 °C. Pd supported BCNT was more selective to CO while Ni favored the formation of CH_4.

Cobalt (1.5–35 wt%) supported on different carbon structures, i.e., nanotubes, nanofibers, low-layered graphite (LGF) as well as carbon nanotube-Nb_2O_5 composites were evaluated and compared with Nb_2O_5 (Tursunov and Tilyabaev 2019). The CNT-Nb_2O_5 composites were synthetized by decomposing methane at 650 °C for 5 h in the presence of Nb_2O_5 which had been previously impregnated with a solution of cobalt nitrate. Overall, the catalysts are 100% selective to CH_4 and CO, the selectivity is dependent on the metal loading. At 3 wt.% Co loading CNT, CNF, and LGF were selective to CO reaching 78%, 23%, and 64%, respectively, while at higher loadings the catalysts offered exclusively CH_4. Also, using specific Co loadings on CNT (3 wt% of Co) and LGF (1.5 wt% of Co), resulted in inactive catalysts apparently due to the amorphous nature of the cobalt particles of those samples. The stability and catalytic performance of the catalysts could be enhanced by incorporation nitrogen atoms into graphene sheets through reaction with the oxygen surface groups of the carbon structures.

The hydrogenation of CO_2 has also been tested under supercritical conditions, enhancing mass transfer and H_2 solubility. For example, Bogdan et al. (2019) evaluated Co and Ni supported on CNT which showed great activity on CO_2 hydrogenation, cobalt being the most active and selective catalyst with 25–26% conversion and 100% selectivity to CH_4 at 300–450 °C. The authors also reported that under supercritical conditions the space–time yield to the product increases significantly (by 5–8 times) compared with the reaction in the gas phase.

Chernyak et al. (2020) used spark plasma sintered Fe/CNT catalysts and evaluated the effect of the sintering temperature on the structure and performance of the catalysts as well as the stability of the materials during supercritical conditions (85 bar and 350 °C) and different CO_2:H_2 ratio. Spark plasma sintering (SPS) is one of the techniques to obtain bulk CNT materials along with hot pressing or pressing with binder. SPS consists of applying high pressure and pulsed current, leading to fast sample heating (up to 1000 °C/min) which allows to preserve the CNT structure. In this work the synthesis starts by producing the multi-walled CNTs by chemical vapor deposition of hexane over the Co–Mo/MgO catalyst at 750 °C. Then a specific equipment for SPS is employed in which the samples were sintered under the pressure of 30 MPa at 800 and 1200 °C under vacuum for 5 min. The activity was similar at both tested sintering temperatures and similar to a catalyst prepared by impregnation and reduction in H_2. Also, the catalysts displayed a quite similar selectivity using a CO_2:H_2 ratio of 1:2, to CO (18–22%) CH_4 (34–36%), C_2–C_4 (33–35%), and C_5 + (9–12%).

The effect of nitrogen doping in CNT on the RWGS and FT synthesis has been reported by Williamson et al. (Williamson et al. 2019) and described that the C–N dipoles in the support when CO_2 and CO were adsorbed were more relevant than the particle-support interactions.

9.3.2 *Electrocatalytic CO_2 Reduction Using Carbon-Based Catalysts*

The CO_2RR has been attracting considerable attention over the last years as it provides a valuable pathway for renewable production of fuels as well as renewable energy storage. Additionally, it offers further advantages, like fully controlled electrochemical process by applying external electrolytes and potentials, the possibility to reduce the whole cost and the consumption of chemicals through the reuse of the electrolytes, and the operation under ambient conditions. In addition, the numerous possibilities for the selection of components of electrocatalytic reaction system combined with the multiple assemblage combinations and the fact that electrochemical setups are generally modular and compact and the scale-up can be simpler (Fernandes et al. 2019).

As for all reactions related to energy like oxygen and hydrogen reactions, here electrocatalysts (ECs) play a critical role in the electro-kinetics, mechanism, and product distribution. The major challenge is the design and successful preparation of highly efficient, low-cost, selective, and stable Ecs capable of addressing the kinetically slow CO_2 process. Various review papers have been published regarding the application of homogeneous and heterogeneous Ecs for the CO_2 electroreduction (Qiao et al. 2014; Kuhl et al. 2012; Feng et al. 2019; Liu et al. 2018a; Zhu et al. 2016; Wang et al. 2017; Xie et al. 2018; Khezri et al. 2017; Abdelkader-Fernandez et al. 2020; Fernandes et al. 2019). Exhaustive reviews addressing the electroreduction of CO_2 from a mechanistic/theoretical perspective have been published recently (Kibria et al. 2019; Handoko et al. 2018; Zhang et al. 2017), even though the CO_2RR complexity involving multiple possible pathways surely will continue to stimulate scientists to look for further insights on this topic.

Nonetheless, here we present an up-to-date review of the recent advances in the last 5 years involving carbon-based materials for the electrochemical transformation of CO_2 into selected fuels (methane, methanol, ethanol, and syngas).

Carbon-based materials are interesting materials for electrocatalysis, mainly owing to their mechanical and physicochemical properties which include electron conductivity, thermal stability, large surface area, and corrosion resistance. Owing to its surface chemistries and tailorable porous structures apart from the properties already referred to above, carbon-based materials have been largely applied both as a support for metal-based electrocatalysts as well as versatile heterogeneous metal-free electrocatalysts itself. Carbon-based materials' electrocatalytic activity is intrinsically related to their structures, chemical surfaces, and defects. Porous carbons display large porous structures, leading to a high transfer of mass and larger surface areas with a greater concentration of active sites per mass of catalyst. Additionally, the incorporation of heteroatoms, in particular nitrogen, portrays a valuable functionalization at a molecular level modulating the electronic and physicochemical for targeted purposes like CO_2 electroreduction.

9.3.2.1 About CO_2 Reduction Reaction (CO_2RR)

The CO_2 molecule is a chemically inert and thermodynamically stable molecule that presents low electron affinity and consists of 2 C=O bonds, each with a bond dissociation energy of 750 kJ mol^{-1} (Ma et al. 2019c). Considering other bond dissociation energy values namely of C–C (336 kJ mol^{-1}), C–H (441 kJ mol^{-1}), and C–O (327 kJ mol^{-1}) is clear why CO_2RR reduction is more difficult to occur compared to the reduction of other organic molecules (Fan et al. 2018; Ma et al. 2019c). In addition, the CO_2RR is more challenging when the electrolyte is water due to the fact that CO_2 is weakly dissolved ($\approx$ 0.33 mol dm^{-3} at 25 °C, 1 atm), (Fan et al. 2018) in acidic and neutral solutions and due to possible occurrence of the hydrogen evolution reaction (HER).

The CO_2RR happens at electrode/electrolyte boundaries and the process involves three main steps. The first comprises the CO_2 chemical adsorption on the surface of the electrocatalyst, and the second is the electron transfer and/or proton migration to break C–O bonds and/or form C–H bonds. The third and last one involves the rearrangement of product species, their desorption from the electrode, and diffusion into the electrolyte.

CO_2RR is a thermodynamically challenging process but it also presents innumerable kinetic problems. Generally, the CO_2RR proceeds thru distinct multi-step paths with the involvement of electrons (2–18) resulting in various products. Parameters like the type of ECs, electrolytes, pH, pressure, temperature, and CO_2 concentration greatly influence the number and type of paths (Mao and Hatton 2015). Various reduction products can be formed such as formic acid (HCOOH, acid medium), carbon monoxide (CO), formate ($HCOO^-$, basic medium), methane (CH_4), methanol (CH_3OH), ethanol (C_2H_5OH), formaldehyde (CH_2O), oxalic acid ($H_2C_2O_4$, acid medium), oxalate ($C_2O_4^{2-}$, basic medium), and ethylene (C_2H_4) among others.

9.3.2.2 Production of Methane

Currently, few examples can be found in the literature regarding electrocatalysts that can maintain high and stable methane selectivity. Among them, transition metals, particularly Cu, exhibit the most excellent catalytic performance on the selectivity of methane (Guo 2021). The protonation of adsorbed CO is the most decisive step considering the overpotential. Therefore, strong bonding with CHO or COH intermediate is critical for the electrocatalyst surface to increase the potential of the protonate CO. Additionally, the competition of HER is a tricky problem due to the similarity of standard potential.

Sun et al. (2016) reported the CO_2RR to methane by N-doped carbon (graphene-like) material (NGMs) coated onto carbon paper electrodes and using ionic liquids as the electrolyte. The authors used distinct bases containing nitrogen (3-pyridinecarbonitrile, 3-hydroxypyridine, 4-dimethylaminopyridine, benzimidazole, and 1-vinylimidazole) to prepare them, which resulted in different percentages of nitrogen species ranging from 4.8 to 1.8%. The $FE_{methane}$ decreased from 93.5% to 20.8% from 3-pyridinecarbonitrile to 1-vinylimidazole. According to the authors, the active sites were pyridinic-N and pyridinic-/pyrrolic N species. The strong interaction between the electrode and the adsorbed CO prevents its escape, facilitating the hydrogenation reaction to yield CH_4. Wang et al. (2018b) developed perfluorinated covalent triazine framework (CTF) derived hybrids and applied them as CO_2RR electrocatalysts. Their studies suggested that the edge-gN and edge-2gN on the CTF hybrid acted as the electrocatalytic sites to obtain CH_4 with a FE of 99.3%. When both F and N-dopants were introduced, F regulated the N to enhance the adsorption of the key intermediate $^*CO$ and $^*CH_2$.

Following their primary studies on CO_2RR on N-doped graphene quantum dots (NGQDs) (Wu et al. 2016), Yakobson and co-workers reported a vast work on computational simulations to identify the underlying mechanisms regulating the process of CO_2RR using this type of materials (Zou et al. 2017). In previous works, the same authors have shown that NGQDs are able to convert electrochemically the CO_2 molecule into different hydrocarbons and oxygenates, like CH_4, C_2H_4, and C_2H_5OH, making them valuable alternatives to Cu for CO_2RR (Wu et al. 2016). The reduction of CO_2–CO is promoted by the N-doped edges on NGQDs which results in the enhancement of their bonding with *COOH (Zou et al. 2017).

9.3.2.3 Production of Methanol

Methanol generation using CO_2RR is tricky due to its 6e- and $6H^+$ transfer processes which require higher selectivity (Fernandes et al. 2019). First, CO_2 is reduced to CO which is adsorbed on the electrocatalyst surface then, the CO is then reduced to CHO^*, CH_2O^*, and CH_3O^*. Finally, the products are eventually desorbed from the electrocatalysts surface to produce methanol (Liang et al. 2021; Yang et al. 2019; Zhao and Lu 2019). Theoretical calculations have suggested that surface geometry could modulate the total energy of metal-free sites via strain effects (Zhou et al.

2019), thus possibly tuning the production of methanol by CO_2RR on curvature carbon-based catalyst surfaces. In fact, Chai and Guo (2016) demonstrated that N-doped CNT with a certain degree of curvature (6.0) could produce methanol whereas the graphene electrocatalysts without curvature produced CO/HCOOH.

Mou et al. (2019) suggested that an increasing difficulty in CH_2O and CO production would lead to CH_3OH production using boron phosphide nanoparticles as electrocatalysts. This catalyst achieved a high $FE_{methanol}$ of 92.0% at −0.5 V versus RHE. DFT calculations disclosed that P and B promote the binding and activation of CO_2 molecules, and the $^*CO + {}^*OH$ to $^*CO + {}^*H_2O$ process is the rate-determining step ($\Delta G = 1.36$ eV). Additionally, CO and CH_2O products are not easily generated on the boron phosphide (111) surface, which is responsible for the high selectivity and activity for methanol production. These studies emphasized that the production of methanol depends on the chemical state of the reactant and on the electrocatalyst surface properties. A proper surface curvature combined with an optimized heteroatom doping configuration will be favorable as the metal-free site for CO_2RR–CH_3OH conversion. Partially oxidized Co nanoparticles on a single-layer nitrogen-doped graphene were reported by Wang and co-workers (Huang et al. 2018a) with a $FE_{methanol}$ of 71.4% at −0.90 V versus. SCE with a current density of 4 mA cm^{-2} and high yield of 1.10 mmol dm^{-3} h^{-1}. Higher faradaic efficiency of 97.0% at −0.98 V versus. SCE was obtained over urchin-like $Co(CO_3)_{0.5}(OH){\cdot}0.11H_2O$ (Huang et al. 2018b). Yang et al. (2019) described the preparation and application of CO_2RR–CH_3OH electrocatalyst of N-doped carbon nanofibers involving Cu single atoms. The electrocatalyst was produced by a scalable 3-step strategy involving the pyrolysis of polyacrylonitrile (PAN) fibers-integrated Cu^{2+}/ZIF-8 nanoparticles. To guarantee only the presence of Cu single atoms (Cu–N_4 moieties) as the main active sites, the material was washed with H_2SO_4 solution after carbonization. This step was important to remove residual Zn atoms and Cu aggregates. The final electrocatalyst was capable of producing methanol with an outstanding FE of 44% at −0.9 V versus RHE. Another MOF-derived carbon obtained thru the carbonization of HKUST-1 ($[Cu_3(BTC)_2(H_2O)_3]_n$, BTC = trimesic acid) was applied as CO_2RR electrocatalyst with high methanol selectivity (Zhao et al. 2017). The oxidation degree and size of carbon-embedded Cu-containing NPs were controlled using different pyrolysis temperatures (900, 1000, and 1100 °C). The authors noticed that the carbon produced at 1000 °C, exhibited the highest proportion of Cu^+ in form of Cu_2O, the lowest average NP size, and the optimal relation between charge and mass transfer resistances, showing the best selectivity toward methanol with FE = 43.2% at −0.3 V versus. RHE and even formed ethanol ($FE_{ethanol}$ = 34.8% at −0.5 V vs. RHE).

9.3.2.4 Production of Ethanol

The reduction of CO_2 to ethanol is highly desirable due to its considerable industrial applications but its 12-electron process, with the equilibrium potential of 0.084 V versus. RHE requires higher selectivity to be achieved (Mohamed et al. 2020). Similar to other multi-carbon (C^{2+}) products, the majority of electrocatalysts reported for the production of C_2H_5OH are based on copper since Cu metal allows the dimerization of CO, which is a vital step for the production of molecules with more than one carbon. However, only a small number of examples can be found in the literature regarding electrocatalysts that are able to produce ethanol with FE > 30%, as other competing C^{2+} products, especially ethylene, are also formed (Hoang et al. 2018; Chen et al. 2018a; Mi et al. 2019; Kim et al. 2021).

Non-metal electrocatalysts have also been reported, with outstanding selectivity for C_2H_5OH production (FE > 90%) when a B and N co-doped nanodiamond (BND) was used (Liu et al. 2017b). B and N-dopants induced reactive sites by regulating the electronic state of the BND. Moreover, the B or N-doped BND presented higher overpotential for HER than graphitic carbon materials, so the dopant-induced active sites can favor CO_2RR over HER in a broader potential window. However, this high selectivity was achieved at a low current density of less than 2 mA cm^{-2} which limits its practical applications. Lu and co-workers (Yuan et al. 2018) have also reported the application of several functionalized graphene oxide (GO) surfaces with different pyridine derivatives (pyridoxine, 4-hydroxypyridine, 4-aminopyridine, 8-hydroxyquionoline, and 5-amino-1,10-phenanthroline) as CO_2RR electrocatalysts. The best results were achieved with GO sheets modified with pyridoxine (GO-VB6) with a 2.32% N-pyridinic content. The GO-VB6 showed optimum CO_2 electro-catalytic activity with an overall faradaic efficiency of 45.8% and a $FE_{ethanol}$ of 37%.

Nitrogen-doped ordered cylindrical mesoporous carbon (c-NC) has also been reported by Song et al. (2017) as a robust metal-free EC for CO_2RR to ethanol. The c-NC (Fig. 9.6a) was prepared by a soft template method via the self-assembly of resol (C precursor), F127 (soft template), and dicyandiamide (N precursor). The c-NC electrocatalyst has distinctive cylindrical channels embedded in the carbon bulk. The authors also prepared the inverse mesoporous N-doped carbon (i-NC) that preserves similar pore parameters and tested it in the same conditions to highlight the advantage of the cylindrical mesoporous structure. It is important to refer that the N content of both materials was strictly controlled to be the same. The c-NC showed 100% selectivity and a high faradaic efficiency of 77% at −0.56 V versus RHE (Fig. 9.6). The HER side reaction was greatly hindered, and CO generation was completely suppressed during CO_2RR in an aqueous solution. Both experimental and theoretical studies demonstrated that the dimerization of key CO^* intermediates and the subsequent proton–electron transfers that led to the superior electrocatalytic performance was facilitated by the synergetic effect of the nitrogen heteroatoms and the cylindrical channel.

Later, the same group reported a hierarchical medium micropore-embedded ordered cylindrical nitrogen-doped mesoporous carbon (MNC) prepared by a similar method (Song et al. 2020). The desolvation induced by the medium micropores uniformly embedded in the mesopore channel wall, led to the accumulation of electrolyte ions, and enabled high local electric potentials. Consequently, the activation of CO_2 molecules and the C–C coupling of the key intermediates were facilitated by surface-active pyridinic and pyrrolic N sites with high electron density. As a result,

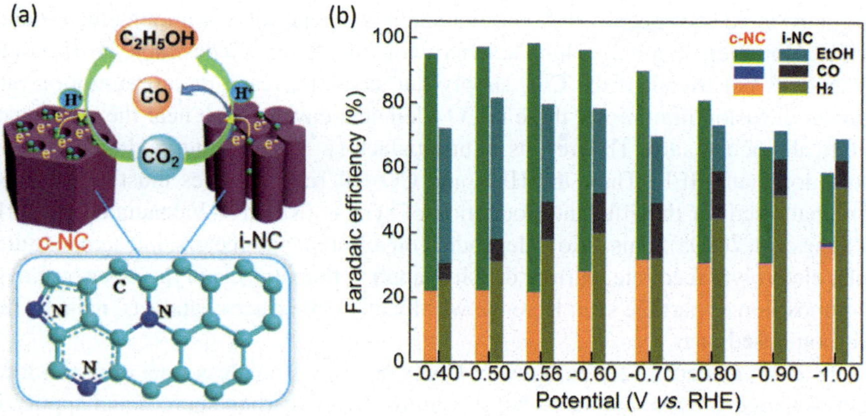

Fig. 9.6 Illustration of c-NC and i-NC for CO_2 electroreduction (**a**) and FEs of CO_2 electroreduction products over c-NC and i-NC electrocatalysts at various applied potentials (**b**). Reproduced with permission from (Song et al. 2017). Copyright 2017 Wiley-VCH

an ethanol production rate as high as 2.3 mmol $gcat^{-1}$ h^{-1} was achieved as well as a FE of 78%.

Recently, Cu single atoms/nanoparticles/clusters have been combined with various carbon matrixes to prepare CO_2RR electrocatalysts. The most commonly used strategy is to use N-doped carbon materials to improve the interactions with the metal ECs. For example, Karapinar et al. (2019) showed that Cu0.5NC produced ethanol with high selectivity of 43% and with a current density of 16.2 mA cm^{-2} at −1.2 V versus RHE if the single-atom Cu sites on N-doped graphene were reversibly converted into Cu nanoparticles. Chen et al. (2020a) have also observed that N-doped graphene quantum dots modified with Cu nanorods (Cu-nr) (NGQ/Cu-nr composite) were able to produce C^{2+} alcohols with FE of 52.4% a partial current density of 147.8 mA cm^{-2} at −0.9 V versus RHE (Chen et al. 2020a). Zhan et al. (2020b) have also reported the CO_2 conversion to ethanol using Cu_2O/Cu nanoparticles supported on vertically ZIF-L coated N-doped graphene. The authors grew the 2D pseudopolymorph of ZIF-8 (ZIF-L), on graphene sheets, introduced a Cu precursor on this nanocomposite and finally, pyrolyzed for 2 h at 1000 °C. The final electrocatalysts resulted in a $FE_{ethanol}$ of 70.52% at −0.87 V versus RHE.

9.3.2.5 Production of Syngas

Syngas (synthesis gas), a highly desirable feedstock for the production of value-added chemicals by post-thermochemical processes, is a gas mixture of carbon monoxide (CO) and hydrogen (H_2). Since the CO_2RR is usually performed in aqueous media, the hydrogen evolution reaction (HER) from the reduction of water or protons (H^+) is a competitive reaction of CO_2RR (Hernandez et al. 2017). A considerable number of attempts have been conducted to design electrocatalysts able to produce syngas with a controllable and specific H_2/CO ratio which is crucial for maximizing the product yield. However, obtaining a high activity while maintaining a desirable H_2/CO ratio (0.5–2) is not trivial. If the CO_2 supply rate does not reach the consumption rate due to diffusion limitations, there is CO_2 deficient environment near the electrocatalytically active sites. This results in unsatisfactory $^*COOH$ intermediates, leading to a dominant HER. Thus, the HER and CO_2RR reaction rates must be adjusted and balanced for the efficient production of syngas (Ramirez-Valencia et al. 2021; Zhang et al. 2020a). This ratio is dependent on different parameters, like temperature, pH, electrolyte feed rate, current density (applied potential), and the electrocatalyst composition or particle size. Here we will focus on the electrocatalysts, namely, the carbon-based ones.

In 2017, Liu et al. (2017a) grew N-CNTs in-situ on stainless steel under an inert atmosphere, using melamine as both the nitrogen and carbon sources and applied it as an electrocatalyst for the conversion of CO_2 into syngas. The integrated electrode displayed efficient, robust, and controllable clean syngas production with a FE_{CO} of 75%. More importantly, the authors were able to tailor the H_2/CO ratio in the clean syngas products in the range between 1:3 and 3:1 by tuning the applied potential or the pyrolysis temperature. Nitrogen-doped tubular carbon foam electrodes were reported by Li et al. (2019) as both, gas diffusers and self-supported electrocatalysts for CO_2 transformation into syngas. The monolithic tubular electrodes were capable of producing syngas over a wide potential range (−0.5 to −1.3 V vs. RHE) and its ratio was tunable from 1:3 to 2:1 through the control of the N defect enrichment being independent of the applied potential. Metal-free porous N-doped carbon prepared using melamine and pentaerythritol as the N and C sources, respectively have also been successfully applied for the production of syngas through CO_2RR showing good selectivity ($FE_{CO} = 74\%$, $FE_{H2} = 18\%$, at −0.6 V) (Chen et al. 2021). More importantly, the authors were able to adjust the syngas ratio from 0.24 to 5.25 with good product selectivity by varying the applied potential from −0.4 to −1.0 V.

Apart from heteroatom-doped a considerable number of noble metal-free carbon-based electrocatalysts have been reported for the syngas production from CO_2RR in the last 5 years. For example, Zhao et al. (2020) reported the generation of syngas through CO_2RR for a sequence of Fe–N–C ECs. These were prepared by calcination at high temperature utilizing transition metal Fe incorporated into urea-formaldehyde resin precursor. The best Fe–N–C electrocatalyst (3%, 950 °C) exhibited a FE close to 100% with no liquid phase product with an attractive activity of CO and H_2 generation ($FE_{CO} = 74\%$, $FE_{H2} = 25\%$) at a low overpotential of −0.6 V. More importantly, the syngas (H_2/CO) ratio from 4:1 to 1:3 can be controlled by adjusting the applied potential with good long-term stability. Zhang et al. (2020a) prepared a series of Ni_xFe_{1-x}–NC single-atom catalysts (SACs), which enabled syngas production with a wide range of CO/H_2 ratio (0.14–10.86). The authors concluded that Fe sites play a crucial role in HER reaction, while the Ni sites in the CO_2RR and that the fine tuning of the Ni/Fe ratio they were able to control the product composition in a large window from high CO end to high H_2 end.

Recently, Atanassov and co-workers (Delafontaine et al. 2020) published an excellent review on the metal-nitrogen-carbon EC for CO_2 reduction toward syngas generation. The authors highlighted the high potential of MNC electrocatalysts due to (i) their versatility and tunable selectivity toward H_2 and CO generation by applying distinct moieties and metal elements, and (ii) their carbonaceous nature which allows a high degree of control of key parameters like hydrophilicity and porosity distributions. Furthermore, they pointed out the durability/robustness under operation as an important aspect of real-like implementation.

9.3.3 Economic Analysis of CO_2-Derived Fuels

Given the early stage of the CO_2 to fuels technology, it is difficult to estimate the future market. Nevertheless, the market will remain low in the short term.

In order to use CO_2, it has to be captured, purified, and transported. Hence, the cost of CO_2-derived fuels will depend on the price of the energy, the technology, as well as that of CO_2 and H_2 and other required materials such as catalysts. Also, while CO_2 supply shortages can increase the costs of CO_2, carbon taxes can encourage the producers of CO_2 to use conversion technologies.

The prices of CO2 capture and purification are different depending on the source of CO_2, fluctuating from 15 to 60 \$/tCO_2 for concentrated CO_2 streams, 40–80 \$/tCO_2 for coal and gas-fired power plants, or above 100 \$/tCO_2 for dilute point sources (e.g., industrial furnaces). Obviously, direct CO_2 capture from the air is the most expensive method, ranging from 94 to 232 \$/tCO_2, due to the required energy to concentrate the stream sources (Keith et al. 2018). Nevertheless, technical developments are expected to reduce costs.

It is also intuitive that the proximity between the source of CO_2 and the point of use of the derived fuels would drastically decrease the costs, however, this is not possible in every situation.

Some techno-economic analyses have been reported. For example, it is estimated that the production costs of methanol and methane from CO_2 are 2–7 times higher compared to fossil sources, electricity being the main associated cost, which represents 40–70% of the total expenses (Agency 2019). Under these conditions, the production of methanol and its derivatives from CO_2 hydrogenation might be realized in areas with low-cost renewable energy and available CO_2, such as in North Africa, Chile, or Iceland where there is already a facility, George Olah, that processes over 5 k ton of CO_2 per year into methanol employing H_2 obtained using renewable electricity.

There have been some recent publications that evaluate the costs of producing gasoline from CO_2. For example, Caballero et al. simulated a capture and one-stage conversion system for CO_2 through Fischer Tropsch (Fernández-Torres et al. 2022). The authors considered different carbon capture (either directly from the air or from other concentrated sources) as well as different hydrogen sources from methane reforming to that renewable wind or solar. The associated costs of the produced gasoline varied from 1.8 to 2 $/l for gas natural derived H_2 to 6–8 $/l for electrolysis derived H_2 using wind or solar renewable energy.

Also, Marchese et al. (2021) evaluated a system consisting of direct air capture and then the conversion of CO_2 through the Fischer-Tropsch mechanism. The authors considered different scenarios where the energy was supplied either by natural gas or renewable sources. The estimated costs were 5–6.3 €/kg of wax depending on the interests rate.

It is expected that production costs of CO_2-derived fuels will decrease in the future mainly due to the availability of low-cost renewable electricity. Nevertheless, in the near future CO_2-derived methane and CO_2-derived liquid fuels will be too expensive unless public taxes are applied. On the contrary CO_2-derived methanol may be competitive depending on the region (Marchese et al. 2021). The main advantages are the climate benefits, hence the price can be linked to it provided public policies are set.

9.4 Conclusion, Future Challenges, and Perspectives

The current challenges and research trends toward efficient carbon-based catalysts for renewable fuels production from biomass/CO_2 were addressed in this review aiming to inspire the broad community of scientists working toward a low-carbon society where biomass renewable resources are the best choice. This review showed the potential of biomass (bio-oil upgrading) and CO_2 conversion as a very promising strategy to produce sustainable and cleaner fuels to reduce the massive dependence on fuel from fossil resources. From this review paper it is possible to conclude that:

- In terms of the use of carbon-based catalysts, the published studies use commonly activated carbon and biochar as support. These carbon materials and specially biochar are easily tunable allowing materials with different physical and chemical properties and exceptional pore structure and specific surface area which results in catalyst better dispersion. Carbon nanotubes and graphene are also interesting materials due to their unique structures (nanomaterials) providing a larger reaction area and excellent electrical and thermal conductivity. In the future, the application of modification methods will improve even more the pore structure and mechanical strength of carbon-based catalysts and their ability to gradually become the protagonist of biomass and CO_2 conversion into fuels. Low-cost carbon supports are key factors to the potential of industrial scale bio-oil/CO_2-based energy in the future. The mechanical strength and the regeneration and recycling capabilities of the catalysts are fundamental requirements for industrialization.
- N-dopants on carbon matrix can be assumed as crucial anchoring sites or defects for particle formation and size reduction.
- HDO is one of the most extensively studied processes for the catalytic upgrading of bio-oil to remove oxygen containing functional groups improving the bio-oil quality including heating value and process efficiency. The development of

cost-competitive non-noble metal catalysts, with high selectivity for C–O bonds cleavage preserving simultaneously the C–C bonds at low H_2 pressures, remains a major challenge. Therefore, novel catalysts, sustainable solvents, and especially the adoption of an "H_2-free" HDO process are key factors to overcome the economic barriers of a potential commercial scale. Future work should also focus on improving the understanding of bio-oil HDO reaction pathways. The commercialization of bio-oil is not yet viable but can be in the near future, after reducing the upgrading costs of bio-oil. Furthermore, some studies have prompted HDO as a versatile route to produce jet fuel hydrocarbons from bio-oil. The development of catalysts to efficiently eliminate oxygen under low levels of H_2/mild conditions avoiding coke formation/catalyst deactivation is crucial to the sustainability of bio-jet fuel production from bio-oil.

- The steam reforming of bio-oil is an environmentally friendly route to produce H_2 or syngas and can be considered an important process in the future although the major challenge of these processes is catalyst deactivation. Carbon materials have been shown to efficiently improve metal dispersion, and water resistance, and limit metal sintering enhancing H_2 production and/or catalyst stability from the bio-oil fraction depending on the process employed.
- CO_2 thermochemical and electrochemical reduction has not only an economic potential due to low-cost feedstocks but also contributes to the mitigation of CO_2 emissions through its utilization. The catalyst and the support play an important role in this conversion. Carbon-based materials as a catalyst or catalyst support showed to be efficient demonstrating that: the N-doped carbon materials favor the CO_2 activation and the electronic properties of graphenic structures may influence by changing the electronic properties of the metal-nanoparticles promoting hydrogen transfer. While there are several works on CO_2 hydrogenation to methanol and hydrocarbons, there is still much work to do on the transformation to formic acid which is relevant to be coupled with intermittent renewable energy generation sites. In a similar way, the CO_2 transformation through electrochemical processes is favored by the presence of nitrogen in the electrocatalysts structure regardless of the presence of metals. Results have shown that N-doping boosts the electrocatalytic activity and that most likely the pyridinic-N sites represent a key factor for this enhanced activity.
- Despite significant advances in CO_2 and biomass conversion into fuels, the low yields, the complexity (multi-step; difficulties in purification) of the processes, the high costs, and the catalysts' deactivation are hurdling the large-scale production and the competitiveness of a CO_2/biomass biorefinery.
- From an overall view of these biomass/CO_2 conversion methodologies, it is clear that a cost-effective and efficient catalytic process has not been developed so far, so there are several opportunities for researchers to develop new catalysts and catalysts supports, design reactors, and find alternative reaction conditions (single-step catalytic conversions) to produce bio-oil biofuels from biomass and CO_2.

- Life cycle assessment of bio-oil production and upgrading technologies and CO_2 (electro)catalytic conversion is essential to bring the vision of the sustainability of the processes and target products at an industry level.

Acknowledgements A. F. P., D. F., and C. F. thanks Fundação para a Ciência e a Tecnologia (FCT/MCTES) funding through the projects UIDB/50006/2020, UIDP/50006/2020 and EXPL/BII-BIO/0436/2021. A.F.P. and D. F. also thanks the work contract under the Scientific Employment Stimulus (2020.01614.CEECIND/CP1596/CT0007 and 2021.00771.CEECIND/CP1662/CT0007). A.B.D. acknowledges financial support from Fundación General CSIC (Programa ComFuturo and iLink project N° 20211 from CSIC (Spain)). E.B acknowledges financial support from ANID-Millennium Science Initiative Program-NCN2021_090.

References

Abdelkader-Fernandez VK, Fernandes DM, Freire C (2020) Carbon-based electrocatalysts for CO_2 electroreduction produced via MOF, biomass, and other precursors carbonization: a review. J Co_2 Utiliz 42:18

Agency IE (2019); IEA (2019) Putting CO_2 to use, IEA, Paris. https://www.iea.org/reports/putting-co2-to-use

Amenaghawon AN, Anyalewechi CL, Okieimen CO, Kusuma HS (2021) Biomass pyrolysis technologies for value-added products: a state-of-the-art review. Environ Dev Sustain

Arora S, Gupta N, Singh V (2020) Improved Pd/Ru metal supported graphene oxide nano-catalysts for hydrodeoxygenation (HDO) of vanillyl alcohol, vanillin and lignin. Green Chem 22:2018–2027

Attia M, Farag S, Chaouki J (2020) Upgrading of oils from biomass and waste: catalytic hydrodeoxygenation. Catalysts 10:1381

Augusto BL, Ribeiro MC, Aires FJCS, da Silva VT, Noronha FB (2020) Hydrogen production by the steam reforming of ethanol over cobalt catalysts supported on different carbon nanostructures. Catal Today 344:66–74

Benzigar MR, Talapaneni SN, Joseph S, Ramadass K, Singh G, Scaranto J, Ravon U, Al-Bahily K, Vinu A (2018) Recent advances in functionalized micro and mesoporous carbon materials: synthesis and applications. Chem Soc Rev 47:2680–2721

Bjelic A, Grilc M, Hus M, Likozar B (2019) Hydrogenation and hydrodeoxygenation of aromatic lignin monomers over Cu/C, Ni/C, Pd/C, Pt/C, Rh/C and Ru/C catalysts: mechanisms, reaction micro-kinetic modelling and quantitative structure-activity relationships. Chem Eng J 359:305–320

Blanco E, Dongil AB, Escalona N (2020a) Synergy between Ni and Co nanoparticles supported on carbon in guaiacol conversion. Nanomaterials 10:2199

Blanco E, Dongil AB, García-Fierro JL, Escalona N (2020b) Insights in supported rhenium carbide catalysts for hydroconversion of lignin-derived compounds. Appl Catal A 599:117600

Blanco E, Sepulveda C, Cruces K, García-Fierro JL, Ghampson IT, Escalona N (2020c) Conversion of guaiacol over metal carbides supported on activated carbon catalysts. Catal Today 356:376–383

Blanco E, Díaz de León JN, García-Fierro JL, Escalona N (2021) Study of supported bimetallic MoRe carbides catalysts for Guaiacol conversion. Catal Today 367:290–296

Bogdan VI, Pokusaeva YA, Koklin AE, Savilov SV, Chernyak SA, Lunin VV, Kustov LM (2019) Carbon dioxide reduction with hydrogen on carbon-nanotube-supported catalysts under supercritical conditions. Energy Technol 7

Cai Z, Wang FM, Zhang XB, Ahishakiye R, Xie Y, Shen Y (2017) Selective hydrodeoxygenation of guaiacol to phenolics over activated carbon supported molybdenum catalysts. Mol Catal 441:28–34

Carrasco JL, Gunukula S, Boateng AA, Mullen CA, Desisto WJ, Wheeler MC (2017) Pyrolysis of forest residues: an approach to techno-economics for bio-fuel production. Fuel 193:477–484

Chai GL, Guo ZX (2016) Highly effective sites and selectivity of nitrogen-doped graphene/CNT catalysts for CO_2 electrochemical reduction. Chem Sci 7:1268–1275

Chen CJ, Sun XF, Lu L, Yang DX, Ma J, Zhu QG, Qian QL, Han BX (2018a) Efficient electroreduction of CO_2–C_2 products over B-doped oxide-derived copper. Green Chem 20:4579–4583

Chen J, Wang M, Wang S, Li X (2018b) Hydrogen production via steam reforming of acetic acid over biochar-supported nickel catalysts. Int J Hydrogen Energy 43:18160–18168

Chen D, Wang W, Liu C (2019) Ni-encapsulated graphene chainmail catalyst for ethanol steam reforming. Int J Hydrog Energy 44:6560–6572

Chen CJ, Yan XP, Liu SJ, Wu YH, Wan Q, Sun XF, Zhu QG, Liu HZ, Ma J, Zheng LR, Wu HH, Han BX (2020a) Highly efficient electroreduction of CO_2 to C2+ alcohols on heterogeneous dual active sites. Angew Chem-Int Edn 59:16459–16464

Chen D, Wang W, Liu C (2020b) Hydrogen production through glycerol steam reforming over beehive-biomimetic graphene-encapsulated nickel catalysts. Renew Energy 145:2647–2657

Chen KY, Deng J, Zhao J, Liu X, Imhanria S, Wang W (2021) Electrocatalytic production of tunable syngas from CO_2 via a metal-free porous nitrogen-doped carbon. Ind Eng Chem Res 60:7739–7745

Chernyak SA, Ivanov AS, Stolbov DN, Maksimov SV, Maslakov KI, Chernavskii PA, Pokusaeva YA, Koklin AE, Bogdan VI, Savilov SV (2020) Sintered Fe/CNT framework catalysts for CO_2 hydrogenation into hydrocarbons. Carbon 168:475–484

Cordero-Lanzac T, Palos R, Arandes JM, Castano P, Rodriguez-Mirasol J, Cordero T, Bilbao J (2017) Stability of an acid activated carbon based bifunctional catalyst for the raw bio-oil hydrodeoxygenation. Appl Catal B-Environ 203:389–399

Cordero-Lanzac T, Rodríguez-Mirasol J, Cordero T, Bilbao J (2021) Advances and challenges in the valorization of bio-oil: hydrodeoxygenation using carbon-supported catalysts. Energy Fuels 35:17008–17031

Cordero-Lanzac T, Hita I, Garcia-Mateos FJ, Castano P, Rodriguez-Mirasol J, Cordero T, Bilbao J (2020) Adaptable kinetic model for the transient and pseudo-steady states in the hydrodeoxygenation of raw bio-oil. Chem Eng J 400

da Silva ALM, den Breejen JP, Mattos LV, Bitter JH, de Jong KP, Noronha FB (2014) Cobalt particle size effects on catalytic performance for ethanol steam reforming—Smaller is better. J Catal 318:67–74

Davda RR, Shabaker JW, Huber GW, Cortright RD, Dumesic JA (2005) A review of catalytic issues and process conditions for renewable hydrogen and alkanes by aqueous-phase reforming of oxygenated hydrocarbons over supported metal catalysts. Appl Catal B 56:171–186

De S, Saha B, Luque R (2015) Hydrodeoxygenation processes: advances on catalytic transformations of biomass-derived platform chemicals into hydrocarbon fuels. Biores Technol 178:108–118

De S, Dokania A, Ramirez A, Gascon J (2020) Advances in the design of heterogeneous catalysts and thermocatalytic processes for CO_2 utilization. ACS Catal 10:14147–14185

Deerattrakul V, Limphirat W, Kongkachuichay P (2017) Influence of reduction time of catalyst on methanol synthesis via CO 2 hydrogenation using Cu–Zn/N-rGO investigated by in situ XANES. J Taiwan Inst Chem Eng 80:495–502

Deerattrakul V, Puengampholsrisook P, Limphirat W, Kongkachuichay P (2018) Characterization of supported Cu-Zn/graphene aerogel catalyst for direct CO_2 hydrogenation to methanol: effect of hydrothermal temperature on graphene aerogel synthesis. Catal Today 314:154–163

Deerattrakul V, Yigit N, Rupprechter G, Kongkachuichay P (2019) The roles of nitrogen species on graphene aerogel supported Cu–Zn as efficient catalysts for CO_2 hydrogenation to methanol. Appl Catal A 580:46–52

Delafontaine L, Asset T, Atanassov P (2020) Metal-nitrogen-carbon electrocatalysts for CO_2 Reduction towards syngas generation. Chemsuschem 13:1688–1698

Dhyani V, Bhaskar T (2018) A comprehensive review on the pyrolysis of lignocellulosic biomass. Renew Energy 129:695–716

De Luna P, Hahn C, Higgins D, Jaffer SA, Jaramillo TF, Sargent EH (2019) What would it take for renewably powered electrosynthesis to displace petrochemical processes? Science 364:eaav3506

di Stasi C, Cortese M, Greco G, Renda S, González B, Palma V, Manyà JJ (2021) Optimization of the operating conditions for steam reforming of slow pyrolysis oil over an activated biochar-supported Ni–Co catalyst. Int J Hydrog Energy 46:26915–26929

Din IU, Shaharun MS, Naeem A, Tasleem S, Ahmad P (2019) Revalorization of CO_2 for methanol production via ZnO promoted carbon nanofibers based Cu–ZrO_2 catalytic hydrogenation. J Energy Chem 39:68–76

Din IU, Shaharun MS, Naeem A, Tasleem S, Rafie Johan M (2018) Carbon nanofibers based copper/zirconia catalysts for carbon dioxide hydrogenation to methanol: effect of copper concentration. Chem Eng J 334:619–629

Dongil AB, Conesa JM, Pastor-Pérez L, Sepúlveda-Escribano A, Guerrero-Ruiz A, Rodríguez-Ramos I (2021) Carbothermally generated copper–molybdenum carbide supported on graphite for the CO_2 hydrogenation to methanol. Catal Sci Technol 11:4051–4059

Elias Bamaca Saquic B, Irmak S, Wilkins M, Smith T (2021) Effect of precursors on graphene supported platinum monometalic catalysts for hydrothermal gasification of biomass compounds to hydrogen. Fuel 290:120079

Esrafili MD, Sharifi F, Dinparast L (2017) Catalytic hydrogenation of CO_2 over Pt- and Ni-doped graphene: a comparative DFT study. J Mol Graph Model 77:143–152

Esteve-Adell I, Bakker N, Primo A, Hensen EJM, García H (2017a) Graphene as metal-free catalyst for aqueous phase reforming of ethylene glycol. ChemistrySelect 2:6338–6343

Esteve-Adell I, Crapart B, Primo A, Serp P, Garcia H (2017b) Aqueous phase reforming of glycerol using doped graphenes as metal-free catalysts. Green Chem 19:3061–3068

Fan Q, Zhang M, Jia M, Liu S, Qiu J, Sun Z (2018) Electrochemical CO_2 reduction to C2+ species: Heterogeneous electrocatalysts, reaction pathways, and optimization strategies. Mater Today Energy 10:280–301

Fasolini A, Cespi D, Tabanelli T, Cucciniello R, Cavani F (2019) Hydrogen from renewables: a case study of glycerol reforming. Catalysts 9

Feitosa LF, Berhault G, Laurenti D, da Silva VT (2019) Effect of the nature of the carbon support on the guaiacol hydrodeoxygenation performance of nickel phosphide: comparison between carbon nanotubes and a mesoporous carbon support. Ind Eng Chem Res 58:16164–16181

Feng DM, Sun Y, Liu ZQ, Zhu YP, Ma TY (2019) Designing nanostructured metal-based CO_2 reduction electrocatalysts. J Nanosci Nanotechnol 19:3079–3096

Fernandes DM, Peixoto AF, Freire C (2019) Nitrogen-doped metal-free carbon catalysts for (electro)chemical CO_2 conversion and valorisation. Dalton Trans 48:13508–13528

Fernández-Torres MJ, Dednam W, Caballero JA (2022) Economic and environmental assessment of directly converting CO_2 into a gasoline fuel. Energy Convers Manag 252:115115

Gawande MB, Fornasiero P, Zbořil R (2020) Carbon-based single-atom catalysts for advanced applications. ACS Catal 10:2231–2259

Gogoi P, Kanna N, Begum P, Deka RC, CV V S, Raja T (2020) Controlling and stabilization of Ru nanoparticles by tuning the nitrogen content of the support for enhanced H_2 production through aqueous-phase reforming of glycerol. ACS Catal 10:2489–2507

Grim RG, Huang Z, Guarnieri MT, Ferrell JR, Tao L, Schaidle JA (2020) Transforming the carbon economy: challenges and opportunities in the convergence of low-cost electricity and reductive CO_2 utilization. Energy Environ Sci 13:472–494

Guo C, Tirumala Venkateswara Rao K, Reyhanitash E, Yuan Z, Rohani S, Xu C, He S (2016) Novel inexpensive transition metal phosphide catalysts for upgrading of pyrolysis oil via hydrodeoxygenation. AIChE J 62:3664–3672

Guo SB (2021) CO_2 electrochemical reduction to methane on transition metal porphyrin nitrogen-doped carbon material M@d-NC: theoretical insight. Theor Chem Accounts 140:16

Guo C, Rao KTV, Yuan ZS, He S, Rohani S, Xu CB (2018) Hydrodeoxygenation of fast pyrolysis oil with novel activated carbon-supported NiP and CoP catalysts. Chem Eng Sci 178:248–259

Handoko AD, Wei FX, Jenndy Yeo BS, Seh ZW (2018) Understanding heterogeneous electrocatalytic carbon dioxide reduction through operando techniques. Nat Catal 1:922–934

Hernandez S, Farkhondehfal MA, Sastre F, Makkee M, Saracco G, Russo N (2017) Syngas production from electrochemical reduction of CO_2: current status and prospective implementation. Green Chem 19:2326–2346

Hita I, Cordero-Lanzac T, Kekalainen T, Okafor O, Rodriguez-Mirasol J, Cordero T, Bilbao J, Janis J, Castano P (2020) In-depth analysis of raw bio-oil and its hydrodeoxygenated products for a comprehensive catalyst performance evaluation. Acs Sustain Chem Eng 8:18433–18445

Hoang TTH, Verma S, Ma SC, Fister TT, Timoshenko J, Frenkel AI, Kenis PJA, Gewirth AA (2018) Nanoporous copper silver alloys by additive-controlled electrodeposition for the selective electroreduction of CO_2 to ethylene and ethanol. J Am Chem Soc 140:5791–5797

Huang JZ, Guo XR, Yue GQ, Hu Q, Wang LS (2018a) Boosting CH_3OH production in electrocatalytic CO_2 reduction over partially oxidized 5 nm cobalt nanoparticles dispersed on single-layer nitrogen-doped graphene. ACS Appl Mater Interfaces 10:44403–44414

Huang JZ, Hu Q, Guo XR, Zeng Q, Wang LS (2018b) Rethinking Co(CO3)(0.5)(OH)center dot 0.11H(2)O: a new property for highly selective electrochemical reduction of carbon dioxide to methanol in aqueous solution. Green Chem 20:2967–2972

Jacobson K, Maheria KC, Dalai AK (2013) Bio-oil valorization: a review. Renew Sustain Energy Rev 23:91–106

Janas D, Kreft SK, Koziol KKK (2015) Steam reforming on reactive carbon nanotube membranes. J Ind Eng Chem 25:222–228

Jeong K-E, Kim H-D, Kim T-W, Kim J-W, Chae H-J, Jeong S-Y, Kim C-U (2014) Hydrogen production by aqueous phase reforming of polyols over nano- and micro-sized mesoporous carbon supported platinum catalysts. Catal Today 232:151–157

Jin W, Pastor-Pérez L, Shen D, Sepúlveda-Escribano A, Gu S, Ramirez Reina T (2019a) Catalytic upgrading of biomass model compounds: novel approaches and lessons learnt from traditional hydrodeoxygenation—A review. ChemCatChem 11:924–960

Jin W, Pastor-Perez L, Villora-Pico JJ, Sepulveda-Escribano A, Gu S, Reina TR (2019b) Investigating new routes for biomass upgrading: "H-2-Free" hydrodeoxygenation using Ni-based catalysts. Acs Sustain Chem Eng 7:16041–16049

Jin W, Pastor-Perez L, Villora-Pico JJ, Pastor-Blas MM, Odriozola JA, Sepulveda-Escribano A, Reina TR (2021) In-situ HDO of guaiacol over nitrogen-doped activated carbon supported nickel nanoparticles. Appl Catal A-General 620

Jurca B, Bucur C, Primo A, Concepción P, Parvulescu VI, García H (2018) N-doped defective graphene from biomass as catalyst for CO_2 hydrogenation to methane. ChemCatChem

Karapinar D, Huan NT, Sahraie NR, Li JK, Wakerley D, Touati N, Zanna S, Taverna D, Tizei LHG, Zitolo A, Jaouen F, Mougel V, Fontecave M (2019) Electroreduction of CO_2 on single-site copper-nitrogen-doped carbon material: selective formation of ethanol and reversible restructuration of the metal sites. Angew Chem-Int Edn 58:15098–15103

Keith DW, Holmes G, Angelo DS, Heidel K (2018) A process for capturing CO_2 from the atmosphere. Joule 2:1573–1594

Khezri B, Fisher AC, Pumera M (2017) CO_2 reduction: the quest for electrocatalytic materials. J Mater Chem A 5:8230–8246

Khosravanipour Mostafazadeh A, Solomatnikova O, Drogui P, Tyagi RD (2018) A review of recent research and developments in fast pyrolysis and bio-oil upgrading. Biomass Convers Biorefinery 8:739–773

Kibria MG, Edwards JP, Gabardo CM, Dinh CT, Seifitokaldani A, Sinton D, Sargent EH (2019) Electrochemical CO_2 reduction into chemical feedstocks: from mechanistic electrocatalysis models to system design. Adv Mater 31:24

Kim JY, Park W, Choi C, Kim G, Cho KM, Lim J, Kim SJ, Al-Saggaf A, Gereige I, Lee H, Jung WB, Jung Y, Jung HT (2021) High facets on nanowrinkled Cu via chemical vapor deposition graphene growth for efficient CO_2 reduction into ethanol. ACS Catal 11:5658–5665

Kondratenko EV, Mul G, Baltrusaitis J, Larrazábal GO, Pérez-Ramírez J (2013) Status and perspectives of CO_2 conversion into fuels and chemicals by catalytic, photocatalytic and electrocatalytic processes. Energy Environ Sci 6:3112–3135

Kordouli E, Kordulis C, Lycourghiotis A, Cole R, Vasudeyan PT, Pawelec B, Fierro JLG (2017) HDO activity of carbon-supported Rh, Ni and Mo–Ni catalysts. Mol Catal 441:209–220

Kuhl KP, Cave ER, Abram DN, Jaramillo TF (2012) New insights into the electrochemical reduction of carbon dioxide on metallic copper surfaces. Energy Environ Sci 5:7050–7059

Kumar B, Atla V, Brian JP, Kumari S, Nguyen TQ, Sunkara M, Spurgeon JM (2017) Reduced SnO_2 porous nanowires with a high density of grain boundaries as catalysts for efficient electrochemical CO_2-into-HCOOH conversion. Angew Chem Int Ed 56:3645–3649

Kumar R, Strezov V (2021) Thermochemical production of bio-oil: a review of downstream processing technologies for bio-oil upgrading, production of hydrogen and high value-added products. Renew Sustain Energ Rev 135:110152

Lam E, Luong JHT (2014) Carbon materials as catalyst supports and catalysts in the transformation of biomass to fuels and chemicals. ACS Catal 4:3393–3410

Lee JD, Wang C, Jin TF, Gorte RJ, Murray CB (2020) Engineering the composition of bimetallic nanocrystals to improve hydrodeoxygenation selectivity for 2-acetylfuran. Appl Catal A-Gener 606

Li J, Zhou Y, Xiao X, Wang W, Wang N, Qian W, Chu W (2018) Regulation of Ni-CNT Interaction on Mn-promoted nickel nanocatalysts supported on oxygenated CNTs for CO_2 selective hydrogenation. ACS Appl Mater Interfaces 10:41224–41236

Li HQ, Xiao N, Wang YW, Li C, Ye X, Guo Z, Pan X, Liu C, Bai JP, Xiao J, Zhang XY, Zhao SJ, Qiu JS (2019) Nitrogen-doped tubular carbon foam electrodes for efficient electroreduction of CO_2 to syngas with potential-independent CO/H-2 ratios. J Mater Chem A 7:18852–18860

Li T, Li H, Li CL (2021) Progress in effects of microenvironment of carbon-based catalysts on hydrodeoxygenation of biomass. ChemCatChem 13:1074–1088

Liang F, Zhang KW, Zhang L, Zhang YJ, Lei Y, Sun XL (2021) Recent development of electrocatalytic CO_2 reduction application to energy conversion. Small 29

Liu KH, Zhong HX, Yang XY, Bao D, Meng FL, Yan JM, Zhang XB (2017a) Composition-tunable synthesis of "clean" syngas via a one-step synthesis of metal-free pyridinic-N-enriched self-supported CNTs: the synergy of electrocatalyst pyrolysis temperature and potential. Green Chem 19:4284–4288

Liu YM, Zhang YJ, Cheng K, Quan X, Fan XF, Su Y, Chen S, Zhao HM, Zhang YB, Yu HT, Hoffmann MR (2017b) Selective electrochemical reduction of carbon dioxide to ethanol on a boron- and nitrogen-co-doped nanodiamond. Angew Chem-Int Edn 56:15607–15611

Liu GY, Thanh TP, Chen HJ, Tricoli A (2018a) A review of metal- and metal-oxide-based heterogeneous catalysts for electroreduction of carbon dioxide. Adv Sustain Syst 2:13

Liu J, Sun Y, Jiang X, Zhang A, Song C, Guo X (2018b) Pyrolyzing ZIF-8 to N-doped porous carbon facilitated by iron and potassium for CO_2 hydrogenation to value-added hydrocarbons. J CO_2 Utiliz 25:120–127

Liu S, Yan Z, Zhang Y, Wang R, Luo S-Z, Jing F, Chu W (2018c) Carbon nanotubes supported nickel as the highly efficient catalyst for hydrogen production through glycerol steam reforming. ACS Sustain Chem Eng 6:14403–14413

Lopez M, Palacio R, Royer S, Mamede AS, Fernandez JJ (2020) Mesostructured CMK-3 carbon supported Ni–ZrO_2 as catalysts for the hydrodeoxygenation of guaiacol. Microporous Mesoporous Mater 292

Lu Y, Zhang Z, Wang H, Wang Y (2021) Toward efficient single-atom catalysts for renewable fuels and chemicals production from biomass and CO_2. Appl Catal B 292:120162

Ma J, Gong H, Zhang T, Yu H, Zhang R, Liu Z, Yang G, Sun H, Tang S, Qiu Y (2019a) Hydrogenation of CO_2 to formic acid on the single atom catalysis Cu/C_2N: a first principles study. Appl Surf Sci 488:1–9

Ma Q, Geng M, Zhang J, Zhang X, Zhao T-S (2019b) Enhanced catalytic performance for CO_2 hydrogenation to methanol over N-doped graphene incorporated Cu–ZnO–Al_2O_3 catalysts. ChemistrySelect 4:78–83

Ma T, Fan Q, Li X, Qiu JS, Wu TB, Sun ZY (2019c) Graphene-based materials for electrochemical CO_2 reduction. J CO_2 Utiliz 30:168–182

Mao XW, Hatton TA (2015) Recent advances in electrocatalytic reduction of carbon dioxide using metal-free catalysts. Ind Eng Chem Res 54:4033–4042

Marchese M, Buffo G, Santarelli M, Lanzini A (2021) CO_2 from direct air capture as carbon feedstock for Fischer-Tropsch chemicals and fuels: energy and economic analysis. J CO_2 Utiliz 46:101487

Mendes FL, Da Silva VT, Pacheco ME, Pinho AD, Henriques CA (2020) Hydrotreating of fast pyrolysis oil: a comparison of carbons and carbon-covered alumina as supports for Ni2P. Fuel 264

Mi YY, Shen SB, Peng XY, Bao HH, Liu XJ, Luo J (2019) Selective electroreduction of CO_2–C_2 products over Cu3N-derived Cu nanowires. ChemElectroChem 6:2393–2397

Mohamed AGA, Huang YY, Xie JF, Borse RA, Parameswaram G, Wang YB (2020) Metal-free sites with multidimensional structure modifications for selective electrochemical CO_2 reduction. Nano Today 33:18

Mortensen PM, Grunwaldt JD, Jensen PA, Knudsen KG, Jensen AD (2011) A review of catalytic upgrading of bio-oil to engine fuels. Appl Catal A-Gener 407:1–19

Mou SY, Wu TW, Xie JF, Zhang Y, Ji L, Huang H, Wang T, Luo YL, Xiong XL, Tang B, Sun XP (2019) Boron phosphide nanoparticles: a nonmetal catalyst for high-selectivity electrochemical reduction of CO_2 to CH_3OH. Adv Mater 31:6

Mukundan S, Konarova M, Atanda L, Ma Q, Beltramini J (2015) Guaiacol hydrodeoxygenation reaction catalyzed by highly dispersed, single layered MoS2/C. Catal Sci Technol 5:4422–4432

Natewong P, Prasongthum N, Mhadmhan S, Reubroycharoen P (2018) Fibrous platelet carbon nanofibers-silica fiber composite supports for a co-based catalyst in the steam reforming of acetic acid. Appl Catal A 560:215–224

Nie R, Peng X, Zhang H, Yu X, Lu X, Zhou D, Xia Q (2017) Transfer hydrogenation of bio-fuel with formic acid over biomass-derived N-doped carbon supported acid-resistant Pd catalyst. Catal Sci Technol 7:627–634

Özkan G, Gök S, Özkan G (2011) Active carbon-supported Ni, Ni/Cu and Ni/Cu/Pd catalysed steam reforming of ethanol for the production of hydrogen. Chem Eng J 171:1270–1275

Pang SS (2019) Advances in thermochemical conversion of woody biomass to energy, fuels and chemicals. Biotechnol Adv 37:589–597

Panwar NL, Paul AS (2021) An overview of recent development in bio-oil upgrading and separation techniques. Environ Eng Res 26:200382

Parrilla-Lahoz S, Jin W, Pastor-Perez L, Carrales-Alvarado D, Odriozola JA, Dongil AB, Reina TR (2021) Guaiacol hydrodeoxygenation in hydrothermal conditions using N-doped reduced graphene oxide (RGO) supported Pt and Ni catalysts: Seeking for economically viable biomass upgrading alternatives. Appl Catal A-Gener 611

Pérez-Mayoral E, Matos I, Bernardo M, Ventura M, Fonseca IM (2020) Carbon-based materials for the development of highly dispersed metal catalysts: towards highly performant catalysts for fine chemical synthesis. Catalysts 10:1407

Prasongthum N, Xiao R, Zhang H, Tsubaki N, Natewong P, Reubroycharoen P (2017) Highly active and stable Ni supported on CNTs-SiO_2 fiber catalysts for steam reforming of ethanol. Fuel Process Technol 160:185–195

Pu C, Zhang J, Chang GG, Xiao YY, Ma XC, Wu J, Luo TT, Huang KX, Ke SC, Li JX, Yang XY (2020) Nitrogen precursor-mediated construction of N-doped hierarchically porous carbon-supported Pd catalysts with controllable morphology and composition. Carbon 159:451–460
Pujro R, García JR, Bertero M, Falco M, Sedran U (2021) Review on reaction pathways in the catalytic upgrading of biomass pyrolysis liquids. Energy Fuels 35:16943–16964
Qiao JL, Liu YY, Hong F, Zhang JJ (2014) A review of catalysts for the electroreduction of carbon dioxide to produce low-carbon fuels. Chem Soc Rev 43:631–675
Qu L, Jiang X, Zhang Z, Zhang X-G, Song G-Y, Wang H-L, Yuan Y-P, Chang Y-L (2021) A review of hydrodeoxygenation of bio-oil: model compounds, catalysts, and equipment. Green Chem 23:9348–9376
Rafiee A, Rajab Khalilpour K, Milani D, Panahi M (2018) Trends in CO2 conversion and utilization: a review from process systems perspective. J Environ Chem Eng 6:5771–5794
Rahman MM (2015) H_2 production from aqueous-phase reforming of glycerol over Cu–Ni bimetallic catalysts supported on carbon nanotubes. Int J Hydrog Energy 40:14833–14844
Ramirez-Valencia LD, Bailon-Garcia E, Carrasco-Marin F, Perez-Cadenas AF (2021) From CO_2 to value-added products: a review about carbon-based materials for electro-chemical CO_2 conversion. Catalysts 11:67
Rautio A-R, Seelam PK, Mäki-Arvela P, Pitkänen O, Huuhtanen M, Keiski RL, Kordas K (2015) Carbon supported catalysts in low temperature steam reforming of ethanol: study of catalyst performance. RSC Adv 5:49487–49492
Remon J, Casales M, Gracia J, Callen MS, Pinilla JL, Suelves I (2021) Sustainable production of liquid biofuels and value-added platform chemicals by hydrodeoxygenation of lignocellulosic bio-oil over a carbon-neutral Mo2C/CNF catalyst. Chem Eng J 405
Ren J, Cao J-P, Zhao X-Y, Liu Y-L (2021) Fundamentals and applications of char in biomass tar reforming. Fuel Process Technol 216:106782
Rivera-Cárcamo C, Scarfiello C, García AB, Tison Y, Martinez H, Baaziz W, Ersen O, Le Berre C, Serp P (2020) Stabilization of metal single atoms on carbon and TiO_2 supports for CO_2 hydrogenation: the importance of regulating charge transfer. Adv Mater Interfaces 8
Roy S, Cherevotan A, Peter SC (2018) Thermochemical CO_2 hydrogenation to single carbon products: scientific and technological challenges. ACS Energy Lett 3:1938–1966
Ruiz HA, Rodríguez-Jasso RM, Fernandes BD, Vicente AA, Teixeira JA (2013) Hydrothermal processing, as an alternative for upgrading agriculture residues and marine biomass according to the biorefinery concept: a review. Renew Sustain Energy Rev 21:35–51
Santos JL, Alda-Onggar M, Fedorov V, Peurla M, Eränen K, Mäki-Arvela P, Centeno MÁ, Murzin DY (2018) Hydrodeoxygenation of vanillin over carbon supported metal catalysts. Appl Catal A 561:137–149
Santos JL, Maki-Arvela P, Monzon A, Murzin DY, Centeno MA (2020) Metal catalysts supported on biochars: part I synthesis and characterization. Appl Catal B-Environ 268
Seelam PK, Huuhtanen M, Sápi A, Szabó M, Kordás K, Turpeinen E, Tóth G, Keiski RL (2010) CNT-based catalysts for H_2 production by ethanol reforming. Int J Hydrog Energy 35:12588–12595
Sharma V, Getahun T, Verma M, Villa A, Gupta N (2020) Carbon based catalysts for the hydrodeoxygenation of lignin and related molecules: a powerful tool for the generation of non-petroleum chemical products including hydrocarbons. Renew Sustain Energy Rev 133:110280
Sikora E, Prekob A, Halasi G, Vanyorek L, Pekker P, Kristaly F, Varga T, Kiss J, Konya Z, Viskolcz B (2018) Development and application of carbon-layer-stabilized, nitrogen-doped, bamboo-like carbon nanotube catalysts in CO_2 hydrogenation. ChemistryOpen 7:789–796
Song YF, Chen W, Zhao CC, Li SG, Wei W, Sun YH (2017) Metal-free nitrogen-doped mesoporous carbon for electroreduction of CO_2 to ethanol. Angew Chem-Int Edn 56:10840–10844
Song YF, Wang SB, Chen W, Li SG, Feng GH, Wei W, Sun YH (2020) Enhanced ethanol production from CO_2 electroreduction at micropores in nitrogen-doped mesoporous carbon. Chemsuschem 13:293–297

Sredojević DN, Šljivančanin Ž, Brothers EN, Belić MR (2018) Formic acid synthesis by CO_2 hydrogenation over single-atom catalysts based on Ru and Cu embedded in graphene. ChemistrySelect 3:2631–2637
Stasi CD, Cortese M, Greco G, González B, Palma V, Manyà JJ (2020) Activated biochar-based metal catalysts for steam reforming of pyrolysis bio-oil model compound. Chem Proc 2:18
Sudarsanam P, Zhong R, van den Bosch S, Coman SM, Parvulescu VI, Sels BF (2018) Functionalised heterogeneous catalysts for sustainable biomass valorisation. Chem Soc Rev 47:8349–8402
Sun XF, Kang XC, Zhu QG, Ma J, Yang GY, Liu ZM, Han BX (2016) Very highly efficient reduction of CO_2–CH_4 using metal-free N-doped carbon electrodes. Chem Sci 7:2883–2887
Tan RS, Tuan Abdullah TA, Johari A, Md Isa K (2020) Catalytic steam reforming of tar for enhancing hydrogen production from biomass gasification: a review. Front Energy 14:545–569
Thompson ST, Lamb HH (2018) Vapor-phase hydrodeoxygenation of guaiacol over carbon-supported Pd, Re and PdRe catalysts. Appl Catal A-Gener 563:105–117
Tursunov O, Tilyabaev Z (2019) Hydrogenation of CO_2 over Co supported on carbon nanotube, carbon nanotube-Nb_2O_5, carbon nanofiber, low-layered graphite fragments and Nb_2O_5. J Energy Inst 92:18–26
Veiga S, Bussi J (2017) Steam reforming of crude glycerol over nickel supported on activated carbon. Energy Convers Manag 141:79–84
Wang X, Li N, Pfefferle LD, Haller GL (2010a) Pt−Co bimetallic catalyst supported on single-walled carbon nanotubes: effect of alloy formation and oxygen containing groups. J Phys Chem C 114:16996–17002
Wang X, Li N, Webb JA, Pfefferle LD, Haller GL (2010b) Effect of surface oxygen containing groups on the catalytic activity of multi-walled carbon nanotube supported Pt catalyst. Appl Catal B 101:21–30
Wang YH, Liu JL, Wang YF, Al-Enizi AM, Zheng GF (2017) Tuning of CO_2 reduction selectivity on metal electrocatalysts. Small 13:15
Wang W, Duong-Viet C, Ba H, Baaziz W, Tuci G, Caporali S, Nguyen-Dinh L, Ersen O, Giambastiani G, Pham-Huu C (2018a) Nickel nanoparticles decorated nitrogen-doped carbon nanotubes (Ni/N-CNT); a robust catalyst for the efficient and selective CO_2 methanation. ACS Appl Energy Mater 2:1111–1120
Wang YS, Chen JX, Wang GX, Li Y, Wen ZH (2018b) Perfluorinated covalent triazine framework derived hybrids for the highly selective electroconversion of carbon dioxide into methane. Angewandte Chemie-International Edition 57:13120–13124
Wang H-H, Zhang S-N, Zhao T-J, Liu Y-X, Liu X, Su J, Li X-H, Chen J-S (2020a) Mild and selective hydrogenation of CO_2 into formic acid over electron-rich MoC nanocatalysts. Sci Bull 65:651–657
Wang S, Wu T, Lin J, Ji Y, Yan S, Pei Y, Xie S, Zong B, Qiao M (2020b) Iron-potassium on single-walled carbon nanotubes as efficient catalyst for CO_2 hydrogenation to heavy olefins. ACS Catal 10:6389–6401
Wang Y, Zhang Z, Zhang S, Wang Y, Hu S, Xiang J, Hu X (2020c) Steam reforming of acetic acid over Ni/biochar catalyst treated with HNO3: Impacts of the treatment on surface properties and catalytic behaviors. Fuel 278:118341
Williamson DL, Herdes C, Torrente-Murciano L, Jones MD, Mattia D (2019) N-doped Fe@CNT for combined RWGS/FT CO_2 hydrogenation. ACS Sustain Chem Eng 7:7395–7402
Witoon T, Numpilai T, Phongamwong T, Donphai W, Boonyuen C, Warakulwit C, Chareonpanich M, Limtrakul J (2018) Enhanced activity, selectivity and stability of a CuO–ZnO–ZrO_2 catalyst by adding graphene oxide for CO_2 hydrogenation to methanol. Chem Eng J 334:1781–1791
Wu JJ, Ma SC, Sun J, Gold JI, Tiwary C, Kim B, Zhu LY, Chopra N, Odeh IN, Vajtai R, Yu AZ, Luo R, Lou J, Ding GQ, Kenis PJA, Ajayan PM (2016) A metal-free electrocatalyst for carbon dioxide reduction to multi-carbon hydrocarbons and oxygenates. Nat Commun 7:6
Wu J, Huang Y, Ye W, Li Y (2017) CO_2 reduction: from the electrochemical to photochemical approach. Adv Sci 4:1700194

Wu T, Lin J, Cheng Y, Tian J, Wang S, Xie S, Pei Y, Yan S, Qiao M, Xu H, Zong B (2018) Porous graphene-confined Fe–K as highly efficient catalyst for CO_2 direct hydrogenation to light olefins. ACS Appl Mater Interfaces 10:23439–23443

Wu YJ, Sun Y, Liang KL, Yang ZG, Tu R, Fan XD, Cheng SC, Yu HP, Jiang EC, Xu XW (2021) Enhancing hydrodeoxygenation of bio-oil via bimetallic Ni-V catalysts modified by cross-surface migrated-carbon from biochar. ACS Appl Mater Interfaces 13:21482–21498

Xie H, Wang TY, Liang JS, Li Q, Sun SH (2018) Cu-based nanocatalysts for electrochemical reduction of CO_2. Nano Today 21:41–54

Xu D, Yang L, Ding K, Zhang Y, Gao W, Huang Y, Sun H, Hu X, Syed-Hassan SSA, Zhang S, Zhang H (2020) Mini-review on char catalysts for tar reforming during biomass gasification: the importance of char structure. Energy Fuels 34:1219–1229

Yang HH, Nie RF, Xia W, Yu XL, Jin DF, Lu XH, Zhou D, Xia QH (2017) Co embedded within biomass-derived mesoporous N-doped carbon as an acid-resistant and chemoselective catalyst for transfer hydrodeoxygenation of biomass with formic acid. Green Chem 19:5714–5722

Yang HP, Wu Y, Li GD, Lin Q, Hu Q, Zhang QL, Liu JH, He C (2019) Scalable production of efficient single-atom copper decorated carbon membranes for CO_2 electroreduction to methanol. J Am Chem Soc 141:12717–12723

Yodsin N, Rungnim C, Tungkamani S, Promarak V, Namuangruk S, Jungsuttiwong S (2019) DFT study of catalytic CO_2 hydrogenation over Pt-decorated carbon nanocones: H_2 dissociation combined with the spillover mechanism. J Phys Chem C 124:1941–1949

Yuan J, Zhi WY, Liu L, Yang MP, Wang H, Lu JX (2018) Electrochemical reduction of CO_2 at metal-free N-functionalized graphene oxide electrodes. Electrochim Acta 282:694–701

Zhang L, Zhao ZJ, Gong JL (2017) Nanostructured materials for heterogeneous electrocatalytic CO_2 reduction and their related reaction mechanisms. Angew Chem-Int Edn 56:11326–11353

Zhang M, Hu Z, Gu L, Zhang QH, Zhang LH, Song Q, Zhou W, Hu S (2020a) Electrochemical conversion of CO_2 to syngas with a wide range of CO/H-2 ratio over Ni/Fe binary single-atom catalysts. Nano Res 13:3206–3211

Zhang YY, Li K, Chen MM, Wang J, Liu JD, Zhang YT (2020b) Cu/Cu_2O Nanoparticles supported on vertically ZIF-L-coated nitrogen-doped graphene nanosheets for electroreduction of CO_2 to ethanol. Acs Appl Nano Mater 3:257

Zhao ZL, Lu G (2019) Cu-based single-atom catalysts boost electroreduction of CO_2–CH_3OH: first-principles predictions. J Phys Chem C 123:4380–4387

Zhao K, Liu YM, Quan X, Chen S, Yu HT (2017) CO2 Electroreduction at low overpotential on oxide-derived cu/carbons fabricated from metal organic framework. ACS Appl Mater Interfaces 9:5302–5311

Zhao J, Deng J, Han J, Imhanria S, Chen KY, Wang W (2020) Effective tunable syngas generation via CO_2 reduction reaction by non-precious Fe-N-C electrocatalyst. Chem Eng J 389:7

Zhou WY, Shen HM, Wang Q, Onoe J, Kawazoe Y, Jena P (2019) N-doped peanut-shaped carbon nanotubes for efficient CO_2 electrocatalytic reduction. Carbon 152:241–246

Zhu DD, Liu JL, Qiao SZ (2016) Recent advances in inorganic heterogeneous electrocatalysts for reduction of carbon dioxide. Adv Mater 28:3423–3452

Zou XL, Liu MJ, Wu JJ, Ajayan PM, Li J, Liu BL, Yakobson BI (2017) How nitrogen-doped graphene quantum dots catalyze electroreduction of CO_2 to hydrocarbons and oxygenates. ACS Catal 7:6245–6250

Chapter 10
Waste-to-Energy: Applications and Perspectives on Sustainable Aviation Fuel Production

Nikolaos C. Kokkinos and Elissavet Emmanouilidou

Abstract Global climate change and depletion of resources are crucial issues for modern societies. Moreover, population growth and rapid industrialization result in increased energy demand. To meet ongoing energy demand, developed and developing countries must adopt sustainable waste management methods. In addition to prevention and recycling, waste-to-energy technologies could certainly be beneficial. Except for heat and electricity generation, biofuels can also be produced from waste. Aviation sector is expanding and its greenhouse gas emissions account for about 2% of total global emissions. Aircraft emissions are more persistent in higher altitudes having a greater environmental impact. Hence, decarbonization of the sector is an ongoing challenge. Sustainable aviation fuels comprise a promising solution over the following years. Utilization of waste materials will contribute to the total production of biojet fuels. To date, eight pathways have been certified by the American Society for Testing Materials standards for blending limits up to 50% with conventional jet fuels. The current chapter highlights sustainable waste management methods focused on waste-to-energy conversion technologies for biojet fuel production from waste materials as feedstocks. Current conversion pathways are further analyzed and discussed toward a "greener" and more sustainable future of aviation industry. Hydrotreated esters and fatty acids pathway is the most mature and promising production method for oleochemical feedstocks, including waste oils. Thermochemical methods and alcohol-to-jet pathway will contribute to biojet fuel production in the following years.

Keywords Waste-to-energy · Biofuels · Biojet · Decarbonization · GHG emissions · Sustainable aviation fuels · Sustainable waste management

---

N. C. Kokkinos (✉) · E. Emmanouilidou
Department of Chemistry, School of Science, International Hellenic University, Ag. Loukas, 654 04 Kavala, Greece
e-mail: nck@chem.ihu.gr

Petroleum Institute, International Hellenic University, Ag. Loukas, 654 04 Kavala, Greece

P. C. Shukla et al. (eds.), *Renewable Fuels for Sustainable Mobility*, Energy, Environment, and Sustainability, https://doi.org/10.1007/978-981-99-1392-3_10

10.1 Introduction

Rapid growth in population and urbanization resulted in higher energy needs and faster consumption of goods. The enormous generation of waste is a crucial environmental concern. Sustainable waste management and energy recovery from waste can minimize environmental pollution and simultaneously promote the energy sector for the production of renewable fuels to meet future energy demand (Brunner and Rechberger 2015; Khan and Kabir 2020).

Waste-to-energy (WtE) technologies include several methods such as incineration, pyrolysis, gasification, fermentation, anaerobic digestion, landfill, and others (Kumar and Ankaram 2019). During the last decades, WtE technologies have undergone rapid development. Different waste management policies and strategies exist around the globe between developing and developed countries (Sun, et al. 2020). Current situation regarding municipal solid waste (MSW) management needs strengthening, especially in developing countries. Major challenges involve improper separation of MSW, lack of advanced technologies, and limited recycling (Shah, et al. 2021).

European targets for sustainable waste management involve the principles of circular economy in terms of renewable energy utilization, energy efficiency improvement, and reduction of dependency on imported sources. Three main policies are included in WtE multi-purpose concept: waste management, energy union, and air quality/climate change (Malinauskaite and Jouhara 2019).

Energy needs in the transportation sector will continue to grow over the next decades. Aviation sector comprises one of the fast-growing transport sectors with a large contribution to the global economy. However, sector's greenhouse gas (GHG) emissions reach approximately 2% in global emissions (Reaching 2021). Aircraft emissions include carbon dioxide (CO_2), water vapor (H_2O), nitrogen oxides (NO_x), carbon monoxide (CO), sulfur oxides (SO_x), soot, and particles. Aviation emissions affect climate change through radiative forcing (RF) and the effects are interconnected and depend on the altitude where they are emitted. Long-term impacts are associated with CO_2 emissions, whereas short-term impacts are related to non-CO_2 emissions. Chemical and particle microphysical properties of the upper atmosphere are affected by aviation emissions and cloud effects, causing warming or cooling (Hashemi Devin and Sabziparvar 2002). In the troposphere, pollutants can be removed, due to water vapor precipitation. On the contrary, the stratosphere, which lies above the atmosphere, lacks water vapors and pollutants remain for years. The exact height of the thin layer between troposphere and stratosphere (tropopause) is affected by several factors including latitude, weather, temperature, and human activity (Sheng et al. 2015).

Decarbonization of aviation sector is a crucial challenge in order to achieve early reduction of emissions by 2030 and deeper ones by 2050 (Reaching 2021; Abrantes 2050). Despite the relatively low contribution of aviation sector to total CO_2 emissions, reducing CO_2 emissions is essential according to the aviation industry's growth projections. In Europe, expected CO_2 emissions will reach 360 million tons by 2023, while fuel demand will amount to 118 Mtoe at the same time (Kousoulidou and Lonza

2016). Toward this direction, sustainable aviation fuels (SAF) present the most viable solution in near future compared to other options such as hydrogen and electrification (Bauen et al. 2020).

In the ground transport sector, in 1894, Rudolf Diesel's self-igniting engine invention quickly rose to prominence in the ground transportation industry. Diesel engines are more efficient, because they have a higher compression ratio and operate with a leaner mixture. However, since stringent regulations have been introduced to control and reduce diesel engine emissions, alternative fuels, such as biodiesel, bio-alcohols, biogas, and bio-oil, need to be developed (Vijayashree and Ganesan 2020). CO_2 reduction of compression ignition engines can be further improved by advanced combustion technologies combined with alternative fuels, such as alcohol fuels (methanol, ethanol). The charge cooling effect lowers the combustion temperature and increases overall efficiency. Moreover, alcohol–diesel fuel blends can reduce particulate matter (PM) and soot emissions (Shamun et al. 2020).

In the air transport sector, military and commercial aircraft engines are mostly powered by gas turbine engines. Despite the fact that gas turbine engines are very efficient, replacement of kerosene-based fuels is essential for the fulfillment of future energy demand needs and mitigating climate change. In order to be compatible with existing aircraft engines, biojet fuels have to meet strict specifications according to American Society for Testing and Materials (ASTM) D7566 standard. Performance characteristics of biojet fuels include low-temperature fluidity, thermal oxidation stability, combustion characteristics, compatibility with current aviation fueling system, fuel volatility, fuel metering, and aircraft range (Chandralingam 2020; Yang et al. 2019). Aircraft fuel efficiency has improved substantially over the last decades and is expected to improve further in the future. Synthetic jet fuels produced by coal or natural gas through Fischer–Tropsch synthesis are not considered sustainable. However, coal-to-liquid (CTL), gas-to-liquid (GTL), and other first-generation synthetic fuels paved the way for biomass-to-liquid biofuels. Alternatives to food plant-derived fuels are the focus of ongoing research in the world. Nevertheless, almost all alternative fuels face implementation challenges compared to kerosene fuels. High energy content per unit of weight and volume is a basic requirement for aircraft fuels. Hydrogen (H_2) may not emit CO_2 and it is also lightweight, but production, handling, infrastructure, and storage are the main drawbacks (Daggett et al. 2006; Marsh 2008).

Utilization of waste materials for biofuel production, including SAF, is of great significance for a sustainable future of aviation industry. In industrialized countries, waste materials such as MSW, forestry, and agricultural residues are considered as renewable feedstock with a high potential to recover energy through different WtE technologies (Sindhu et al. 2019; Montoya Sánchez et al. 2022). In Fig. 10.1, an illustrative flowchart connecting waste generation and its utilization for energy recovery in the fast-growing aviation sector is shown.

Hydroprocessing of oleochemical/lipid feedstocks or Hydroprocessed Esters and Fatty Acids (HEFA) pathway is the most widely used technology for SAF production. Currently, this is the only fully commercialized method. Nevertheless, future demand is not going to be fulfilled only from these kinds of feedstocks, but lignocellulosic

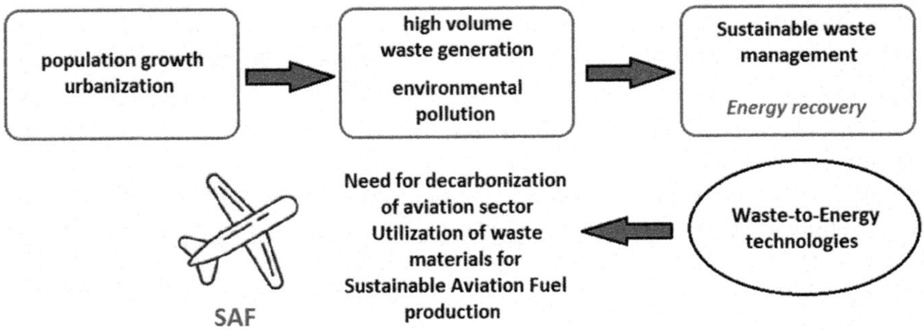

Fig. 10.1 Waste generation and utilization as feedstock through WtE technologies for SAF production

residues and MSW could contribute to aviation biofuel production (Zhang, et al. 2022; Monteiro, et al. 2022).

Catalytic conversion of alcohols as intermediate (alcohol-to-jet) technology as well as thermochemical methods, such as gasification-Fischer-Tropsch synthesis, are rising in technological readiness level. However, techno-economic models reveal that without policy support measurements, none of the approved pathways could be economically viable. Financial viability could be possible through the cumulative impact of multiple policies (Dahal 2021; Machineni 2019; Launay et al. 1998). Biofuels are evaluated against the broader sustainability claims that include: (a) climate security and appropriate land use, (b) social issues, (c) economic viability in existing sustainable environment, (d) technological advances combined with sustainable standards, and (e) governments encouraging plans to provide a smooth legislative agenda (Thangaraja et al. 2020).

10.2 Historical Overview—Environmental Legislation

First attempts on solid waste management in Western cultures date back to Greek Era in 500 BC. Municipal dumps had been organized in order to prevent garbage throwing in the streets. Similarly, European and American cities adopted primitive methods in the nineteenth century for the disposal of waste such as landfilling, burning, or utilization as fertilizers (Louis 2004). In the United States, the Federal Congress in 1970 legislated comprehensive laws for solid waste planning, energy conservation, recycling, etc. (Kovacs 1993).

Rapid urbanization and humans' population growth resulted in a rapid increase of MSW. Responsible authorities were pushed in order to take action and develop new strategies. There was realization that landfilling could not be efficient. First MSW incinerators without energy recovery were constructed in UK and US in 1970 and 1885, respectively. Hence, since the late nineteenth century, the development of MSW incinerators has been significant. Volumetric reduction of waste, optimal heat, and materials recovery and prevention of emissions were the main targets for successful implementation of incinerators (Makarichi et al. 2018).

In fact, utilization of energy was introduced as a technical requirement, when air pollution started to be systematically controlled. Meanwhile, first incineration plants were constructed with regard to human's health and environmental protection. During 1950s and 1960s filters were installed inside incinerators in order to mitigate air pollution. However, due to the compositional differentiation in the increased MSW components further technological improvement was required (Brunner and Rechberger 2015).

Improper management of MSW can cause air, water, soil, and aesthetic contamination at different scales. Open dumping of waste, which is a common practice in poor economies, results in water bodies' pollution and threatens public health, due to the release of organic and inorganic contaminants. Leaching of toxic pollutants affects surface and groundwater systems. Moreover, incineration of waste emits persistent organic pollutants, whereas landfills can generate methane (GHG) when organic materials decompose (Vergara and Tchobanoglous 2012; Das et al. 2019).

Environmental protection and public health were the main drivers to eliminate uncontrolled waste disposal. Technical standards started to gradually develop. Moreover, waste hierarchy emerged as an important driver in Europe in 1977, where more sustainable options such as reduction, reuse, recycling, and energy recovery from waste were introduced (Wilson 2007). Environmental Protection Agency of US (EPA) established the Resource Conservation and Recovery Act (RCRA) regulations for non-hazardous and hazardous waste as well as for used oil management and standards or underground storage tanks (Bajpai and Tyagi 2006). The RCRA law covers both hazardous and solid waste. With RCRA, US EPA was able to address hazardous waste throughout the entire life cycle of wastes (known as "cradle-to-grave"). Moreover, the Comprehensive Environmental Response, Compensation, and Liability Act (CERCLA), also known as Superfund, was enacted by US Congress in 1980. This specific act created a tax for petroleum and industrial industries and provided federal authority for facing releases of hazardous substances to the environment. In 1986, CERCLA act was amended by the Superfund Amendments and Reauthorization Act (SARA). Innovative technologies for cleaning up of hazardous waste sites and revision of the Hazard Ranking System (HRS) were regulated (Vallero 2019).

During 2015, the introduced "environmental" tax obligated manufactures and importers of recyclable goods to pay non-tax revenues of the federal budget. Regulation of MSW treatment activities was introduced in 2016 inside Federal Law No. 89 as an independent chapter, where regional operators have been nominated for several activities such as accumulation, transportation, treatment, and disposal (Kortov et al. 2016).

In Korea, waste management legislation was enacted in 1960. By the establishment of the Waste Management Law in the middle of 1980s, waste management involved not only waste minimization, but also suitable treatment methods and maximization of materials recycling (Yang et al. 2014).

According to EU policy on waste management (Directive 2008/98/EC of the European Parliament, Article 4), the following waste hierarchy should be applied

Fig. 10.2 Waste management hierarchy

(Gopinath et al. 2010): (1) prevention; (2) preparing for reuse; (3) recycling; and (4) other recovery, including energy recovery and (5) disposal (Fig. 10.2).

Member States shall take all the appropriate measures for waste hierarchy application in order to deliver the best environmental outcome. Moreover, general environmental protection principles such as precaution, sustainability, technical, and economic feasibility as well as humans' health and socioeconomic impacts have to be taken into consideration. Even if prevention is the first priority in the Waste Frame Directive (WFD), different waste management operations are essential to regulate landfill and incineration (Aziz and Maulood 2015; Thapa et al. 1970).

10.3 Energy Supply and Demand

Global human's population is expected to reach 9.6 billion by 2050. Energy demand will be definitely increased and WtE markets are going to be expanded by 2023. Potential energy of 13 GW could be generated globally from WtE facilities (Mubeen and Buekens 2019).

According to International Energy Agency (IEA), global oil demand is projected to reach 101.6 mb/d in 2023. In order to meet the expected growing demand, producers from Middle East will dominate in global oil supply market. However, demand is going to peak earlier compared to pre-pandemic forecast if governments adopt stronger policies and shift on renewable energy (Calhoun 1963).

Currently, bioenergy accounts for one-tenth of the global primary energy supply. Biofuel consumption increased 5% between 2010 and 2019. IEA's Net Zero Scenario estimates that biofuels derived from waste and non-edible energy crops could contribute around 45% to biofuel consumption instead of 7% in 2020. Nevertheless, advanced technologies for these types of feedstocks will need to be commercialized (Neamah 2014).

Global biofuel demand is projected to be increased by 28% by 2021–2026. The main driver for biofuel demand is government policies; however, other parameters such as total transport fuel demand, costs, and appropriate policy design could affect demand's fulfilling (IEA 2021).

Mitigation of climate change and GHG emissions are ongoing concerns that have shifted governments to regulate new policies during last two decades. Toward this direction, bioenergy supply chain has to be designed efficiently. Stricter regulations can potentially increase the share of renewables in total energy consumption (Daneshmandi et al. 2022).

10.4 Sustainable Waste Management

Waste is considered as the result of inadequate contemplation. Waste classification is a difficult task, because the composition of waste is different around the globe; whereas most countries have established their own classification systems (Wen et al. 2013).

Waste management is "set of interacting units or elements that form an integrated whole intended to perform some function" (Clark 1978). Conventional approach for waste management encountered waste generation, collection, and disposal methods as independent processes. Nevertheless, these procedures are definitely interlinked. Hence, in order to shift to a sustainable waste management, several parameters regarding system's efficiency have to be taken under serious consideration (Seadon 2010).

Sustainability is an internationally leading topic that is continuously analyzed by researchers, policymakers, and society members. Conservation of resources with regard to environmental protection and human health is the main priority of sustainable waste management (Fig. 10.3) (Gören 2014).

Key challenges for sustainable waste-to-energy improvement in developing countries that have to be addressed include: establishment of regulations and technical standards, business model design, technical localization, labor training, and fly ash treatment (Yan and 2020).

"***Zero waste***" concept is a visionary waste management system to deal with waste issues in the twenty-first century. Proper waste management policies, plans, and initiatives are of great significance in local and international levels. Zero landfilling and maximum resource recovery from waste are the main goals of "*zero waste*" approach. Identification of priority of waste strategy fields and development of national "*zero waste*" guidelines can contribute to "*zero waste*" concept application widely (Zaman 2015). Nevertheless, "*zero waste*" achievement in urban centers remains utopian, unless a good understanding of inputs and outputs is well established. Reducing waste could be possible if there was expansion in the life of products;

Fig. 10.3 Sustainable waste management priorities

however, financial solutions are required. Circular systems need a cooperative effort from all stakeholders and policy support is essential (Awasthi et al. 2021).

Several models for waste management have been developed over the years. During the 1980s, earlier models covered MSW management focused on the relationship between each factor. Main intention was to minimize the cost of mixed waste management and to a lesser extent the recycling. Nevertheless, the term sustainable waste management was not used. On the contrary, during the 1990s, recycling and other waste management methods were included in most models. Sustainability was inserted as term and the main categories of models were cost–benefit analysis models, life-cycle inventory models, and multi-criteria models (Morrissey and Browne 2004).

Solid waste management (SWM) has the potential to be more effective by the establishment of environmental strategies based on "reduce", "reuse", and "recycle" (3R) concepts. Sophisticated techniques should be applied in global level, where developed nations can share the knowledge and new methodologies, setting the good paradigm for developing countries. Of course, economic feasibility is essential to be evaluated in the long run and life-cycle assessment (LCA) models should be improved (Das et al. 2019).

Integrated approach for solid waste management includes the combination of several methods for both developed and developing countries. Different collection and treatment options as well as stakeholders' involvement and development of relevant technologies related to products' design are crucial (Joseph 2006).

Toward a circular economy, waste reduction and material reusability are of great significance. Waste materials from tires, glass, and plastic are generated in large volumes in many countries and undergo mainly landfilling. New strategies and economical solutions are vital to extend products' life and to improve utilization of recycled materials (Ferdous et al. 2021). Moreover, direct and indirect emissions can be reduced by utilization of waste recycling and recovery. Bio-waste, plastics recycling, and energy recovery are key sustainability factors for waste management systems to protect human health and to minimize environmental impact (Maria et al. 2020).

10.4.1 Waste-to-Energy Conversion Technologies

WtE comprises a rapidly growing and eco-friendly approach to waste treatment. The term WtE is used interchangeably and involves a variety of processes for several streams conversion to energy. Promising conversion technologies include thermal conversion methods (incineration, pyrolysis, and gasification), biochemical methods (anaerobic digestion and alcohol fermentation), and landfill gas recovery (Beyene et al. 2018; Foster 2021).

In general, waste can be classified as urban, industrial, biomass, and biomedical waste. Utilization of WtE strategies can achieve sustainable waste management.

Potential renewable resources include agricultural, industrial, and municipal solid waste (Rajesh and Majid 2019).

MSW has either solid or semisolid form and is generated in municipal areas. Basically, MSW presents heterogeneous nature and its composition varies throughout the world. Major components are organic matter, paper, metal, glass, plastic, textile, wood, etc. Currently, major technologies for MSW management are incineration, composting, and landfill (Rao et al. 2017).

Improvement of WtE technologies for MSW management can reduce GHG emissions especially in developed countries, where WtE facilities are more applicable. In the US, there are 86 WtE facilities, mainly in New York and Florida. WtE facilities include mass burn and refuse-derived fuel technologies. However, presently only 13% of MSW is utilized for energy recovery, whereas 53% is landfilled. Landfilling and anaerobic digestion with gas recovery are the most widely used WtE methods for MSW management. Future technologies include plasma with fluidized bed gasification, plasma-assisted gasification, gasification with pyrolysis, etc. (Chand Malav 2020). In Europe, energy recovery from MSW is mainly applied in Germany, France, and UK. Nevertheless, incineration capacity of 18 countries is less than a quarter compared to MSW generation, whereas in 10 countries incineration capacities do not exist at all. In 2016, there were 512 WtE plants in Europe, while 330 new plants present the potential to be built for energy recovery from waste with a total capacity of about 50 million tons (Scarlat et al. 2018a). Generally, advanced waste management technologies aim to reduce the final disposal of waste. Incineration is mostly applied and if this option is not possible, waste is landfilled (Mubeen and Buekens 2019). Future trends in WtE technologies include biological hydrogen production, dark fermentation, and bioelectrochemical processes, such as microbial fuel cells (MFC) and microbial electrolysis cells (MEC). MFC and MEC are considered the most eco-friendly for conversion of MSW to energy (Beyene et al. 2018). In Fig. 10.4, current and future trends in WtE technologies are illustrated.

In developing countries, WtE plants have been constructed in some cases; however, the development of WtE technologies requires financial support and government regulations. Main challenges include sorting and handling of enormous volumes of MSW, economic instability, and lack of sustainable MSW management strategies (Kumar and Samadder 2017; Khan et al. 2022).

Biomass waste-to-energy technologies can also be applied for biofuel production. Liquid biofuel is an important alternative to fossil fuel in both air and ground transportation. The aviation sector is a fast-developing transport sector and its contribution to GHG emission reduction is very significant. Development of SAF can be achieved based on materials and production methods used in ground transportation. SAF properties must meet specific standards regarding engine operating conditions, storage, environmental, and safety concerns (Gil 2022; Yilmaz and Atmanli 2017). Production of SAF from biomass sources is an integration of several technologies such as thermochemical and biochemical conversions. Second-generation (2G) biomass sources, which include non-edible crops, lignocellulosic materials, and organic waste (such as MSW), are rich in fatty acid components and can undergo both thermochemical and biochemical conversion methods (Wang et al. 2019; Anjani et al. 2021).

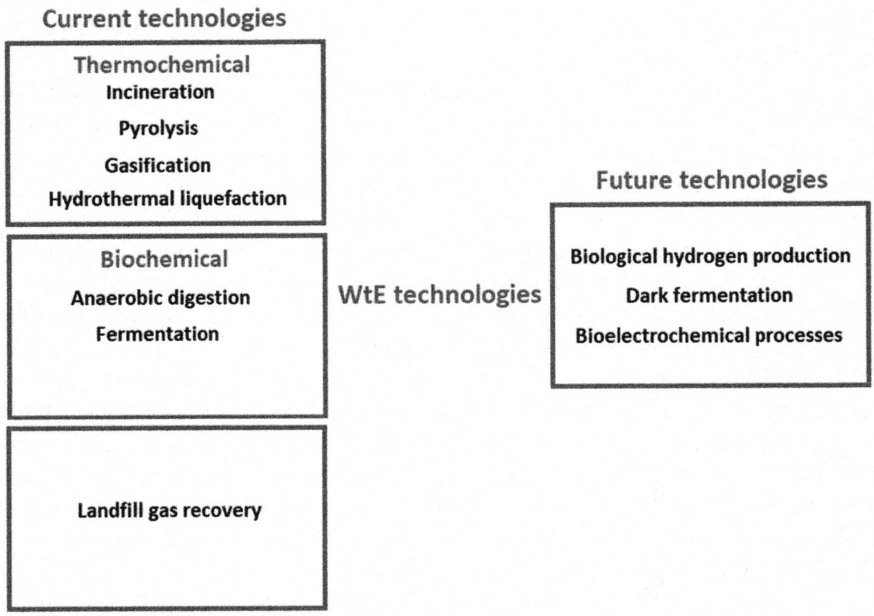

Fig. 10.4 Current and future WtE technologies

10.4.1.1 Thermochemical Conversion Technologies

Incineration is a thermal process that is applied for degradation and deconstruction of organic matter in a furnace by monitoring burning at high temperatures between 750 and 1100 °C in the presence of oxygen. It is considered as one of the most effective and mature methods, as it decreases the volume of solid waste up to 90%. Main stages of incineration process are drying and degassing, pyrolysis and gasification and oxidation. However, additional treatment of flue gas system is required (Makarichi et al. 2018; Beyene et al. 2018; Saini and Saini 2021), due to the release of gaseous pollutants and heavy metals. Hence, careful waste sorting is essential as some materials are non-combustible or hazardous and should not be incinerated. Pretreatment methods are also needed in order to reduce the moisture content in cases such as food waste (Jeevahan et al. 2018).

Pyrolysis is a thermochemical process where waste materials are combusted in the absence of oxygen at high temperatures of 500–800 °C. Char, oil, and gas are the main products. This method is the best in case of MSW management; however, it presents high operational and maintenance costs when applied at commercial level. Common types of pyrolysis are: conventional, fast, and flash pyrolysis. Waste separation is not required and environmental issues are limited. The gas produced during pyrolysis is called syngas and consists mainly of methane, hydrogen, carbon monoxide, and carbon dioxide. Syngas can be utilized in several energy applications such as engines, boilers, and biofuels production (Mubeen and Buekens 2019; Beyene et al. 2018; Saini and Saini 2021).

Gasification is also a thermochemical method that involves the conversion of organic matter into syngas in a controlled oxidizing environment at high temperatures of 800–1200 °C) without combustion. The volume and the mass of waste can be reduced by 90% and 70%, respectively. Solid and semisolid waste are utilized in this method, where in case of MSW gasification is a well-established process for renewable energy production (Mubeen and Buekens 2019; Foster 2021; Moya et al. 2017).

Hydrothermal liquefaction (HTL) is a thermochemical process for both dry and wet biomass conversion into crude bio-oil at high pressures (50–250 bar) and temperatures (250–400 °C) that is further upgraded and refined into liquid biofuels. Wet biomass is used without any drying requirement. However, due to higher water and oxygen content of bio-oil more research is needed for its utilization as drop-in fuel (Foster 2021; Perkins 2019).

10.4.1.2 Biochemical Conversion Technologies

Anaerobic digestion (AD) is a biological process that is used for the decomposition of organic matter from bacteria in the absence of oxygen. Biogas is produced and consists mainly of methane and carbon dioxide. Feedstocks include MSW, biosolids, manure, etc. Power, heat, and biofuels can be obtained from biogas; however, compared to other WtE technologies, AD is still considered as not economical (Kumar and Ankaram 2019; Mubeen and Buekens 2019; Beyene et al. 2018).

Fermentation is a biochemical reaction that involves the conversion of organic materials into an alcohol and an acid in the absence of oxygen. Bio-ethanol has been produced in large quantities from food waste after pretreatment due to the complex nature of lignocellulosic components. Any material that contains sugar can be converted into ethanol. Three main types of feedstocks are sugar-, starch-, and cellulose-based materials. Starch should be firstly hydrolyzed by enzymes in order to generate fermentable sugars, whereas in case of cellulose depolymerization of carbohydrates is also required (Mubeen and Buekens 2019; Beyene et al. 2018; Jeevahan et al. 2018).

10.4.2 Landfill Gas Recovery

Sanitary landfilling is a WtE technology that involves the controlled disposal of waste on land in order to eliminate environmental contamination through biogas recovery and leachate management. A promising approach for energy recovery is the construction of biogas production plants from landfill gas recovery. In US, there were 654 biogas recovery plants from landfills in 2017, where the energy potential of biogas reached 8.0 billion m^3 of biogas/year. In Europe, landfill biogas is mainly produced in Italy, UK, France, and Spain. Also, it dominates the market in Portugal, Greece, Estonia, and Ireland (Scarlat et al. 2018b).

Nevertheless, environmental impact is higher compared to other disposal methods, since, especially in most developing countries, unsanitary landfilling is applied as a simpler practice. Hence, restrictions for MSW disposal on landfill sites motivated governments to adopt more efficient methods for disposal (Kumar and Samadder 2017). Although waste landfilling is currently considered as a generally accepted low-cost method, regular monitoring is required. EU Directive aims to reduce the waste in landfills, even if in many European countries, landfilling is still the main waste management technology (Vaverková 2019).

10.5 Waste Materials as Biojet Fuel Feedstock

10.5.1 Decarbonizing Aviation Industry

Aviation sector is one of the most rapidly growing transportation sectors and plays a key role in global economy. Sector's emissions contribute around 2% of total carbon dioxide (CO_2) emissions globally. COVID-19 pandemic has certainly affected sector's emissions. However, pre-COVID levels are projected to be reached soon. Hence, decarbonization of aviation industry is an ongoing challenge and contribution of SAF is of great significance to achieve early and deeper CO_2 reduction emissions by 2030 and 2050, respectively (Abrantes 2050; Gollakota and Shu 2022).

The aviation industry has set ambitious targets to reduce its environmental impact while contributing to economic growth. In 2009, the International Air Transport Association (IATA) aimed to stabilize CO_2 emissions after 2020 by implementing a combination of factors including fleet renewal, infrastructure measures, use of renewable fuels, etc. Compared to other solutions that has been proposed, biojet fuel utilization is currently the most attractive option to operate in the existing aircraft engines as blends with conventional jet fuels (Kousoulidou and Lonza 2016).

Europe's Biofuel Flight Path Initiative was introduced in 2011 by European Commission in partnership with airline and biofuel producers to support the use of biojet fuels (Deane and Pye 2018). Aviation emissions were inserted in EU emission trading scheme since 2012, while in 2016 the Carbon Offsetting and Reduction Scheme for International Aviation (CORSIA) was introduced in the International Civil Aviation Organization (ICAO) (Gray et al. 2021).

The International Energy Agency (IEA) stated that in order to achieve a 2 °C target, aviation emissions should not exceed 1000 Gt of CO_2 from 2011 onward (Stupp et al. 2012). In 2019, aviation sector was responsible for million tons (Mt) of CO_2 emissions. Top emitter of GHG emissions was USA. GHG emissions came from aviation bunkers and amounted to 179 Mt of CO_2. Together with USA, other counties such as China, UK, Japan, and Germany contributed to 40% of global GHG emission from aviation bunkers. Since aviation sector's emissions are expected to increase in the future, negative-carbon technologies have to be developed along with bioenergy production (Hasan et al. 2021). In Europe, the Renewable Energy Directive (RED) motivated EU members to improve the production of biofuels from several biomass sources. Nevertheless, utilization of food crops for biofuels production remains an ongoing concern regarding GHG emissions due to the indirect land use (Aracil et al. 2017).

In order to achieve a sustainable energy system and meet long-term GHG emissions target, EU has adopted ambitious energy and climate regulations in its 2030 framework. Research and development of innovative eco-friendly technologies could contribute to sector's emissions reduction, according to the Aviation Strategy for Europe. EU target policy for 2020 does not specify the exact targets for transport sector, but emphasizes on low-carbon technologies application (O'Connell et al. 2019).

Biojet fuel market is entirely connected with stringent requirements of GHG emission standards that are stated by ICAO and aviation industries ought to pay credits in case of not satisfying the regulations. Due to high safety requirements, only drop-in SAF with excellent performance in jet engines can be used according to approved American Society for Testing Material (ASTM) D7566 standard (Reaching 2021; Zhang et al. 2020).

10.5.2 Waste Material Feedstock

Biofuels, including SAF, can be produced by a plethora of feedstocks such as MSW, agricultural and crop residues, energy crops, sewage sludge, livestock manure, and other organic materials. Biomass is a renewable organic material that can be converted into biofuels and chemicals. First-generation biomass mainly includes edible plant materials, whereas second-generation biomass consists of non-edible plant residues and third-generation biomass of micro- and macro-algae (Nanda et al. 2018).

In general, second-generation biofuels are more suitable for biofuels production as they do not compromise food security and direct land use. Non-edible feedstock, lignocellulosic biomass, waste cooking oils, and organic have overcome the limitations of first-generation feedstock and can be converted into biofuels by both thermochemical and biochemical methods. Moreover, agricultural wastes are still underutilized and can be potentially used as an alternative feedstock for biofuels and value-added chemicals production in near future (Anjani et al. 2021; Osman et al. 2021). Waste biomass feedstocks (Fig. 10.5) that can be utilized for SAF production are further reported.

MSW provides an alternative feedstock for SAF production in agreement with circular economy concept. According to European waste hierarchy, only the non-recycled fraction of MSW can be used for energy recovery, including biofuel production (Aracil et al. 2017).

Food waste consists mainly of highly degradable compounds such as lipids, protein, and carbohydrates. It is generated from food processing plants, kitchen,

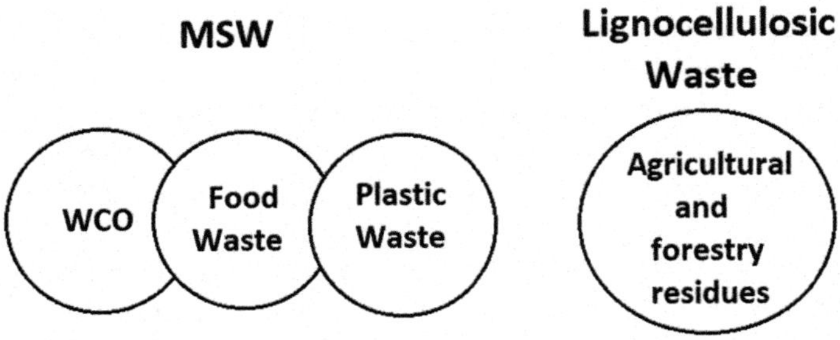

Fig. 10.5 Waste feedstocks for SAF production

restaurants, industries, agricultural field, etc. and accounts around one-third of MSW. Biofuels such as bio-oils, syngas, and bioethanol can be produced as main products through pyrolysis or biochemical conversion (Yukesh Kannah, et al. 2020; Sridhar 2021).

Waste triglyceride feedstock includes waste cooking oil (WCO), animal fat, oil extracted from industrial residues, and bio-oil from pyrolysis. WCO is oil residue that is produced after cooking. It is considered as a serious environmental hazard if disposed improperly. It is mainly generated from households, restaurants, and food industries and contains triglycerides and high amounts of free fatty acids (FFAs) and water. Utilization of WCO as an alternative feedstock for biofuels including SAF could be really promising for future energy needs (Foo et al. 2021; Goh 2020).

Lignocellulosic feedstock includes mainly agricultural and forestry residues as well as textile and solid urban waste. Due to the abundance of this renewable type of feedstock, lignocellulosic waste materials are an attractive option for SAF production. Their contribution in total SAF production will be really beneficial (Moreno-Gómez et al. 2020). Nevertheless, due to highly recalcitrant nature of lignocellulosic biomass, deconstruction of lignocellulosic matrix into its components (cellulose, hemicellulose, and lignin) is required. Deconstruction can be achieved through several pretreatment methods including physical, chemical, biological, physico-chemical, or combination of them. The selection of the most appropriate pretreatment method depends on the downstream applications of lignocellulosic biomass (Ashokkumar et al. 2022; Costa et al. 2020).

Finally, plastic production in recent years raised the interest for their potential utilization as feedstocks for biofuel production. Due to their slow degradation and limited recycling rate after disposal, plastics can be used as potential feedstock for SAF production through thermal processes (Zhang et al. 2022; Fahim et al. 2021).

Waste feedstock is a renewable raw material for SAF production and due to its abundance, it can guarantee the supply of production process. Moreover, its utilization as feedstock minimizes waste disposal issues and the cost of waste materials is much lower than other feedstock. For oleochemical feedstock, WCOs have been mostly investigated for SAF production, whereas for lignocellulosic-based materials agricultural and forestry residues gained the attention (Moreno-Gómez et al. 2020). Neste has identified 30 million tons of waste feedstock that could be used for biojet fuel production. However, since there is a significant gap between biojet fuel price and convention jet fuel, effective policies are essential to encourage the production and use of biojet fuel in the short-to-medium term. Technology push and market pull policies can promote development and market penetration of biojet fuel (Reaching 2021).

10.6 Sustainable Pathways for Greener Biojet Fuel Production

To date, eight pathways for SAF production have received certification by ASTM to be used as blends with conventional jet fuels in order to meet the specific requirements in existing aircraft engines. In Table 10.1, ASTM-certified pathways and the year of receiving certification are presented, while further down detailed description of each production route is given (Reaching 2021; Wang and Tao 2016).

The most widely used route for SAF production is Hydroprocessed Esters and Fatty Acids Synthetic Paraffinic Kerosene (HEFA-SPK) or catalytic hydroprocessing. This pathway is fully commercialized and involves biojet fuel production through hydroprocessing of vegetable oils, animal fats, waste grease, and algal oil (oleochemical feedstocks). ASTM certification was received in 2011 for maximum blend level of biojet fuel of 50% with conventional jet fuels. Hydroprocessing of lipid feedstocks includes hydrotreating and hydrocracking reactions. In first case, carbon double bonds undergo hydrogenation followed under mild conditions by several catalysts such as Pd, Pt, NiMo, and CoMo over γ-Al_2O_3 or activated carbon. The majority of aviation biofuels is currently produced by HEFA pathway from oleochemical feedstocks and this technology is going to dominate over the next years (Monteiro et al. 2022; Chong and Chemmangattuvalappil 2021).

Chen and Wang (2019) investigated WCO conversion through hydroprocessing over Pd/C catalyst. High concentrations of C15–C18 normal alkanes were produced. Moreover, Xu et al. (2019) used a two-step catalytic conversion and hydrogenation of waste triglycerides into aviation range hydrocarbons with a maximum yield of 60%.

Table 10.1 ASTM-certified pathways for SAF production

Pathway	Year of certification
FT	2011
HEFA	2011
SIP	2014
SPK/A	2015
ATJ	2016/2018
Co-processing	2018/2020
CHJ	2020
HC-HEFA	2020

Main components of produced biojet fuel were found compatible to conventional aviation fuels. Asiedu et al. (2021) investigated catalytic transfer hydrogenation of WCO over a fixed bed of granular activated carbon with a biojet fuel fraction yield of approximately 52%. In addition, Shah et al. (2019) studied catalytic pyrolysis hydrogenation over NiMo catalyst of agricultural wastes (eucalyptus sawdust) and waste frying oil. Obtained upgraded bio-oil exhibited similar properties with conventional aviation fuels.

Another promising technology based on gasification and further upgrading of biomass products is Fischer-Tropsch Synthetic Paraffinic Kerosene (FT-SPK) pathway. Gasification combined with FT synthesis received certification under ASTM D7566 standard on 2009 and covers any type of feedstock allowing a maximum blend limit of 50% biojet fuel with conventional jet fuels. Syngas is produced by gasification of biomass feedstocks. It primarily consists of CO and H_2, which are the building blocks for FT liquid hydrocarbons synthesis (Wang and Tao 2016; Ng et al. 2021; Richter et al. 2018). The length of final products depends on catalyst, pressure, and temperature. Several catalysts have been reported such as Fe, Co, and Ru based on supported materials (Martínez del Monte et al. 2019; Liuzzi et al. 2021).

Several thermal processes can be applied to convert plastics into biofuels that have unlimited applications in airline industry, as well as in transportation and power generation industries (Fahim et al. 2021; Erdoğan 2020). Among them, catalytic pyrolysis is a promising conversion method. Improvement of bio-oil quality under the most economic viable way is an ongoing concern for researchers (Saha et al. 2019). Sarker et al. (2012) investigated the production of aviation range hydrocarbons from polystyrene (PS) waste plastics through thermal degradation and fractional distillation. Biofuel yield reached 23% having a carbon range of C_6–$C_{16.}$ Moreover, Ali et al. (2021) proposed pyrolysis of waste plastic feedstock over graphite as catalyst and an output of 80% of biojet fuel was observed having similar chemical properties with conventional aviation hydrocarbons. In addition, according to Liu (2020), catalytic hydrocracking of polyethylene (PE) over bifunctional catalyst Pt/Al/MCM-48 produced biojet fuel range hydrocarbons with a yield of 85.9%.

Synthesized paraffinic kerosene with aromatics (SPK/A) pathway has been certified by ASTM in 2015 allowing a blend limit of 50%. This is also a thermochemical production method based on FT synthesis and involves the addition of light aromatic compounds. Potential feedstocks for biojet fuels production include coal, natural gas, and biomass-based materials. However, it is still in demonstration stage based on fuel readiness level (FRL) (Reaching 2021; Abrantes 2050).

Alcohol-to-Jet Synthetic Paraffinic Kerosene (ATJ-SPK) pathway involves dehydration and oligomerization of alcohols into biofuel hydrocarbons. Isobiobutanol AJT pathway is approved by ASTM in 2016 allowing 30% blend of biojet with jet fuels, whereas ethanol AJT pathway received ASTM approval in 2018 with a maximum blend limit of biojet fuel 50%. Feedstocks include sugars from starch-rich or from lignocellulosic biomass. An ongoing challenge of this pathway is the high market value of alcohols as intermediates, as well as feedstock sustainability and availability (Reaching 2021; Bauen et al. 2020; Susan van Dyk 2021).

Synthesized iso-paraffins from hydroprocessed fermented sugars (SIP-SPK) pathway or also known as direct sugars to hydrocarbon conversion (DSHC) was certified by ASTM in 2014. Sugar feedstocks are fermented into C15 hydrocarbon compounds which are further upgraded to biojet fuel hydrocarbons. Maximum blend limit is 10% of biojet fuel with conventional jet fuels. However, due to the low energy input, this type of fuel is more viable for value-added chemical production (Anjani et al. 2021; Susan van Dyk 2021).

More recent ASTM-certified pathways include: hydrocarbon-HEFA-synthesized paraffinic kerosene (HEFA/SPK) pathway, catalytic hydrothermolysis jet (CHJ-SPK) pathway, and co-processing technologies. HEFA/SPK route was certified by ASTM in 2020 and bio-derived lipid feedstock is derived from microalgal species, Botryococcus braunii. Maximum blend limit is 10% and currently this method is under pilot to demonstration at technological level (Reaching 2021; Goh 2022). Catalytic hydrothermolysis jet (CHJ-SPK) pathway received ASTM certification in 2020 for blending limits up to 50%. It involves conversion of lipid feedstocks to bio-crude oil and further upgrading through hydrothermal liquefaction technology. However, it is still in pilot stage of development (Reaching 2021; Susan van Dyk 2021). Finally, co-processing technologies involve lipid and FT liquid conversion through hydrotreating processes in existing petroleum refineries and in both cases a blending limit of 5% is allowed. Lipids as intermediates received ASTM 1655 certification in 2018, whereas FT liquids in 2020 (Susan van Dyk 2021; Van Dyk 2019).

Minimum jet fuel selling price (MJFSP) estimation results reveal that there is a wide range in prices between ASTM-certified production pathways, depending on feedstock type and production cost. In most of biojet fuels, MJFSP is higher compared to the selling price of conventional jet fuel (approximately 18 $/GJ). In fact, recent literature data indicate a wide range of 23–310 $/GJ for HEFA pathway, 4–215 $/GJ for ATJ pathway, 31–108 $/GJ for CHJ pathway, 34–82 $/GJ for FT pathway, and 37–60 $/GJ for SIP pathway (Dahal 2021). HEFA and ATJ pathways are more sensitive to feedstock cost, whereas gasification technologies present greater sensitivity to capital cost. The price gap between biojet and conventional jet fuels could be bridged by airlines as (a) for lower blends (5%) a small increase in ticket price could be paid by passengers, (b) passengers may have the opportunity to voluntarily pay a further fee, and (c) airports may offer initiatives for biojet fuel usage (Reaching 2021).

In terms of CO_2 emissions, SAF is responsible for the generation of less than 80% of CO_2 compared to their fossil-based counterparts. It is estimated that CO_2 will be reduced around 11% in 2050, by increasing SAF uptake (Hasan et al. 2021). Gasification and FT pathways for biojet fuel production present lower emissions than hydroprocessing or combined pyrolysis and hydroprocessing. In case of lignocellulosic feedstock, forestry residues perform the lowest GHG emissions (O'Connell et al. 2019).

10.7 Conclusions

Through suitable WtE technologies, waste materials such as MSW as well as agricultural and industrial residues can be converted into proper energy forms. Besides electricity and heat production from waste, biomass-based waste materials can be used as feedstocks for biofuel production, including biojet fuels.

Decarbonization of the transportation sector is an ongoing concern. Aviation industry contributes to approximately 2% of global GHG emissions. Since aviation sector will continue to grow over the following years, alternative kerosene drop-in fuels provide a promising solution to existing aircraft infrastructure. Sustainable aviation fuels can be produced from various waste feedstocks such as MSW, waste cooking oil, and lignocellulosic materials. Conversion methods include thermochemical and biochemical processes. Nevertheless, to meet ASTM standard requirements, biojet fuels can be used only as blends with conventional jet fuels up to a maximum blending limit of 50%.

Currently, eight pathways have been certified by ASTM for SAF production. HEFA pathway is the most mature technology and is fully commercialized. It involves catalytic hydrogenation of lipid feedstocks, including WCO. This conversion route is projected to dominate over the following years, although feedstock availability and cost remain challenging. Moreover, lignocellulosic feedstocks can potentially contribute to biojet fuel production in the medium to long term. Advanced thermochemical methods such as gasification-FT synthesis and HTL as well as biochemical ATJ process can be applied for a variety of waste feedstocks.

Waste-to-energy conversion methods can be utilized in biojet fuel production for a sustainable aviation industry. Proper solid waste management can contribute to efficient waste material recovery in order to meet the ongoing energy demand in aviation sector. Nevertheless, the commercialization of advanced biojet fuel production methods requires further research and development. The availability of raw materials, high production cost, and lack of policy support remain ongoing challenges for advanced methods.

References

Abrantes I et al (2021) Sustainable aviation fuels and imminent technologies—CO_2 emissions evolution towards 2050. J Clean Prod 313

Ali Z et al (2021) Jet fuel produced from waste plastic with graphite as a catalyst. Mater Today: Proc

Anjani R, Thandlam A, Shu CM (2021) Biomass to bio jet fuels: a take off to the aviation industry, pp 183–213

Aracil C et al (2017) Proving the climate benefit in the production of biofuels from municipal solid waste refuse in Europe. J Clean Prod 142:2887–2900

Ashokkumar V et al (2022) Recent advances in lignocellulosic biomass for biofuels and value-added bioproducts—A critical review. Bioresour Technol 344(Pt B):126195

Asiedu A et al (2021) Waste cooking oil to jet-diesel fuel range using 2-propanol via catalytic transfer hydrogenation reactions. Biofuels 12(6):723–736

Awasthi AK et al (2021) Zero waste approach towards a sustainable waste management. Resour Environ Sustain 3

Aziz SQ, Maulood YI (2015) Contamination valuation of soil and groundwater source at anaerobic municipal solid waste landfill site. Environ Monit Assess 187(12):755

Bajpai D, Tyagi VK (2006) Biodiesel: source, production, composition, properties and its benefits. J Oleo Sci 55(10):487–502

Bauen A et al (2020) Sustainable aviation fuels: status, challenges and prospects of drop-in liquid fuels, hydrogen and electrification in aviation. Johnson Matthey Technol Rev 64(3):263–278

Beyene HD, Werkneh AA, Ambaye TG (2018) Current updates on waste to energy (WtE) technologies: a review. Renew Energy Focus 24:1–11

Brunner PH, Rechberger H (2015) Waste to energy-key element for sustainable waste management. Waste Manag 37:3–12

Calhoun JC Jr (1963) A definition of petroleum engineering. J Petrol Technol 15(07):725–727

Chand Malav L et al (2020) A review on municipal solid waste as a renewable source for waste-to-energy project in India: current practices, challenges, and future opportunities. J Clean Prod 277

Chandralingam R (2020) Lowest emission sustainable aviation biofuels as the potential replacement for the Jet-A fuels. Aircr Eng Aerosp Technol. ahead-of-print

Chen R-X, Wang W-C (2019) The production of renewable aviation fuel from waste cooking oil. Part I: bio-alkane conversion through hydro-processing of oil. Renew Energy 135:819–835

Chong JW, Chemmangattuvalappil NG (2021) Aviation biofuels: conversion routes and challenges. J Oil Palm Environ Health 12:69–85

Clark RM (1978) Analysis of urban solid waste services: a systems approach. Ann Arbor Mich: Ann Arbor Sci

Costa FF et al (2020) Lignocellulosics to biofuels: an overview of recent and relevant advances. Curr Opin Green Sustain Chem 24:21–25

Daggett D, Hadaller O, Hendricks R, Walther R (2006) Alternative fuels and their potential impact on aviation. In: 25th international congress of the aeronautical sciences

Dahal K et al (2021) Techno-economic review of alternative fuels and propulsion systems for the aviation sector. Renew Sustain Energy Rev 151

Daneshmandi M, Sahebi H, Ashayeri J (2022) The incorporated environmental policies and regulations into bioenergy supply chain management: a literature review. Sci Total Environ 820:153202

Das S et al (2019) Solid waste management: Scope and the challenge of sustainability. J Clean Prod 228:658–678

Deane JP, Pye S (2018) Europe's ambition for biofuels in aviation—A strategic review of challenges and opportunities. Energ Strat Rev 20:1–5

Di Maria F et al (2020) Is the policy of the European Union in waste management sustainable? An assessment of the Italian context. Waste Manag 103:437–448

Erdoğan S (2020) Recycling of waste plastics into pyrolytic fuels and their use in IC engines

Fahim I, Mohsen O, ElKayaly D (2021) Production of fuel from plastic waste: a feasible business. Polymers (Basel) 13(6)

Ferdous W et al (2021) Recycling of landfill wastes (tyres, plastics and glass) in construction—A review on global waste generation, performance, application and future opportunities. Resour Conserv Recycl 173

Foo WH et al (2021) The conundrum of waste cooking oil: transforming hazard into energy. J Hazard Mater 417:126129

Foster W et al (2021) Waste-to-energy conversion technologies in the UK: processes and barriers—A review. Renew Sustain Energy Rev 135

Gil A (2022) Challenges on waste-to-energy for the valorization of industrial wastes: electricity, heat and cold, bioliquids and biofuels. Environ Nanotechnol Monit Manag 17

Goh BHH et al (2020) Progress in utilisation of waste cooking oil for sustainable biodiesel and biojet fuel production. Energy Convers Manag 223

Goh BHH et al (2022) Recent advancements in catalytic conversion pathways for synthetic jet fuel produced from bioresources. Energy Convers Manag 251

Gollakota ARK, Shu CM (2022) Covid-19 and energy sector—Unique opportunity for switching to clean energy. Gondwana Res

Gopinath A, Puhan S, Nagarajan G (2010) Effect of unsaturated fatty acid esters of biodiesel fuels on combustion, performance and emission characteristics of a DI diesel engine. Int J Energy Environ 1(3):411–430

Gören S, Akküçük U (2014) Sustainable waste management, p 253

Gray N et al (2021) Decarbonising ships, planes and trucks: an analysis of suitable low-carbon fuels for the maritime, aviation and haulage sectors. Adv Appl Energy 1

Hasan MA et al (2021) Climate change mitigation pathways for the aviation sector. Sustainability 13(7)

Hashemi Devin M, Sabziparvar A-A (2002) Aviation and global climate change

IEA (2021) Renewables 2021, Paris. https://www.iea.org/reports/renewables-2021. Accessed 23 June 2022

IRENA (2021) Reaching zero with renweables: biojet fuels. https://www.irena.org/-/media/Files/IRENA/Agency/Publication/2020/Sep/IRENA_Reaching_zero_2020.pdf

Jeevahan J et al (2018) Waste into energy conversion technologies and conversion of food wastes into the potential products: a review. Int J Ambient Energy 42(9):1083–1101

Joseph K (2006) Stakeholder participation for sustainable waste management. Habitat Int 30(4):863–871

Khan I, Kabir Z (2020) Waste-to-energy generation technologies and the developing economies: a multi-criteria analysis for sustainability assessment. Renew Energy 150:320–333

Khan AH et al (2022) Current solid waste management strategies and energy recovery in developing countries—State of art review. Chemosphere 291(Pt 3):133088

Kortov S et al (2016) Municipal solid waste management in a new legislation: comprehensive approach. In: E3S web of conferences 6

Kousoulidou M, Lonza L (2016) Biofuels in aviation: fuel demand and CO_2 emissions evolution in Europe toward 2030. Transp Res Part D: Transp Environ 46:166–181

Kovacs WL (1993) Solid waste management: historical and future perspectives. Resour Conserv Recycl 8:113–130

Kumar A, Samadder SR (2017) A review on technological options of waste to energy for effective management of municipal solid waste. Waste Manag 69:407–422

Kumar S, Ankaram S (2019) Waste-to-energy model/tool presentation. In: Current developments in biotechnology and bioengineering, pp 239–258

Launay F, Roucoux A, Patin H (1998) Ruthenium colloids: a new catalyst for alkane oxidation by tBHP in a biphasic water-organic phase system. Tetrahedron Lett 39(11):1353–1356

Liu Y (2020) Hydrocracking of polyethylene to jet fuel range hydrocarbons over bifunctional catalysts containing Pt- and Al-modified MCM-48. Reactions 1(2):195–209

Liuzzi D, Pérez-Alonso FJ, Rojas S (2021) Ru-M (M = Fe or Co) catalysts with high ru surface concentration for fischer-tropsch synthesis. Fuel 293

Louis GE (2004) A historical context of municipal solid waste management in the United States. Waste Manag Res 22(4):306–322

Machineni L (2019) Lignocellulosic biofuel production: review of alternatives. Biomass Convers Biorefinery 10(3):779–791

Makarichi L, Jutidamrongphan W, Techato K-A (2018) The evolution of waste-to-energy incineration: a review. Renew Sustain Energy Rev 91:812–821

Malinauskaite J, Jouhara H (2019) The trilemma of waste-to-energy: a multi-purpose solution. Energy Policy 129:636–645

Marsh G (2008) Biofuels: aviation alternative? Renew Energy Focus 9(4):48–51

Martínez del Monte D et al (2019) Effect of K, Co and Mo addition in Fe-based catalysts for aviation biofuels production by Fischer-Tropsch synthesis. Fuel Process Technol 194

Monteiro RRC et al (2022) Production of jet biofuels by catalytic hydroprocessing of esters and fatty acids: a review. Catalysts 12(2)

Montoya Sánchez N et al (2022) Conversion of waste to sustainable aviation fuel via fischer-tropsch synthesis: front-end design decisions. Energy Sci Eng

Moreno-Gómez A et al (2020a) Production of biojet fuel from waste raw materials. In: Process systems engineering for biofuels development, pp 149–171

Morrissey AJ, Browne J (2004) Waste management models and their application to sustainable waste management. Waste Manag 24(3):297–308

Moya D et al (2017) Municipal solid waste as a valuable renewable energy resource: a worldwide opportunity of energy recovery by using waste-to-energy technologies. Energy Procedia 134:286–295

Mubeen I, Buekens A (2019) Energy from waste. In: Current developments in biotechnology and bioengineering, pp 283–305

Nanda S et al (2018) A broad introduction to first-, second-, and third-generation biofuels. In: Recent advancements in biofuels and bioenergy utilization, pp 1–25

Neamah A (2014) Separation of the petroleum system

Ng KS, Farooq D, Yang A (2021) Global biorenewable development strategies for sustainable aviation fuel production. Renew Sustain Energy Rev 150

O'Connell A et al (2019) Considerations on GHG emissions and energy balances of promising aviation biofuel pathways. Renew Sustain Energy Rev 101:504–515

Osman AI et al (2021) Conversion of biomass to biofuels and life cycle assessment: a review. Environ Chem Lett 19(6):4075–4118

Perkins G et al (2019) Recent advances in liquefaction technologies for production of liquid hydrocarbon fuels from biomass and carbonaceous wastes. Renew Sustain Energy Rev 115

Rajesh C, Majid M (2019) Sustainable waste management through waste to energy technologies in India-opportunities and environmental impacts. Int J Renew Energy Res 9:309–342

Rao MN, Sultana R, Kota SH (2017) Municipal solid waste. In: Solid and hazardous waste management, pp 3–120

Richter S et al (2018) Paths to alternative fuels for aviation. CEAS Aeronaut J 9

Saha S et al (2019) Bio-plastics and biofuel. In: Plastics to energy, pp 365–376

Saini K, Saini K (2021)

Sarker M et al (2012) Polystyrene (PS) waste plastic conversion into aviation/kerosene category of fuel by using fractional column distillation process. Int J Energy Environ 3:871–880

Scarlat N, Fahl F, Dallemand J-F (2018a) Status and opportunities for energy recovery from municipal solid waste in Europe. Waste Biomass Valoriz 10(9):2425–2444

Scarlat N, Dallemand J-F, Fahl F (2018b) Biogas: developments and perspectives in Europe. Renew Energy 129:457–472

Seadon JK (2010) Sustainable waste management systems. J Clean Prod 18(16–17):1639–1651

Shah Z et al (2019) Preparation of jet engine range fuel from biomass pyrolysis oil through hydrogenation and its comparison with aviation kerosene. Int J Green Energy 16(4):350–360

Shah AV et al (2021) Municipal solid waste as a sustainable resource for energy production: state-of-the-art review. J Environ Chem Eng 9(4)

Shamun S, Belgiorno G, Di Blasio G (2020) Engine parameters assessment for alcohols fuels application in compression ignition engines. In: Singh AP et al (eds) Alternative fuels and their utilization strategies in internal combustion engines. Springer Singapore, Singapore, pp 125–139

Sheng H, Marais K, Landry S (2015) Assessment of stratospheric fuel burn by civil commercial aviation. Transp Res Part D: Transp Environ 34:1–15

Sindhu R et al (2019) Biofuel production from biomass. In: Current developments in biotechnology and bioengineering, pp 79–92

Sridhar A et al (2021) Conversion of food waste to energy: a focus on sustainability and life cycle assessment. Fuel 302

Stupp D et al (2012) Gasoline ether oxygenate occurrence in Europe, and a review of their fate and transport characteristics in the environment. CONCAWE Rep (4)
Sun L et al (2020) An overview of waste-to-energy: feedstocks, technologies and implementations. In: Waste-to-energy, pp 1–22
Susan van Dyk JS (2021) Progress in commercialization of biojet/sustainable aviation fuels (SAF): technologies, potential and challenges. https://www.ieabioenergy.com/wp-content/uploads/2021/06/IEA-Bioenergy-Task-39-Progress-in-the-commercialisation-of-biojet-fuels-May-2021-1.pdf
Thangaraja J, Sivaramakrishna A, Desikan R (2020) Biofuels from renewable biomass resources: an overview of technologies for production, environmental and economic impacts. In: Singh AP et al (eds) Alternative fuels and their utilization strategies in internal combustion engines. Springer Singapore, Singapore, pp 25–47
Thapa B, Kc AK, Ghimire A (1970) A Review on bioremediation of petroleum hydrocarbon contaminants in soil. Kathmandu Univ J Sci Eng Technol 8(1):164–170
Van Dyk J et al (2019) Potential synergies of drop-in biofuel production with further co-processing at oil refineries. Biofuels Bioprod Biorefining 13
Vallero DA (2019) Regulation of wastes. In: Waste, pp 33–66
Vaverková (2019) Landfill impacts on the environment—Review. Geosciences 9(10)
Vergara SE, Tchobanoglous G (2012) Municipal solid waste and the environment: a global perspective. Annu Rev Environ Resour 37(1):277–309
Vijayashree P, Ganesan V (2020) A comprehensive review on oxygenated fuel additive options for unregulated emission reduction from diesel engines. In: Singh AP et al (eds) Alternative fuels and their utilization strategies in internal combustion engines. Springer Singapore, Singapore, pp 141–165
Wang W-C, Tao L (2016) Bio-jet fuel conversion technologies. Renew Sustain Energy Rev 53:801–822
Wang M et al (2019) Biomass-derived aviation fuels: challenges and perspective. Prog Energy Combust Sci 74:31–49
Wen X et al (2013) Comparison research on waste classification between China and the EU, Japan, and the USA. J Mater Cycles Waste Manag 16(2):321–334
Wilson DC (2007) Development drivers for waste management. Waste Manag Res 25(3):198–207
Xu J et al (2019) Integrated catalytic conversion of waste triglycerides to liquid hydrocarbons for aviation biofuels. J Clean Prod 222:784–792
Yan M, Waluyo J (2020) Challenges for sustainable development of waste to energy in developing countries. Waste Manag Res 38(3):229–231
Yang W-S et al (2014) Past, present and future of waste management in Korea. J Mater Cycles Waste Manag 17(2):207–217
Yang J et al (2019) An overview on performance characteristics of bio-jet fuels. Fuel 237:916–936
Yilmaz N, Atmanli A (2017) Sustainable alternative fuels in aviation. Energy 140:1378–1386
Yukesh Kannah R et al (2020) Food waste valorization: biofuels and value added product recovery. Bioresour Technol Rep 11
Zaman AU (2015) A comprehensive review of the development of zero waste management: lessons learned and guidelines. J Clean Prod 91:12–25
Zhang F et al (2022) From trash to treasure: Chemical recycling and upcycling of commodity plastic waste to fuels, high-valued chemicals and advanced materials. J Energy Chem 69:369–388
Zhang L, Butler TL, Yang* B (2020) Recent trends, opportunities and challenges of sustainable aviation fuel. In: Green energy to sustainability, pp 85–110
Zhang Y et al (2022) A review of aviation oil production from organic wastes through thermochemical technologies. Appl Energy Combust Sci 9

Part IV
Miscellaneous

Chapter 11
Feasibility Study of Laser Plasma-Assisted Stratified Combustion and Spray Investigations in a Constant Volume Chamber

Aaishi Ashirbad, Dhananjay Kumar, and Avinash Kumar Agarwal

Abstract Due to their great efficiency and fuel economy, gasoline engines, particularly Direct Injection Spark Ignition (DISI) engines, have been extensively developed and employed in passenger vehicles. Stratified and homogeneous modes are the two basic operating modes of the DISI engines. Stratified mode offers superior efficiency at part-load engine operation. Laser ignition coupled with the stratified mode of operation can increase combustion efficiency by enabling multipoint laser ignition. However, there are several challenges associated with optics and laser beam delivery that must be taken into consideration. Since fuel is injected directly, it requires adequate fuel–air mixing, which dictates the combustion characteristics. Hence, spray characterisation is required to assess the potential of laser ignition in the stratified mode operation. For the fundamental spray and combustion investigations, optical techniques such as Shadowgraphy, Schlieren imaging, Laser-Induced Fluorescence (LIF), Mie scattering and Phase Doppler Interferometry (PDI) are used, which are discussed along with their principles. Different spray chamber types are also discussed, along with design considerations and a demonstration of the horizontal cylindrical combustion chamber. This chapter also covers the fundamentals of laser ignition and associated challenges for a stratified mode DISI engine operation.

Keywords Laser ignition · Optical investigations · Stratified mode combustion · DISI

Abbreviations

PFI	Port fuel injection
SI	Spark ignition
CI	Compression ignition
NO_x	Oxides of nitrogen

A. Ashirbad · D. Kumar · A. K. Agarwal (✉)
Department of Mechanical Engineering, Indian Institute of Technology Kanpur, Kanpur, Uttar Pradesh 208016, India
e-mail: akag@iitk.ac.in

P. C. Shukla et al. (eds.), *Renewable Fuels for Sustainable Mobility*, Energy, Environment, and Sustainability, https://doi.org/10.1007/978-981-99-1392-3_11

GDI	Gasoline direct injection
DISI	Direct injection spark ignition
BSFC	Brake-specific fuel consumption
TDC	Top Dead Centre
CO_2	Carbon dioxide
THC	Total hydrocarbon
LIF	Laser-induced fluorescence
PDI	Phase Doppler interferometry
CVC	Constant volume chamber
CVCC	Constant volume combustion chamber
FoS	Factor of safety
SG	Spray guided

11.1 Introduction

Spark-ignition (SI) engines have dominated the passenger vehicle segment of the global transport sector for a long time. An increase in demand for petroleum products and deteriorating environmental conditions motivate the advancements in engine technology and improvements in emission control technologies. Most engine research and development activities focus on reducing fuel consumption and emissions and maximising engine performance and thermal efficiency (Kumar and Agarwal 2022). These factors are influenced by the fuel–air mixture formation processes, either internal or external. In the latest gasoline direct injection (GDI) engine technology, fuel–air mixing occurs in the combustion chamber by delivering high-pressure gasoline directly into the in-cylinder air. However, fuel–air mixing occurs in the intake manifold, i.e. externally, in the port fuel injection (PFI) engine technology, which powers most gasoline engines today. PFI systems have evolved extensively over the decades. However, they have some limitations. In PFI systems, throttled engine operation leads to pumping losses, which are significant at part loads. PFI engines knock at higher loads/lower speeds. Hence, they generally have lower compression ratios. Their higher in-cylinder combustion temperature increases the NO_x emissions because the PFI systems mainly use a stoichiometric mixture. Charge in the crevices of the PFI engines lead to higher hydrocarbon (HC) emissions. In PFI, the injector sprays fuel into the intake manifold, mixing it with the air inducted (Kumar and Ashok 2021). The time lag between the intake valve opening and fuel injection leads to considerable cylinder wall wetting, specifically closer to the intake valve (Kumar et al. 2022). Direct Injection Spark ignition (DISI) technology offers superiority in performance and emission controls over the PFI. DISI engines deliver better fuel economy, lower BSFC, compliance with strict emission standards, reduced engine noise, wider speed limit, improved start-ability and lower NO_x than PFI engines. DISI engines tend to achieve BSFC closer to diesel engines while maintaining higher specific power output closer to gasoline PFI engines. DISI engines operate in stratified lean mode at part loads, resulting in lower BSFC than PFI engines (Duronio et al. 2020). Simultaneously, the DISI engines deliver higher power output than PFI engines in homogenous mode at high-load conditions. Fuel injection in the combustion chamber during the intake stroke can effectively cool the incoming charge due to fuel vaporisation. Charge cooling increases the mixture density, enabling the engine to induct more air, leading to higher volumetric efficiency and more power output. Charge cooling also decreases the in-cylinder temperature at the TDC, reducing the possibility of knocking and allowing a higher compression ratio, improving engine efficiency. DISI engines combined with turbochargers or superchargers in the homogeneous mode accomplish significant engine downsizing without compromising the power output. Hence, summarising the advantages of DISI engines includes lower pumping losses, lower heat losses and higher compression ratio on account of charge cooling, increased volumetric efficiency, superior control over the air–fuel ratios, ease of cold starting with lesser fuel enrichment and lower CO_2 emissions. Although DISI engines have a few drawbacks, namely, difficulty in managing the stratified charge conditions, increased injector deposits, higher THC emissions at lower loads, and higher soot and NO_x emissions at higher loads. Further in research and development will focus on lowering BSFC, meeting emission standards and transitioning between stratified and homogeneous charge modes.

11.1.1 DISI Engines

DISI engines, also known as GDI engines, operate mainly in stratified and homogenous modes. Fuel is injected into the combustion chamber early in the intake stroke in homogenous charge mode. Early fuel injection offers adequate time for mixing fuel and air; hence, a homogenous charge is formed. This mode is used at high engine loads and speeds. While in stratified mode, fuel is injected later in the compression stroke to make an ignitable charge in the vicinity of the spark plug. Stratified mode is used at lower engine loads and speeds; therefore, less fuel is injected in stratified mode than in homogenous mode. Stratified charge mode fulfills low-speed and low-load requirements without any acceleration. However, the combustion of the lean mixture results in higher NO_x emissions due to the ineffectiveness of the catalytic converter. Therefore, the stratified charge mode is augmented with EGR to limit NO_x emissions. Charge stratification is not feasible at higher speeds due to increased in-cylinder turbulence. Homogenous charge mode is used at high-load and high-speed conditions and during acceleration since this mode produces higher torque and power. While transiting from stratified-to-homogenous charge mode, the homogeneous lean mode also caters to medium speeds and loads. In this mode, fuel is injected early in the intake stroke, keeping the mixture slightly leaner. EGR is not required in the stoichiometric mode as the catalytic converter can easily control NO_x emissions at $\lambda = 1$. Homogenous stratified mode is operated under low-speed, high-load conditions. In this mode, split injections are done, in which a portion of the fuel is injected early during the intake stroke, and the remaining fuel is injected later to achieve the charge stratification. This injection strategy is adopted during acceleration when a large fuel quantity is injected. Homogeneous mode is achieved by injecting fuel early in the intake stroke, allowing it to evaporate and mix uniformly with ambient air. The amount of injected fuel and air intake controls the engine load in each cycle. The intake air mass must be monitored to maintain the air–fuel mixture within the ignitable range. A throttle is generally used to limit the intake airflow. Engines typically operate at a stoichiometric air–fuel ratio. The intake airflow restriction causes significant pumping losses, increasing fuel consumption. There is a need to reduce these pumping losses. One possible way is to allow more air than necessary to enter the combustion chamber, thereby increasing the air–fuel ratio. However, this can cause the mixture to go beyond the ignition limit, leading to partial burns or misfires. Injecting the specified amount of fuel at such lean air–fuel ratios with the same injection timing as in homogeneous mode would result in fuel over-dilution and misfire. Hence, a concentrated charge cloud in the spark plug region enables successful ignition. This is accomplished by adding fuel later in the compression stroke and igniting the cloud before it becomes too dilute. It produces an ignitable charge cloud near the spark plug, surrounded by ambient air or recirculated exhaust gas.

The significant advantage of stratified combustion mode over homogenous combustion mode includes reduced pumping and wall heat transfer losses and increased volumetric efficiency (Huegel et al. 2015). Engines operating in stratified mode attain a higher specific heat ratio ($k = 1.4$) than those operating in homogeneous mode ($k = 1.3$). Stratified operation eliminates the need for an air restrictor to limit air intake, reducing pumping losses and fuel consumption. In stratified mixture combustion, the flame does not reach cylinder walls directly due to the insulating layer of air surrounding the charge cloud. This also reduces the cylinder wall heat transfer losses and improves the engine efficiency. As the fuel is injected later, after the closure of the intake valve, the volumetric efficiency is increased by maximising the air inducted. Moreover, charge cooling occurs as the heat of the air inside the cylinder enables fuel evaporation, reducing the knocking tendency of the engine.

Stratified charge mode enables the engine to run at a higher compression ratio because lean end gas increases the knock limit, leading to higher efficiency. In the homogeneous mode, the mixture cannot be leaned any further. This condition shows an increase in fuel efficiency in the stratified mode compared to the homogenous mode. Three types of combustion systems are used for generating charge stratification in DISI engines: wall-guided, air-guided and spray-guided (SG), as shown in Fig. 11.1 (Baumgarten 2006). These three systems employ different ways of transporting fuel spray in the vicinity of spark plug.

In a spray-guided combustion system, charge stratification occurs mainly due to the spray dynamics, with little contribution from the interactions between bulk air charge motion and piston cavity profile. In this combustion system, the spark plug is located near the injector, so the fuel injected in the vicinity of spark plug gets ignited easily. In a wall-guided combustion system, stratification is obtained from the spray's interaction with the piston cavity. The spark plug is generally located in central position, while the injector is on the side. In an air-guided system, charge stratification results from the interactions between fuel spray and air charge motion induced in the

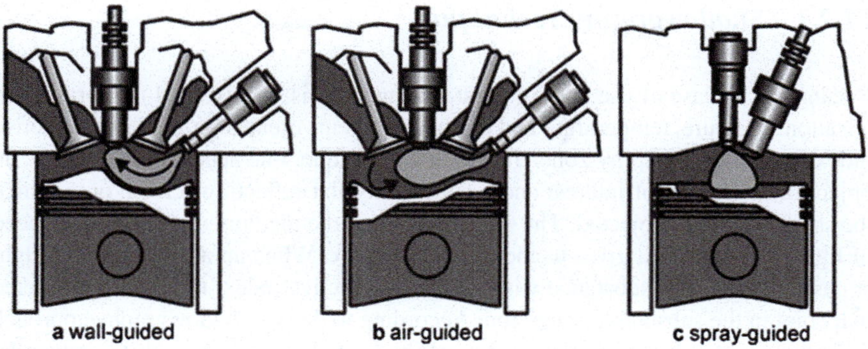

Fig. 11.1 Different combustion system designs for stratified mode engine operation (Baumgarten 2006)

combustion chamber. The spark plug and fuel injector configurations are like wall-guided combustion system. Despite the advantages of lower fuel consumption and improved thermal efficiency, stratified combustion has drawbacks. The most significant downsides of stratified combustion include combustion instability, higher particulate emissions and interference with exhaust gas after-treatment systems (Whitaker et al. 2011; Farron et al. 2011). This chapter focusses on investigating the parameters which affect combustion instability. Successful stratified combustion depends on various factors. In stratified mode, considerable air–fuel mixing occurs before ignition. The ignition and fuel–air mixing systems control the combustion predominantly. Fuel–air mixing can be controlled by injection duration, ignition delay, fuel type, fuel injection pressure, air motion and fuel properties. The proximity of ignition and injection can lead to adverse conditions at the spark gap, such as fuel vapour fluctuations, gradients in air–fuel ratio and wetting of the spark plug electrodes due to liquid fuel droplets (Piock et al. 2010). These factors may lead to poor ignition and unstable flame kernel evolution, eventually leading to combustion instability due to misfire or partially burnt cycles. Therefore, it is necessary to understand the ignition and early flame evolution processes to improve combustion stability.

11.2 Optical Techniques for Combustion and Spray Investigations

Several optical techniques are used to investigate the spray and combustion in the constant volume chambers. Broadly, these optical techniques can be (i) direct visualisation and (ii) laser based. Some of these techniques and their working principles are discussed in the following sub-sections.

11.2.1 Shadowgraphy Technique

Because of its ease of use and low cost, this method is widely used for spray visualisation. Pressure, temperature and mixture gradients create non-uniform refractive index patches in the spray zone. To use this technique, one needs a light source, an object in the region of interest and a sufficiently flat reflecting screen, onto which the shadow can be projected. The speed of light in the medium divided by the speed of light in the vacuum gives us the refractive index. When an angled beam of light is passing through a substance with a higher refractive index, the light is refracted to a point in the substance's direction. According to theory, shadowgraphy approach finds the second derivative of density, while the Schlieren approach identifies the first derivative of the density (Settles 2001).

11.2.2 Schlieren Imaging Technique

Besides placing a sharp edge stop in front of the camera at the focal point of the optics, the Schlieren imaging technique shares the same optical configuration as the shadowgraphy. Schlieren imaging technique offers benefit that cut-off by sharp edge may be readily altered. In contrast, optical path length primarily determines the sensitivity of shadowgraphy. Vapour phase sprays are often seen using the Schlieren method. This approach involves refracted light beams, which convert density changes into light intensity variations. Bright and dark zones are created in the camera. An intense pointed light source, two concave mirrors, a sharp edge and a fast camera make up the Schlieren setup. Two spherical focussing mirrors are the main components of this Schlieren imaging technique experiment. A schematic of the experimental setup is shown in Fig. 11.2.

The reliability of the Schlieren system relies on the larger focal lengths of the two concave mirrors. This technique helps understand the spray and combustion and can be applied to gaseous and liquid fuels. Lee et al. (2021) analysed the behaviour of hydrogen spray at different ambient pressures via a hollow cone injector. They reported that spray collapsed severely at high ambient pressures due to the substantial pressure differential between inner and outer surfaces of hollow-cone-shaped spray. Images of the spray boundaries and computed spray cone angles are shown in Fig. 11.3. Spray cone angle is reduced with increasing ambient pressure. The aerodynamic action at the nozzle region caused a reduction in static ambient gas pressure spray's inner and outer parts as flow motion was generated. Beyond 2 MPa, the centroid position shifted in the other way as pressure rose.

Because of the extreme spray contraction at the injector, the spray plume collided instead of mixing. In this way, injected hydrogen lost its velocity and spreads outward. Spraying hydrogen in its gaseous state offers a similar feel to spraying gasoline in its liquid form. However, aerodynamic effects are more likely to be felt by the spray. Since hydrogen has a weak spray stiffness, it is essential to regulate and design the

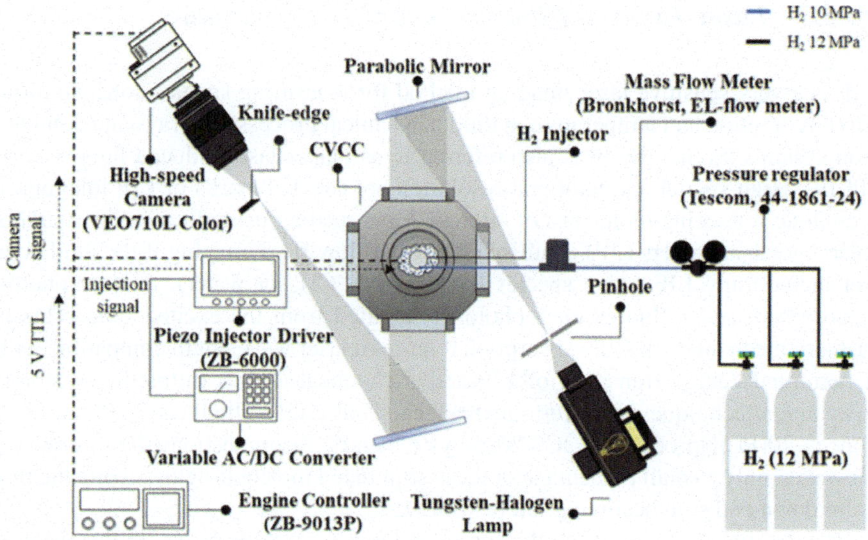

Fig. 11.2 Z-type two-mirror Schlieren imaging system for visualising high-pressure hydrogen spray (Lee et al. 2021)

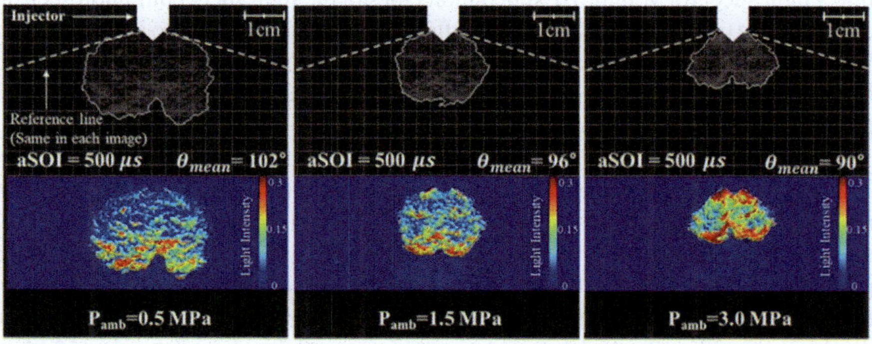

Fig. 11.3 Spray cone angle at different ambient pressures (Top: spray images with boundary detection in the same background, Bottom: grey scaled light intensity images) (Lee et al. 2021)

spray while considering environmental factors carefully. Combustion efficiency may decrease if hydrogen accumulates near the injector and remains in a rich mixture. To counteract the reduction in spray area caused by hydrogen's shrinking atomisation, the nozzle's angle should be widened. Hydrogen dispersion may be solved in several ways, including changing the injector's geometry and lowering the surrounding pressure. Since the injection evolution characteristics are known, a hydrogen-injection method can be designed.

11.2.3 Laser-Induced Fluorescence (LIF) Technique

LIF is a very sensitive laser imaging method for measuring species concentration, mixture fraction and temperature in fluid mechanical processes, sprays and combustion systems. Sometimes it is also referred to as planer laser-induced fluorescence (PLIF). High spatial and temporal resolution are two hallmarks of LIF imaging, a technique for seeing molecules of interest. Flow seeding using fluorescent markers (also known as tracers LIF) is utilised for scalar flow-field imaging if the fluid does not include any LIF-active species (such as N_2, CH_4 or water). A laser photon is absorbed, and a fluorescence photon is emitted from the excited state. This is primary mechanism of LIF. Absorption requires a laser wavelength compatible with a permitted energy transition of LIF-active chemical (atom) (https://www.smart-piv.com/en/techniques/lif-plif/#:~:text=Planar%20Laser%20Induced%20Fluorescence%20(PLIF,processes%2C%20sprays%20and%20combustion%20systems). However, only a small percentage of these stimulated molecules glow, while the rest calm down and stop becoming fluorescent.

It can differentiate between the sprays in liquid to vapour phase. In the absence of oxygen, quenching has a negligible effect on fluorescence intensity and may be disregarded. A typical schematic is shown in Fig. 11.4 (Chang et al. 2020). The fluid region of interest, such as flames or sprays, is intersected by a laser beam created into a light sheet (or volume). Excited molecules in the light sheet emit fluorescence, captured by a time-gated digital camera after passing through a filter to isolate specific wavelengths. The LIF signals are amplified using an image intensifier for pulsed UV LIF applications. Calibration measurements are the basis for transforming LIF pictures into interpretable concentration or temperature fields.

Feng et al. (2021) used LIF imaging in a constant volume spray chamber to examine the differences between gasoline and diesel sprays produced by a common rail direct injection (CRDI) system (Feng et al. 2021). In evaporating conditions, fuel injection has minimal effect on the liquid phase penetration. In contrast, vapour phase penetration steadily increased due to the development of a branch-like structure in the downstream vapour phase. The first important observation is that gasoline spray displays a significantly shorter liquid penetration length by approximately 45% than diesel spray. A minor difference between the vapour penetration lengths of gasoline and diesel sprays is observed. It indicates that the difference in physical properties has a small impact on the development of the vapour phase. As for the maximum equivalence ratio of sprays, physical properties have a particular influence on them. The fuel-rich regions in gasoline sprays are likely to generate high soot emissions. This is probably one of the reasons that high soot emissions can be observed when gasoline PPC mode is used at high loads (Manente et al. 2010). Figure 11.5 displays the spray evolution concerning the liquid fluorescence intensity, equivalence ratio and vapour temperature at $P_i = 60$ MPa, $t_i = 1.25$ ms, $T_a = 773$ K and $P_a = 2$ MPa.

The injector nozzle tip is centred on the left side of each image. The fuel is horizontally injected to the right. The region where fuel is in its liquid form is somewhat tiny. The liquid phase disappeared after just a few centimetres of spray penetration.

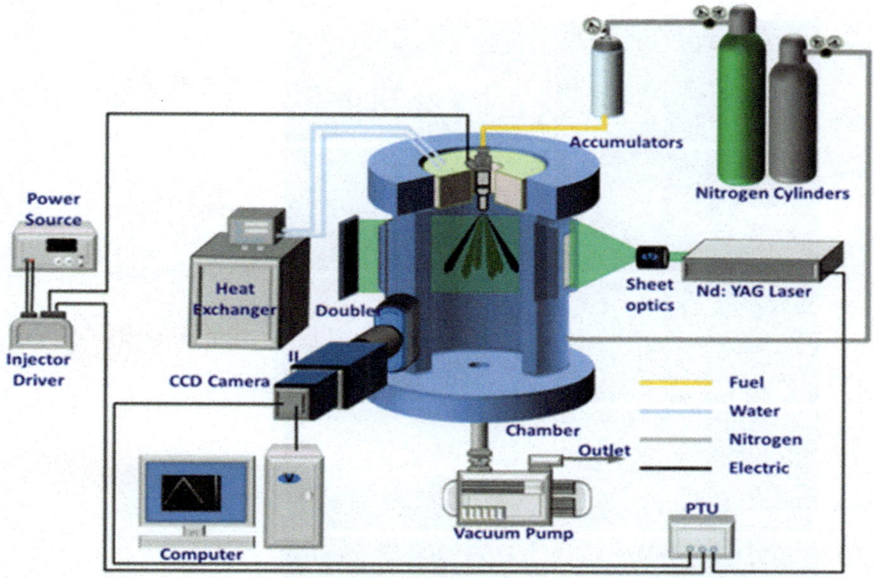

Fig. 11.4 Typical experimental setup for LIF imaging (Chang et al. 2020)

The time it takes for spray velocity and evaporation to equalise is negligible, thanks to the fuel. Hence, there is a slight variation in liquid phase penetration duration over time. They also observed that branch-like structure appeared in the downward liquid phase, highlighted in the white-dotted circle in Fig. 11.5. Locally uneven mixtures, caused by the branched structure formed by shear stress at the interface between liquid phase and surrounding gas, may significantly impact the combustion, performance and emissions (Kosaka et al. 1992).

11.2.4 Mie Scattering Technique

The light elastically scattered by particles with a diameter equal to or bigger than the wavelength of incoming light is called 'Mie scattering'. Mie scattering is directly proportional to the square of exposed particle diameter. Since Mie scattering is much more powerful than Rayleigh scattering, it may act as a source of interference for the latter, which scatters light at a considerably lower efficiency. For practical Mie imaging studies, it is vital to consider the considerable angular dependence of the scattered intensity, which is most pronounced for smaller particles. Mie scattering is a common Particle Image Velocimetry (PIV) method for determining flow speeds. Zhau et al. investigated the spray behaviour in flash boiling conditions for blended alkanes (n-pentane, iso-octane and n-decane) using the Mie scattering technique. A typical schematic of the experimental setup is shown in Fig. 11.6.

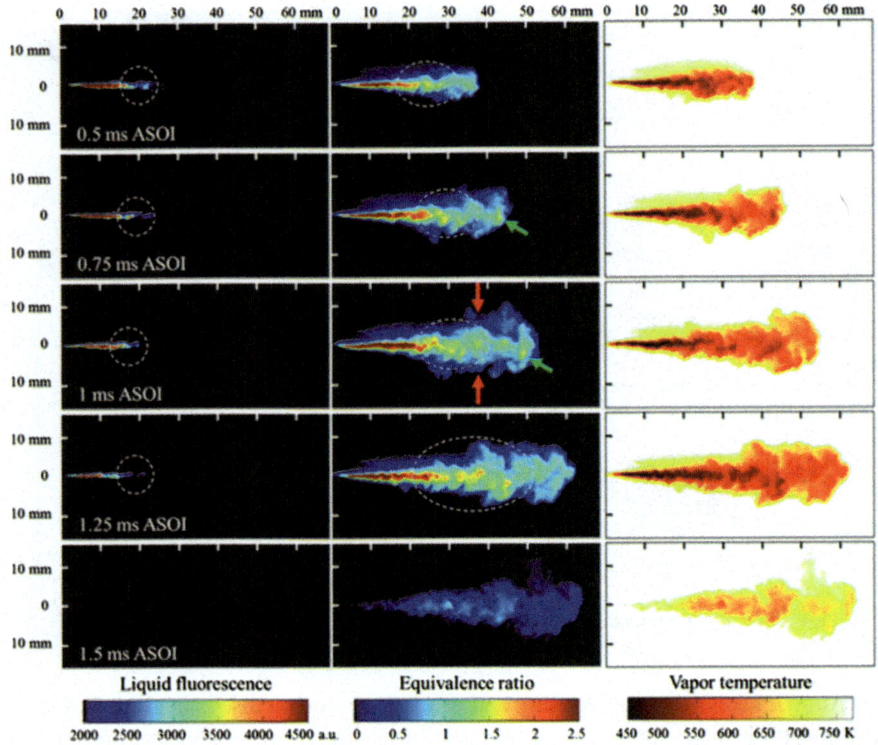

Fig. 11.5 Fluorescence intensity, equivalence ratio and vapour temperature distributions over time for a variety of liquid sprays, while conditions are $P_i = 60$ MPa, $t_i = 1.25$ ms, $T_a = 773$ K and $P_a = 2$ MPa (Feng et al. 2021)

They systematically analyse the macro-spray characteristics for the mixture containing n-pentane, iso-octane and n-decane in different superheated conditions, creating flash boiling phenomena. They also suggested a new superheated index of multicomponent fuel and its impact on the spray structure. Figure 11.7 represents the spray structure of different fuels in ternary plot at various fuel temperatures and 1.0 ms after the start of injection. Subcooled spray refers specifically to spray that mainly retains a single liquid phase. For this scenario, the primary factor in fuel plume separation is the injector orifice's orientation, which determines the flow of fuel plumes. This allows us to differentiate between spray plumes by looking at their cross sections. Figure 11.7 shows that at a fuel temperature of 30 °C, the spray cross patterns of all 15 test fuels are differentiated.

The component percentages of fuel do not heavily impact the spray cross-pattern geometry. Still, n-pentane stands out from the rest of test fuels because its spray plumes are visibly larger and closer together. Although the spray plumes of all test fuels tend to widen and get closer as the fuel temperature rises from 30 to

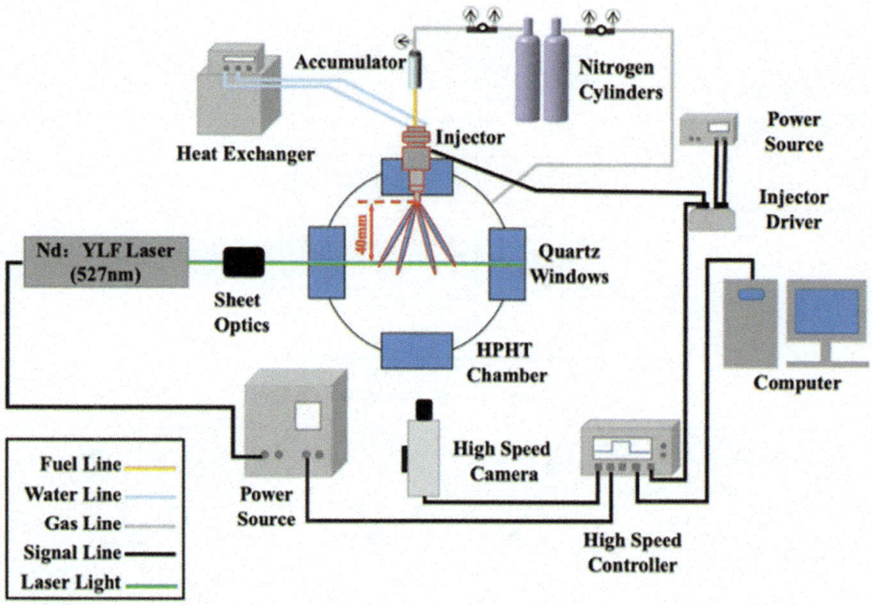

Fig. 11.6 Typical experimental setup for spray imaging using Mie scattering (Zhou et al. 2021)

50 °C, the spray cross pattern of n-pentane goes from being separated to being joined considerably more clearly than other fuels.

11.2.5 Phase Doppler Interferometry (PDI) Technique

PDI is an advanced technique for the microscopic characterisation of liquid sprays. It enables measuring the droplet size and velocity distribution, and volume flux at a point within the spray plume. Characterising each droplet that travels through the probe volume and constructing precise ensemble statistics provides the most comprehensive data on the sprays. High spatial resolution measurements may be taken with little disruption owing to the probe volume generated by two (or four) laser beams crossing at a single point. Moving the instrument or nozzle to gather data at many predetermined places is one way to characterise a spray pattern. Data is combined to show spray pattern's overall droplet size and velocity distribution or volume flow. PDI device contains a laser transmitter that sends out two laser beams of same wavelength with a phase difference to intersect at a small point called 'focal volume'. The constructive and destructive interference of two lasers at the intersection creates an interference pattern of a particular frequency. A typical schematic of 3D-PDI is shown in Fig. 11.8.

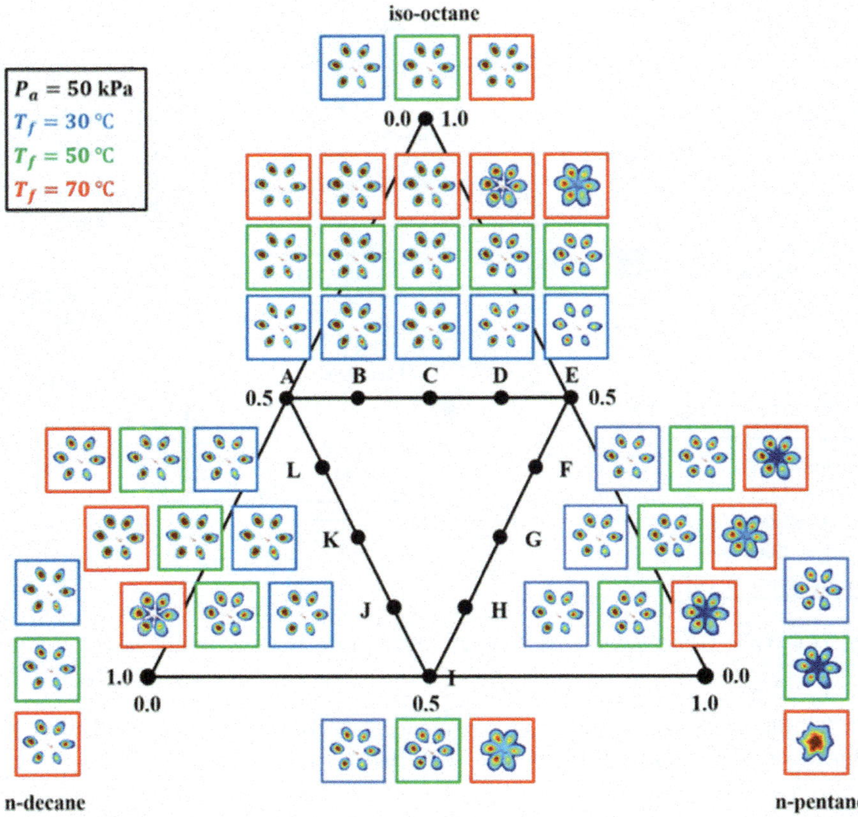

Fig. 11.7 Spray structures of different fuels in ternary plot at various fuel temperatures and 1.0 ms after the start of injection (Zhou et al. 2021)

The PDI instrument's receiver unit detects and analyses the refracted interference pattern caused by a droplet acting as a prism. The high-powered lasers enable measurements within a light-to-medium level of spray droplet density and enable the characterisation of points within spray pattern without losing accuracy due to droplets entering the laser beams outside the laser-intersection region. Extensive data processing is sometimes needed to turn the findings into usable data. Fdida et al. investigated the primary atomisation of cryogenic LOX/nitrogen and LOX/helium sprays and measures droplet size and velocity distributions near the injector (Fdida et al. 2019). Shadowgraphy is generally performed before PDI to visualise the droplet locations and helps to select the region of interest.

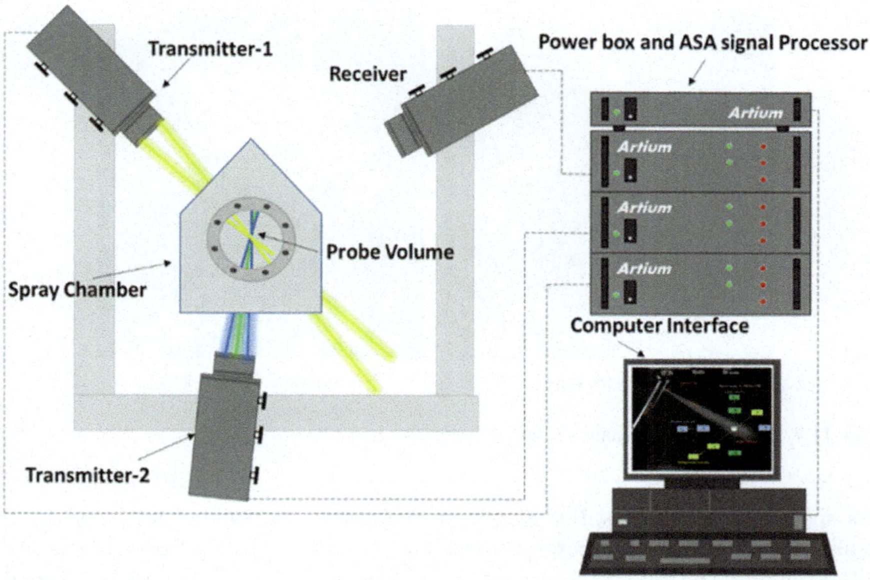

Fig. 11.8 A typical schematic of PDI setup for spray characterisation

11.3 Constant Volume Combustion/Spray Chamber Development

CVC for spray and combustion experiments in a DISI engine under simulated conditions was designed and instrumented at the Engine Research Laboratory (ERL) at IIT Kanpur. The following sections report various CVC's design, development and experimental activities.

11.3.1 General Design Considerations

For spray and combustion fundamental research investigations, constant volume chambers with optical access are preferred. The main objective of designing these chambers is to create environmental conditions similar to actual engines. For instance, the chambers should be capable of producing and withstanding the high pressure-temperature environment similar to engine running at TDC. Therefore, it is crucial to consider high pressure and temperature while designing a constant volume chamber (CVC). Depending on the experimental requirements, CVC might be spherical, cubical, pentagonal or cylindrical. Cylindrical chambers are further classified into horizontal and vertical, depending on orientation. The cylinder and fuel injector axes are parallel in vertical CVC instead of perpendicular in horizontal CVC. The chamber selection also depends on application and optical techniques employed

(a) Isometric view-1 (a) Isometric view-2

Fig. 11.9 Image of the constant volume chamber

for specific investigations. For spray and combustion study the chamber should be equipped with ignition sources, specifically if it simulates the stratified combustion of a DISI engine. The following sub-section will demonstrate the design of CVC capable of simulating stratified and homogeneous combustion of gasoline spray using laser and spark ignition.

11.3.2 Design of Horizontal CVC

CVC was designed to have a volume of 2.98 L, with an inner diameter of 130 mm and a length of 225 mm. It has three optical windows, one for laser ignition (providing access to the laser beam) and the other for visualising the flames. The clear aperture of an optical window is 50 mm, and its thickness is 25 mm. Figure 11.9 shows the image of the developed CVC. Toughened glass is used for optical access to CVC because of its mechanical and thermal strength and low cost. The CVC had four holes machined in its body for housing the pressure sensor, gas filling, spark plug and injector.

11.3.3 Material Selection and Structural Analysis of CVC

Stainless steel is used to manufacture the CVC due to its high corrosion resistance, rust resistance and low strain. Theoretically, spherical vessels are considered superior due to their uniformability to distribute fluid pressure. However, cylindrical vessels are chosen over spherical vessels because cylindrical vessels are easy to fabricate and are economical. A simulation was performed on the CVC model to test its strength under limiting conditions, which helped to determine the window material.

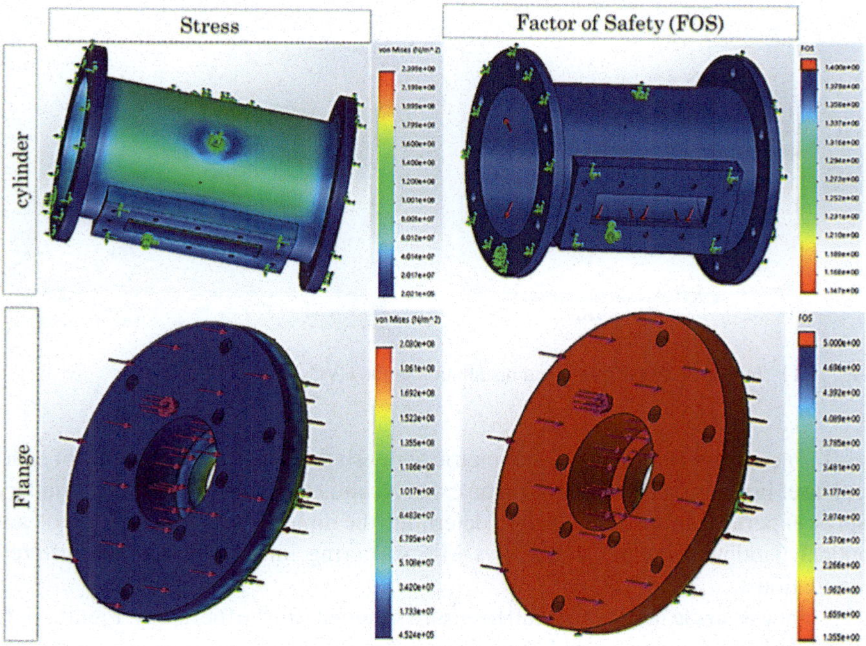

Fig. 11.10 Stress and FoS of main cylinder and associated flanges for side mounting of optical window

The minimum factor of safety (FoS) is set as the deciding parameter in choosing a suitable window material. The minimum FoS for the cylinder and the flange is chosen to be 1.2 and 1.45, respectively. Simulations determined that toughened glass has a reasonable FoS, close to quartz and sapphire, but at a significantly lower cost. Therefore, toughened glass windows are chosen because they offer good mechanical and thermal strength at a reasonable cost. Internal pressure, temperature and bolting stresses are incorporated in determining the structural limits of CVC components. Figure 11.10 shows the CVC component's stress analysis, depicting the points experiencing maximum stress.

11.3.4 Injector and Spark-Plug Assembly

For a spray-guided multi-hole DISI injector, an injector holder is designed, manufactured and installed on CVC. A pneumatic high-pressure pump (Maximator; HTPU-M1-450 SS) is used to inject gasoline at high fuel injection pressure (FIP). Figure 11.11 shows the injector holder and its installation on the CVC. The upper part acts as the common rail for DISI injector.

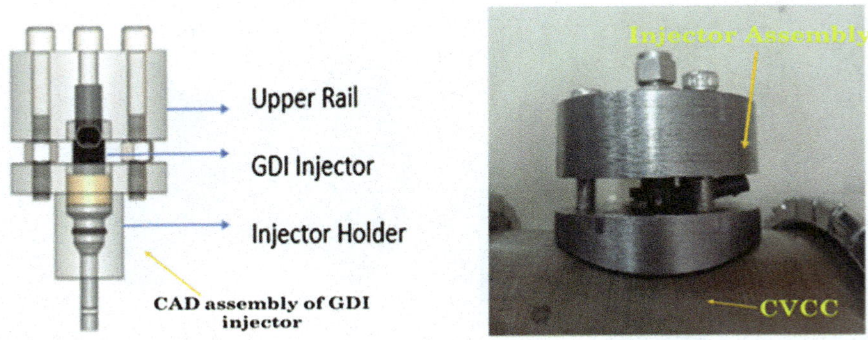

Fig. 11.11 Injector CAD drawing and installation in the CVC

It is necessary to generate a symmetric spray so that comparative effects of spark and laser ignition can be observed at an exact location within the fuel spray boundary. Spray experiments are conducted to determine the dimensions of injector's axis and optical window. Figure 11.12 shows Mie scattering images of spray in different orientations.

A slot was made in the upper rail to ensure symmetry in the fuel spray. Figure 11.13 shows injector orientation and optical view depicting symmetry w.r.t. laser and the spark plug. The spark plug mount is designed, manufactured and installed on the CVC.

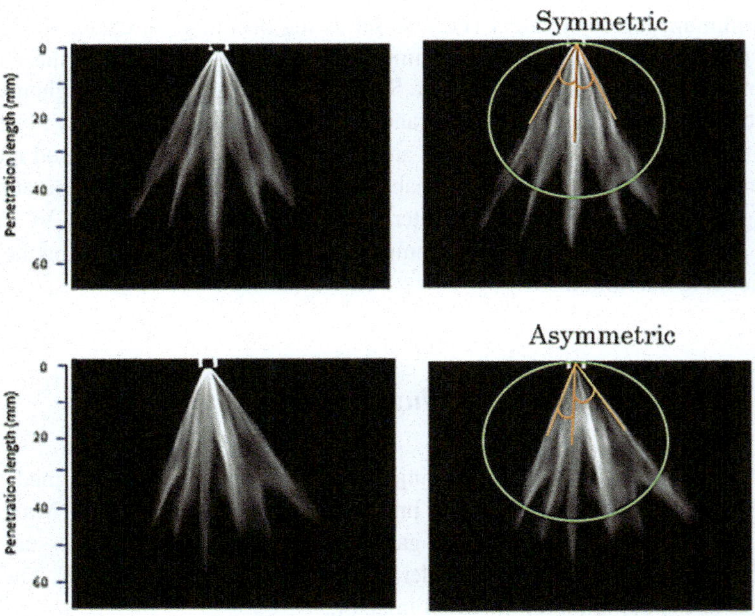

Fig. 11.12 Mie scattering imaging from the front (top row) and side view (bottom row)

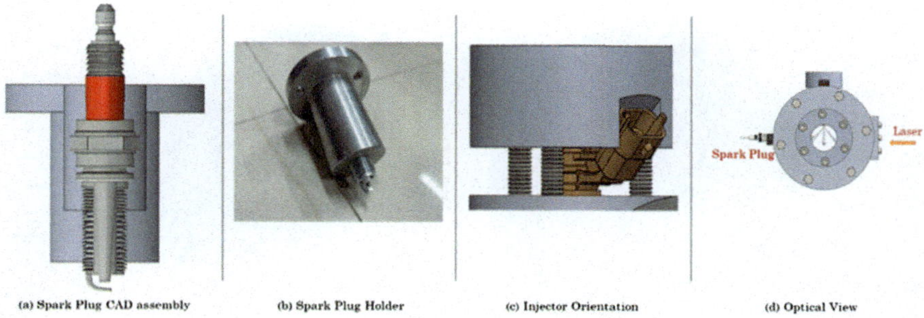

Fig. 11.13 Installation of injector and spark plug in CVC

11.3.5 Structural and Thermal Analyses of Optical Windows

Optical windows provide access to the laser beam and allows flame visualisation. High-pressure gasoline spray penetration length is the primary criterion for selecting the dimensions of optical windows. Figure 11.14 shows the structural analysis of optical windows incorporating clamping forces, internal pressure and chamber temperature.

11.3.6 Manufacturing, Assembly and Testing of CVC

Different sub-systems of CVC, such as injector, spark plug and optical window assembly, are designed, manufactured and assembled. Figure 11.15 shows different views of CAD assembly of CVC.

The CVC is tested at different operating pressures and temperatures. The CVC can be heated up to 200 °C using two-band heaters of 575 W rating. A thermocouple is mounted on CVC to measure the chamber temperature and control it with a precision of ±5 °C. Figure 11.16 shows the temperature measurement and control circuit schematic. CVC is tested to withstand up to 60 bar static pressure. A piezoelectric pressure transducer is mounted onto the CVC, transferring the signal to the high-speed data acquisition system via a charge amplifier.

11.4 Ignition Systems

One of the significant challenges for stratified combustion is to prepare a repeatable fuel–air mixture at the ignition location. This must be done to minimise partial burns or misfires while maximising the stratified combustion efficiency. Another critical factor is the ignition system.

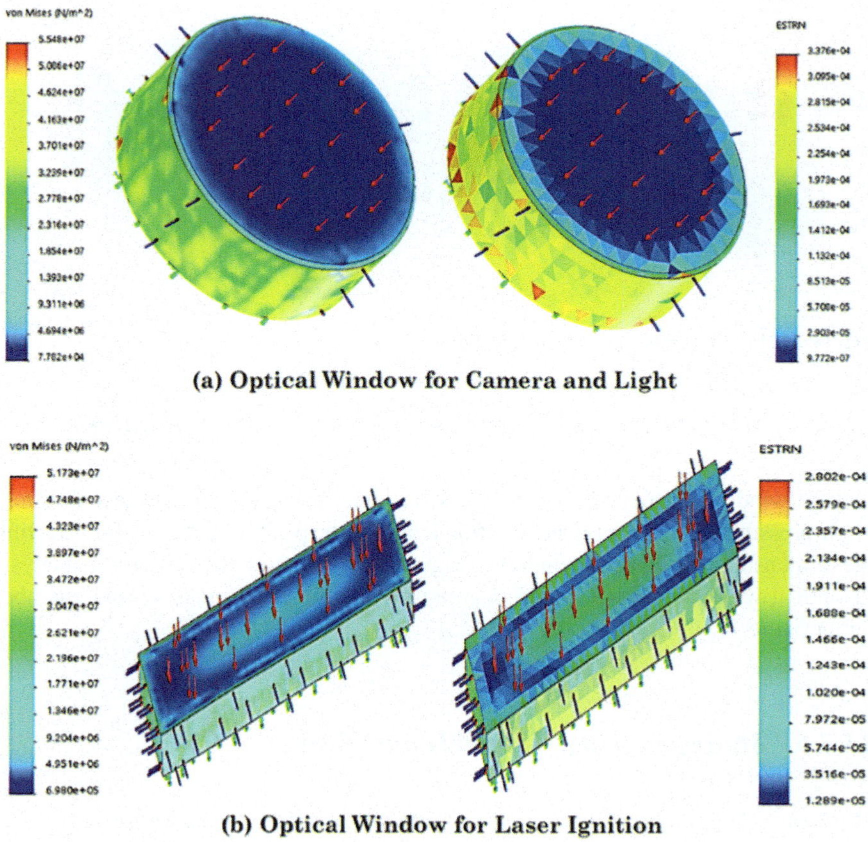

Fig. 11.14 Structural analysis of different windows used for this chamber

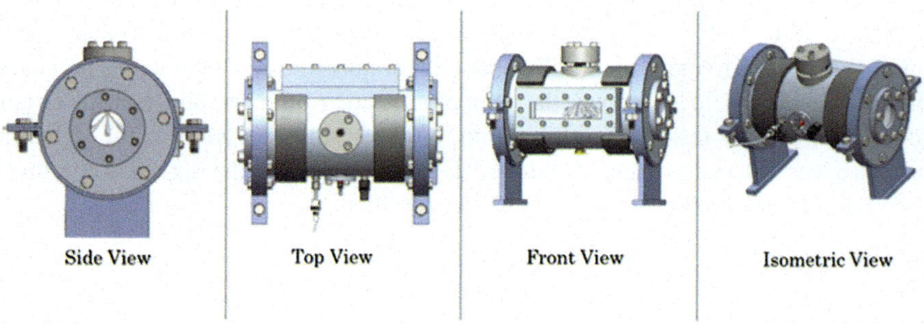

Fig. 11.15 CAD assembly of the CVC

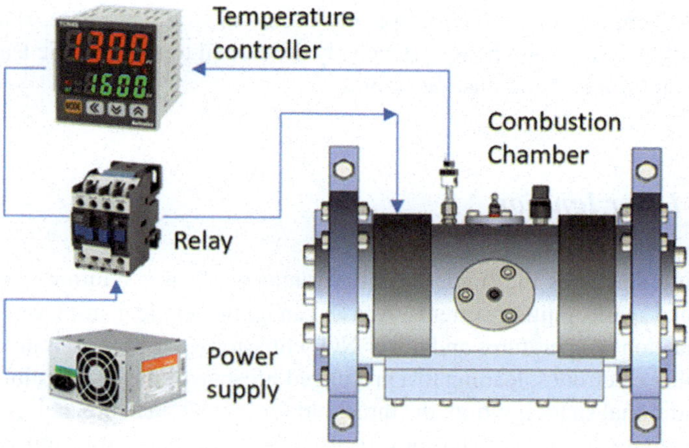

Fig. 11.16 Schematic of temperature measurement and control circuit

11.4.1 Limitations of Spark Ignition

The combustible charge is ionised between the spark plug electrodes in a conventional electric SI system. The spark electrode surface interacts with plasma and may face issues such as erosion of spark plug electrodes. Several factors contribute to the erosion of the spark plug electrodes. These factors include electrode gap, electrode shape and size, fuel type, electrode material, combustion temperature and spark energy. In DISI engines, EGR, globally lean air–fuel mixture, and boost are used to attain high combustion efficiency and low NO_x emissions (Kneifel et al. 2009). These conditions can cause misfiring and combustion instability in electric spark-ignition systems (Piock et al. 2010). In spray-guided combustion systems, the spray is in the proximity of the electrodes, velocity and equivalence ratio distribution gradients are higher at the spark gap and therefore charge can quickly go beyond the ignitability range (Drake et al. 2005; Dahms et al. 2009).

Furthermore, extended spark intervals up to 3 ms and high-velocity variations at the spark gap can cause the spark channel to grow wider (Dahms et al. 2011; Martin et al. 2010), thereby incorporating a wider range of velocity and equivalence ratios. This can cause an early flame development to have an asymmetric shape with substantial cycle-to-cycle variations (Matsumoto et al. 2010). Also, initial flame kernel development and convective motion of the spark channel have a significant role in the misfiring tendency. For successful flame propagation, movement of the flame kernel in the downward and outward direction from the spark plug is necessary (Peterson et al. 2011). The spark plug electrodes can further influence the primary kernel incidents, early flame quenching, upsetting in-chamber flows and promoting irregular flames (Febler et al. 2008; Orlandini et al. 2009). The electrodes are further dependent on spray wetting/choking and enhance the residue formation (Weinberg and Wilson 1971). Moreover, the spark-ignition systems are not very effective in igniting the lean mixture, which frequently occurs in stratified mode. Hence, alternative ignition systems are being extensively developed to improve the lean air–fuel mixtures' ignition and combustion stability.

11.4.2 Laser Ignition

This section discusses the fundamental advantages of laser ignition systems over conventional spark-ignition systems. A laser can ignite very lean mixtures, leading to low combustion temperatures and lower NO_x emissions. There is no issue of erosion of spark plug electrodes, leading to a prolonged lifespan of the laser ignition system. Unlike traditional spark ignition, the threshold for optical breakdown decreases with increasing ambient pressure (Bradley et al. 2004). Laser ignition supports ignition pressures as high as 35 bar, leading to improve the engine efficiency. There is flexibility in choosing the ignition location, which can be used to focus on the combustion location to reduce the maximum flame travel distance. This may lead to higher efficiency, specifically in very lean mixtures. A single laser source may be converted to a multipoint ignition system to enhance the combustion of a lean mixture with greater efficiency (Dale et al. 1978; Ryu et al. 2009; Phuoc 2000; Morsy and Chung 2003; Morsy et al. 2001). Ignition timing can be precisely controlled for high engine efficiency.

Compared to the electric spark, a short ignition delay has been observed for laser ignition (Grob et al. 2009). There is a heat loss while transferring the spark energy through electrodes to the flame kernel. Laser ignition permits the maximum energy transfer to the flame region by eliminating the heat loss from the electrodes. Also, it was observed that flames produced by a laser source could travel up to 20 times faster than unstretched laminar flames (Bradley et al. 2004). Rapid flame growth and quicker ignition can reduce cyclic variations and combustion duration. Since the laser ignition system includes smaller components, it eliminates the need for extra space in the cylinder head, leading to the feasibility of having larger valves, hence contributing to even higher efficiency engine development. Laser ignition offers significant prospects for improving engine combustion (Dearden and Shenton 2013). Different aspects must be fully exploited to develop an optimised laser ignition system control, such as laser delivery to the focal point, optical sensing and the development of inexpensive high-energy lasers. The opportunity for feedback control and optical sensing makes laser ignition exciting in the following areas:

(a) Ignition passage through the cylinder head can be easily optimised for laser ignition since fewer critical components than in conventional spark-ignition systems. This may lead to increased intake and exhaust valve sizes. The laser beam's cylinder entry part could be smaller than a spark plug.
(b) By focussing the laser beam via optical elements to the appropriate point, the ignition location can be dynamically changed, allowing a deeper ignition location than a conventional spark plug. The condensed ceramic lasers can deliver

multipoint ignition at predetermined locations, and the provision that the pump energy variations in the laser array can control the pulse timing (Tsunekane et al. 2011). Using a single laser beam, diffused optical elements can form plasma at multiple locations in the combustion chamber. Evolving technologies, such as liquid lenses, can automatically change their focus when sufficient voltage is applied. The liquid lenses provide sharp images by changing the focal point in a fraction of a second. Further evolution of these techniques could change the focal position to incorporate various engine operating conditions. This could be beneficial in DISI engines where optimised ignition location can be chosen for stratified or homogenous engine operation.

(c) With recent progress in high-frequency lasers such as fibre lasers, laser ignition systems of the future have the potential to deliver multiple pulses during a combustion cycle. Likewise, multiple spark discharges have been extensively researched for stratified and homogenous combustion. A high-frequency laser system precisely handles several pulses and pulse timings, and delivers quick sparks as required.

(d) The laser ignition system potentially offers an optical route in and out of the combustion chamber because of its self-cleaning nature. This visual pathway will monitor the light produced from the combustion event and analyse different combustion parameters through pattern recognition, signal processing and optical sensing methods. LIF and LII are being explored, but their complexity and high costs are significant factors for slower progress. In theory, all these methods can be used for real-time, online feedback control of the combustion using laser ignition in IC engines.

Despite numerous advantages, laser ignition has several challenges and drawbacks. The lasers have a high upfront cost. Mass production and the evolution of inexpensive lasers can make laser delivery systems cheaper. Therefore, it could be used commercially in practical engines. The robustness of the optical windows via which the beam is directed into the chamber is essential (Dearden and Shenton 2013; Ranner et al. 2007). The optical windows should sustain high pressure and temperature (Phuoc 2005). Transmittivity of the laser beam reduces due to carbon residue formation on the optical surfaces, which might lead to engine misfires.

11.5 Spray and CVCC Studies

Successful stratified engine operation depends heavily on the injection and ignition systems. It is challenging to produce a repeatable charge cloud at the ignition point. This section focuses on improving ignition systems for a successful stratified operation. Spark energy can be increased to ensure successful ignition, but surroundings in the vicinity of spark plug pose many challenges, such as high air velocities and wet electrodes. These lead to flames getting extinguished. The residual energy in a single coil ignition system is low, due to which this system cannot re-ignite the extinguished flames.

11.5.1 Spark-Ignited Direct-Injected Gasoline Spray

Spark breakdown occurs within the spark plug electrodes, which triggers the combustion. The high voltage generated between the spark plug electrodes creates an electric field that produces a highly conductive thin plasma after the breakdown. This conductive film allows free electrons to move and collide with air molecules in the spark plug gap. Thus, ionising them and generating more free electrons lead to an exponential growth of electrons, called an avalanche. Then localised charge strength controls the laminar burning speed that defines flame kernel evolution. This localised charge composition governs the premixed combustion, whereas the remaining fuel-lean charge supports the flame growth. Several challenges exist in the spark-ignited DI, preventing the success of stratified combustion (Peterson et al. 2011). Airflow motion has a severe impact on the ignition breakdown.

11.5.2 Laser-Ignited Direct-Injected Gasoline Spray

It is challenging to ignite stratified charge successfully using a laser in DISI engines. When moving to spray-guided operation, Grob et al. (2009) and Febler et al. (2008) discovered similar combustion stability and ignition delay for laser and spark plug-induced ignition, even slightly quicker ignition and flame development with laser. The similarities of these ignition systems emerge as plasma in each case combines with rich charge where the ignition delay is short. Furthermore, intermittent misfire is reported with LI, whereas no misfire observed in spark ignition. The rationale behind these occasional misfires is unknown, although they thought it to be caused by the dynamics of laser-driven ignition compared to turbulence-stratified fuel spray mixing. A high-energy Q-switched laser pulse will be ~10 ns and concentrated into a diameter of a few microns to attain the plasma breakdown. In contrast, an electric spark discharge can persist up to 2–3 ms (Dahms et al. 2009; Sick et al. 2010), with the spark channel spanning up to 10 mm in a turbulent flow. Air–fuel equivalence ratio and velocity distribution near the spark-ignition point will likely be outside the ignitability range.

This is because of the superimposition of stratified flow's turbulence and the condensed space–time scales. Although condensed space–time scales have hampered laser ignition in stratified operated engines, these parameters only helped to understand the conditions affecting successful combustion. The spark channel can last up to 24 CAD for a spark time interval of 2 ms at 2000 rpm. Localised airflow conditions are bound to change dramatically throughout this time, making it difficult to comprehend the impact of localised equivalence ratio, turbulence and air motion in the combustion processes. Soot and NO_x emissions can be improved along with the accurate monitoring of ignition timing. This provides precise charge formation, improving combustion. Due to its short spark duration, Grob et al. (2009) found incorporating laser ignition in DISI engines challenging. This is because the spark kernel generated by the laser is generally located in either too-rich or too-lean mixture regions outside the ignition range. Optimal laser spark location and injection parameters significantly enhance the ignition. Laser-induced ignition is currently not a viable alternative to traditional electrical ignition because of its short spark duration. Due to its proximity to the spray, mixture enrichment at the spark location has a negative impact on the heat release. Grob et al. (2009) investigated that laser spark position significantly affected combustion. Ignition timings vary according to variations in ignition location to achieve successful stratified combustion. Since the spark interval is short, cycle-to-cycle variations cannot be controlled even in a conventional SI system. A computation model showed that even high laser energy could not reduce the possibility of a misfire beyond a specific limit. The strain rate in the in-cylinder air flow affecting the LI might be the reason. Other laser variations, such as using a double spark, should be investigated to improve the possibility of laser ignition in DISI engines. Several optical approaches could explore the feasibility of combustion using the laser. This will entail researching the laser spark's lifespan and initial stages of flame kernel growth. A computer model is developed to understand the transport mechanisms and chemical reactions involved in laser-induced combustion. The model helps us understand the flame front growth of self-sustaining reactions produced by laser-induced plasma. The ignitability range could be calculated as a function of various characteristics using parametric analyses. The calculated ignition maps demonstrated that laser-induced ignition is extremely sensitive to the equivalence ratio and the degree of heat dissipative effects that transfer the particles out of the hot laser plasma. Ignition failure is more likely in lean mixtures than in rich ones. Some studies observed random misfires, even in homogenous charge operation, attributed to localised velocity fluctuations. Calculating ignition limits for turbulent flow will be one of the next steps in the research. The impact of kernel shape on subsequent flame development will also be examined. This job will need two-dimensional calculations, allowing for a more detailed examination of the effect of non-homogeneous flow fields around the hot kernel. Genzale et al. (2011) studied laser ignition at various timings of stratified gasoline spray (Genzale et al. 2011). Experiments were conducted in a cubical CVC under replicated DISI engine conditions. A Bosch automotive DISI injector is used for high injection spray.

Figure 11.17 shows the laser ignition at 340 μs AEOI at an injection duration of 1.2 ms. The green and white curves represent the liquid and axial vapour boundaries. The liquid droplets move downstream of the ignition location at nearly 300 μs, suggesting that vaporised air–fuel mixture is present at the ignition location. The laser plasma becomes visible at 340 μs, and the flame propagates away from the injector tip towards a rich charge. They summarised that laser ignition is impossible until 250 μs AEOI. Before 250 μs, the plasma gets formed and extinguishes quickly. The rationale behind this is the presence of a rich mixture that leads to unwanted flame evolution. Ignition is possible when the charge becomes leaner AEOI, and the

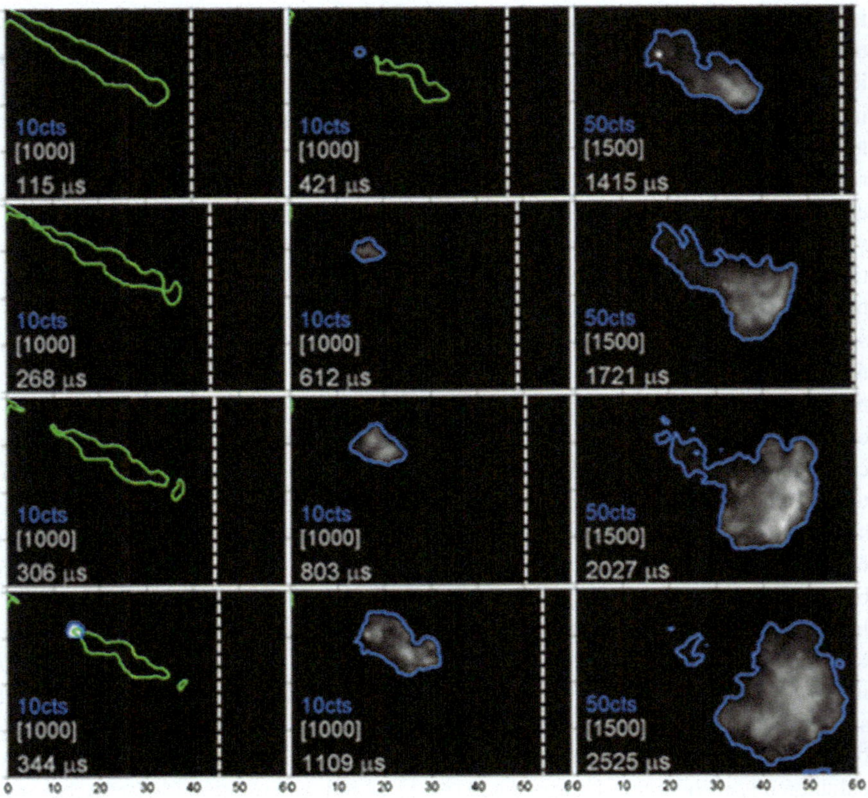

Fig. 11.17 Flame kernel images of gasoline spray with laser ignition (Genzale et al. 2011)

fuel droplets evaporate. However, a reduction in combustion efficiency is observed with more delay in the laser timings. The main charge gets away from the ignition location, and the charge becomes leaner at the ignition location, which might be a reason for the reduction in combustion efficiency.

11.6 Conclusions and Future Directions

With the advancements in engine technology, several methodologies are being adopted by the research community to fulfill the energy demand with reduced emissions. DISI technology has proven to improve engine performance with fuel economy. Stratified and homogeneous modes are the two basic operating modes of DISI engine. In the stratified mode operating DISI engine, laser ignition can extend the lean limit further. Fuel is delivered early in the intake stroke in homogeneous charge mode. Early fuel injection allows appropriate fuel–air mixing, resulting in the formation of a homogeneous charge. This mode is employed when the engine runs at high speeds and loads. Fuel is injected later in the compression stroke when the engine is in stratified mode to provide an ignitable charge close to the spark plug. Less fuel is fed into the engine in the stratified mode compared to the homogeneous mode since it is employed at lower loads and speeds. However, since fuel is injected directly, it requires adequate fuel–air mixing, which dictates the combustion characteristics. Hence, spray characterisation is required to assess the potential of laser ignition in the stratified mode operation. Various optical techniques are discussed for the fundamental spray and combustion investigations. It was recommended that for assessing the charge distribution using microscopic characteristics, the shadowgraphy technique is quite useful, requiring very few optical components. However, the Schlieren imaging technique is preferable for capturing the blue flames and minor density gradients. Different types of CVCs are discussed. A case study of a horizontal cylinder CVC design is also included. By enabling the ignition of prolonged lean fuel–air combinations, combustion gets enhanced. The choice of an ignition site within the CVC remains flexible using an LI system. By using converge lenses of various focal lengths or moving the converge lens within the plug in direction of laser beam, the location of ignition point may be adjusted. Laser ignition coupled with the stratified mode of operation certainly can increase the combustion efficiency by enabling multipoint laser ignition. However, several challenges associated with optics and laser beam delivery must be considered.

References

Baumgarten C (2006) Mixture formation in internal combustion engines. Springer Science & Business Media. https://doi.org/10.1007/3-540-30836-9_6

Bradley D, Sheppard CGW, Suardjaja IM, Woolley R (2004) Fundamentals of high-energy spark ignition with lasers. Combust Flame. https://doi.org/10.1016/J.COMBUSTFLAME.2004.04.002

Chang M, Lee Z, Park S, Park S (2020) Characteristics of flash boiling and its effects on spray behaviour in gasoline direct injection injectors: a review. Fuel. https://doi.org/10.1016/j.fuel.2020.117600

Dahms R, Fansler TD, Drake MC, Kuo TW, Lippert AM, Peters N (2009) Modeling ignition phenomena in spray-guided spark-ignited engines. Proc Combust Inst. https://doi.org/10.1016/J.PROCI.2008.05.052

Dahms RN, Drake MC, Fansler TD, Kuo TW, Peters N (2011) Understanding ignition processes in spray-guided gasoline engines using high-speed imaging and the extended spark-ignition model SparkCIMM. Part A: Spark channel processes and the turbulent flame front propagation. Combust Flame. https://doi.org/10.1016/j.combustflame.2011.03.012

Dale JD, Smy PR, Clements RM (1978) Laser ignited internal combustion engine—An experimental study. SAE Technical Papers. https://doi.org/10.4271/780329

Dearden G, Shenton T (2013) Laser ignited engines: progress, challenges and prospects. Opt Express. https://doi.org/10.1364/OE.21.0A1113

Drake MC, Fansler TD, Lippert AM (2005) Stratified-charge combustion: modelling and imaging of a spray-guided direct-injection spark-ignition engine. Proc Combust Inst. https://doi.org/10.1016/J.PROCI.2004.07.028

Duronio F, De Vita A, Allocca L, Anatone M (2020) Gasoline direct injection engines–A review of latest technologies and trends. Part 1: Spray breakup process. Fuel. https://doi.org/10.1016/j.fuel.2019.116947

Farron C, Matthias N, Foster D, Andrie M, Krieger R, Najt P et al (2011) Particulate characteristics for varying engine operation in a gasoline spark ignited, direct injection engine. In: SAE world congress and exhibition. https://doi.org/10.4271/2011-01-1220

Fdida N, Mauriot Y, Vingert L, Ristori A, Theron M (2019) Characterising primary atomisation of cryogenic LOX/Nitrogen and LOX/Helium sprays by visualisations coupled to Phase Doppler Interferometry. Acta Astronaut. https://doi.org/10.1016/j.actaastro.2018.12.029

Febler M, Wetzel M, Schenk M, Rottengruber H, Huhn W (2008) Determination of thermodynamic potentials for alternative ignition systems in spray-guided combustion. In: Proceedings of the 8th international symposium on internal combustion engine diagnostics

Feng L, Sun X, Pan X, Yi W, Cui Y, Wang Y, Wen M, Ming Z, Liu H, Yao M (2021) Gasoline spray characteristics using a high-pressure common rail diesel injection system by the method of laser-induced exciplex fluorescence. Fuel. https://doi.org/10.1016/j.fuel.2021.121174

Laser Induced Fluorescence (Lavision). https://www.smart-piv.com/en/techniques/lif-plif/#:~:text=Planar%20Laser%20Induced%20Fluorescence%20(PLIF,processes%2C%20sprays%20and%20combustion%20systems. Accessed 10th Nov 2022

Genzale CL, Pickett LM, Hoops AA, Headrick JM (2011) Laser ignition of multi-injection gasoline sprays. SAE World Congress and Exhibition. https://doi.org/10.4271/2011-01-0659

Grob V, Kubach H, Spicher U, Schiebl R, Maas U (2009) Influence of laser-induced ignition on spray-guided combustion—Experimental results and numerical simulation of ignition processes. SAE Technical Papers. https://doi.org/10.4271/2009-01-2623

Huegel P, Kubach H, Koch T, Velji A (2015) Investigations on the heat transfer in a single cylinder research SI engine with gasoline direct injection. SAE Int J Engines. https://doi.org/10.4271/2015-01-0782

Kneifel A, Buri S, Velji A, Spicher U, Pape J, Sens M (2009) Investigations on supercharging stratified part load in a spray-guided DI SI engine. SAE Int J Engines. https://doi.org/10.4271/2008-01-0143

Kosaka H, Won YH, Kamimoto T (1992) A study of the structure of diesel sprays using 2-D imaging techniques. SAE Trans. https://doi.org/10.4271/920107

Kumar TS, Ashok B (2021) Critical review on combustion phenomena of low carbon alcohols in SI engine with its challenges and future directions. Renew Sustain Energy Rev. https://doi.org/10.1016/j.rser.2021.111702

Kumar D, Agarwal AK (2022) Fundamentals, evolution, and modeling of ignition systems for spark ignition engines. In: Agarwal AK, Kumar D, Sharma N, Sonawane U (eds) Engine modeling and simulation. energy, environment, and sustainability. Springer, Singapore. https://doi.org/10.1007/978-981-16-8618-4_9

Kumar D, Sonawane U, Chandra K, Agarwal AK (2022) Experimental investigations of methanol fumigation via port fuel injection in preheated intake air in a single-cylinder dual-fuel diesel engine. Fuel. https://doi.org/10.1016/j.fuel.2022.124340

Lee S, Kim G, Bae C (2021) Behavior of hydrogen hollow-cone spray depending on the ambient pressure. Int J Hydrog Energy. https://doi.org/10.1016/j.ijhydene.2020.11.001

Manente V, Tunestal P, Johansson B, Cannella WJ (2010) Effects of ethanol and different type of gasoline fuels on partially premixed combustion from low to high load. In: SAE World Congress and Exhibition. 10.4271/2010-01-0871

Martin H, Helmut T, Tobias B, Frank A, Harald W (2010) Influence of a multi-spark ignition system on the inflammation in a spray-guided combustion process. SAE Int J Fuels Lubr. https://doi.org/10.4271/2009-24-0117

Matsumoto A, Zheng Y, Xie X bin, Lai MC, Moore W (2010) Characterization of multi-hole spray and mixing of ethanol and gasoline fuels under di engine conditions. SAE Technical Papers. https://doi.org/10.4271/2010-01-2151

Morsy MH, Chung SH (2003) Laser-induced multipoint ignition with a single-shot laser using two conical cavities for hydrogen/air mixture. Exp Thermal Fluid Sci. https://doi.org/10.1016/S0894-1777(02)00252-2

Morsy MH, Ko YS, Chung SH, Cho P (2001) Laser-induced two-point ignition of premixture with a single-shot laser. Combust Flame. https://doi.org/10.1016/S0010-2180(00)00218-2

Orlandini I, Gartung K, Schlerfer J (2009) Application of 3D-CFD simulations in the development of spark plugs. SAE Technical Papers. https://doi.org/10.4271/2009-01-0706

Peterson B, Reuss DL, Sick V (2011) High-speed imaging analysis of misfires in a spray-guided direct injection engine. Proc Combust Inst. https://doi.org/10.1016/J.PROCI.2010.07.079

Phuoc TX (2000) Single point versus multipoint laser ignition: experimental measurements of combustion times and pressures. Combust Flame. https://doi.org/10.1016/S0010-2180(00)00137-1

Phuoc TX (2005) A comparative study of the photon pressure force, the photophoretic force, and the adhesion van der Waals force. Optics Commun. https://doi.org/10.1016/J.OPTCOM.2004.10.047

Piock WF, Weyand P, Wolf E, Heise V (2010) Ignition systems for spray-guided stratified combustion. SAE Int J Engines. https://doi.org/10.4271/2010-01-0598

Ranner H, Tewari PK, Kofler H, Lackner M, Wintner E, Agarwal AK et al (2007) Laser cleaning of optical windows in internal combustion engines. https://doi.org/10.1117/1.2793704

Ryu SK, Won SH, Chung SH (2009) Laser-induced multipoint ignition with single-shot laser using conical cavities and prechamber with jet holes. Proc Combust Inst. https://doi.org/10.1016/J.PROCI.2008.05.080

Settles GS (2001) Shadowgraph techniques. In: Schlieren and shadowgraph techniques. Experimental fluid mechanics. Springer, Berlin, Heidelberg. https://doi.org/10.1007/978-3-642-56640-0_6.

Sick V, Drake MC, Fansler TD (2010) High-speed imaging for direct-injection gasoline engine research and development. Exp Fluids. https://doi.org/10.1007/S00348-010-0891-3/FIGURES/15

Tsunekane M, Pavel N, Taira T (2011) Composite, all-ceramics, high-peak power Nd:YAG/Cr4+:YAG monolithic micro-laser with multiple-beam output for engine ignition. Opt Express. https://doi.org/10.1364/OE.19.009378

Weinberg FJ, Wilson JR (1971) A preliminary investigation of the use of focused laser beams for minimum ignition energy studies. Proc R Soc London. Math Phys Sci. https://doi.org/10.1098/RSPA.1971.0012

Whitaker P, Kapus P, Ogris M, Hollerer P (2011) Measures to reduce particulate emissions from gasoline DI engines. SAE Int J Engines. https://doi.org/10.4271/2011-01-1219

Zhou X, Zhai Q, Hung DL, Li X, Xu M (2021) Study of component proportion effects on flash boiling atomisation with ternary-alkane fuel mixtures. Fuel. https://doi.org/10.1016/j.fuel.2021.120798

Chapter 12
Understanding Combustion in CI Engines for Adoption of Renewable Fuels

Ashutosh Jena and Avinash Kumar Agarwal

Abstract The combustion in diesel engines is a complex phenomenon involving two-phase flow, fluid–geometry interactions and fluid–fluid interactions. Diesel engines have seen numerous designs of the combustion chamber and injection system over the years since their invention. The initial developments were primarily based on experimental results and predictive theoretical models. However, recent developments in optical diagnostics have enabled researchers to understand in-cylinder combustion in a much more refined way. The hybrid approaches of computational simulations and optical investigations have contributed immensely to developing cleaner and more efficient diesel engines to meet emission norms. However, adopting renewable fuels would require subtle design and strategic changes in modern diesel engines. Extension of current technologies for adopting renewable fuels having significantly different physiochemical properties requires in-depth knowledge of diesel combustion. The chapter overviews the state-of-the-art optical diagnostics for in-cylinder combustion visualisation in diesel engines. The differences between optical and all-metal engines have been highlighted, and various methods to minimise the gaps have been discussed. A comprehensive literature review on the spray flames in diesel engines has been done to summarise the current understanding of diesel combustion. Various parameters affecting flame evolution have been discussed. The effect of fuel properties and the combustion of biodiesel and methanol have been discussed based on flame visualisation studies.

Keywords Flames · Optical diagnostics · CDC · Biofuels

12.1 Introduction

Diesel engines have been under continuous scrutiny since the 90s. Since then, diesel engine technology has undergone significant changes. Design optimisation and the introduction of electronic control systems have resulted in a more than 70% reduction

A. Jena · A. K. Agarwal (✉)
Department of Mechanical Engineering, Indian Institute of Technology Kanpur, Kanpur, Uttar Pradesh 208016, India
e-mail: akag@iitk.ac.in

P. C. Shukla et al. (eds.), *Renewable Fuels for Sustainable Mobility*, Energy, Environment, and Sustainability, https://doi.org/10.1007/978-981-99-1392-3_12

in the emission of oxides of nitrogen. In contrast, CO and HC emissions have been cut by >80% during the same period. However, a sustainable diesel engine needs to achieve a bigger goal of overall CO_2 reduction. Decarbonisation of propulsion systems is the prime motive of the combustion community. However, this long-term goal requires significant research efforts. Sustainable diesel alternative fuels have also shown promise in emission legislation compliance. Therefore, diesel engine research efforts must focus on developing highly efficient combustion systems and technologies suitable for alternative and renewable fuels, considering the existing fuelling infrastructure.

Sustainable renewable fuels will play a pivotal role in determining the future of IC engines. Modern engines have been optimised for conventional diesel to meet prevailing emission norms. However, the physiochemical properties of alternative fuels are quite different from diesel; therefore, research and development are required for each alternative fuel candidate. Biofuels have gained prominence recently due to their sustainability, carbon neutrality and low carbon footprint. Extensively investigated biofuels are classified into three generations based on their feedstocks (Puricelli et al. 2021). Apart from these categories, biofuel production via genetic engineering, CO_2 and thermochemical processes is also covered in the literature (Liew et al. 2014; Alalwan et al. 2019; Awudu and Zhang 2012). Alcohol and biodiesel are two prominent diesel alternatives. Alcohols are manufactured from bio-based feedstocks. The fuel-bound oxygen in alcohols is beneficial in reducing soot emissions from an engine. The soot reduction mechanism can be explained based on three different theories: (1) a reduction in the production of aromatics due to the substitution of alcohols (Duraisamy et al. 2020); (2) a lower cetane number increases the ignition delay, promoting premixed combustion, leading to lower soot formation (Wei et al. 2020; Gong et al. 2020) and (3) improved combustion due to availability of fuel oxygen (Ren et al. 2008; Wang et al. 2012).

Moreover, increased aliphatic hydrogens result in a higher number of radical sites on the soot surface, accelerating the oxidation processes (Russo et al. 2015). An increased number of active sites can sustain a higher oxidation rate at lower temperatures; therefore, soot oxidation can be effectively improved during the late combustion phase (Guo et al. 2020). The effect of methanol blending on soot formation and structure has interested researchers. However, poor miscibility of methanol with diesel limits the methanol fraction in the methanol–diesel blends. Mixture stirring (Wei et al. 2020), homogenising machine (Lin 2003) and magnetic stirring (Jamrozik 2017) are a few methods which were used to achieve stable blending with no or little stabilising additives, with limited success. Unlike alcohols, biodiesel can be blended with diesel without any additives or sophisticated methods (Agarwal et al. 2017).

Combined studies, including experiments in all-metal and optical engines, have gained attention for engine development. Based on the literature review and fundamental theories, strategies for advanced combustion can be developed. Experiments can be conducted over extensive load and speed envelope in all metal engines. The results from the metal engine can be analysed. Based on the trends of the results, experiments/simulations can be planned, as shown in Fig. 12.1. A study in a constant-volume combustion chamber (CVCC) provides enhanced optical access to the spray flames under engine-like thermodynamic conditions without the issues of engine dynamics. The experimental conditions can be easily controlled; hence, fundamental analysis of the effect of various parameters, such as oxygen concentration, and fuel properties, can be studied with greater accuracy. However, these experiments are performed in quiescent conditions; therefore, the effect of the spray–swirl interactions cannot be analysed in CVCC experiments. In-cylinder, PIV studies under cold-flow (non-reacting) conditions shed light on the additional effects of air motion on spray evolution. Further, combustion visualisation (reacting condition) experiments reveal the fundamental reasons for the trends observed in the metallic engine.

This chapter focusses on the development and state-of-the-art combustion diagnostics research in diesel engines. In the next sections, a brief history and important innovations are discussed. In the subsequent sections, limitations of optical diagnostics in its current state of the art have been highlighted. Finally, the combustion in conventional diesel engines has been discussed elaborately based on optical engine experiments and fundamental investigations. The effect of fuel properties and research in biofuels has also been included in this chapter to give a comprehensive picture of sustainable diesel engines.

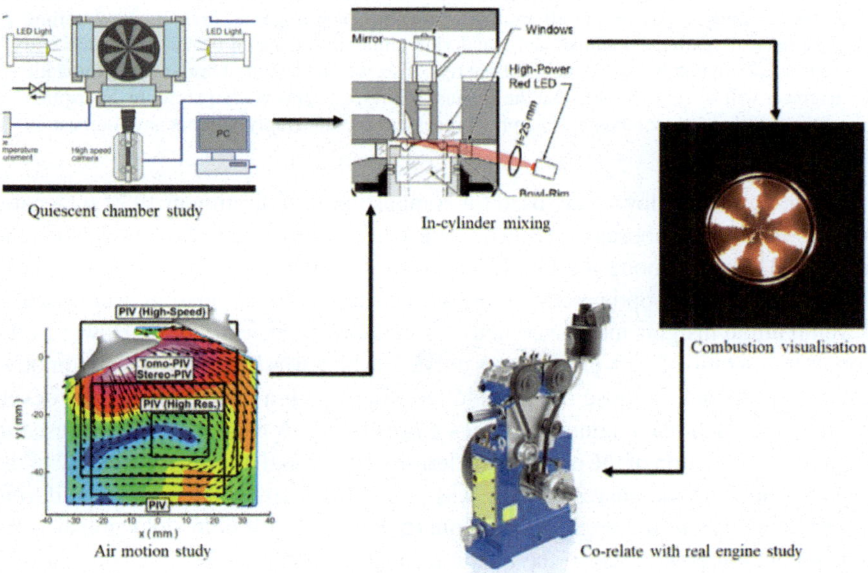

Fig. 12.1 Hybrid approach for diesel engine development

12.2 Historical Evolution

Combustion visualisation has played a crucial role in the evolution of diesel engines. The first ever description of the complex phenomenon occurring in a conventional diesel engine combustion chamber was narrated by Riccardo, the Father of Engines, perhaps based on his imagination, in 1931. He said,

> 'Let us imagine ourselves seated comfortably on top of the piston... about the end of the compression stroke. We are in complete darkness. The atmosphere is a trifle oppressive, for the temperature is well over 500 °C. It is (so) draughty that... we should be blown off our perch and hurled about like leaves in a gale. Suddenly, above our heads... a rainstorm of fuel begins to descend.... But the velocity of the droplets approaches much more nearly that of rifle bullets than of raindrops.... Then... a brilliant gleam of light appears. In an instant, this is followed by myriad others all around us,... until on all sides... the space is filled with a merry blaze of moving lights. From time to time, the smaller lights wink and go out, while the larger ones develop fiery tails like comets. Occasionally they strike the walls, but... they merely bounce off like drops of water spilled on a red-hot plate.... Looking round again, we see the lights are growing yellower ... here and there, they are crowding together in dense nebulae, and these are burning now in a sickly smoky flame, half suffocating for want of oxygen. Now...overhead,...we see that what at first was a cold rain falling through utter darkness has given place to a cascade of fire as from a rocket. For a while, this continues, then ceases abruptly.... Above and all around us are still some lingering fireballs, now trailing long tails of sparks and smoke and wandering aimlessly in search of oxygen which will consume them finally and set their souls at rest. If so, well and good; if not, some unromantic engineer will merely grumble that the exhaust is dirty. ...'. This was an accurate description of the diesel engine combustion chamber processes and, amazingly, without any combustion visualisation.

Later, quartz windows were used for combustion visualisation in 1930 at General Motors Research Laboratory (Bowditch 1961). Withrow and Boyd (Withrow and Boyd 1931) were among the first to visualise the flame front in a gasoline L-head engine. The L-head optical engine design and its modified versions were used to develop initial insights into the combustion chamber in the early days. However, the compression ratio of such engines was only ~7, which was lesser than the automotive engines used back in those days. A slotted piston with a quartz crown was developed to approach real engine conditions more closely. A canted mirror was used to capture the reflection of the combustion luminosity through the quartz window of the piston. Attaching the quartz window was one of the many challenges faced during the development of the engine. A flat quartz window with both sides polished for transparency was attached to the piston through an adhesive (Shell Chemical Co.; Epon 828). The cylinder head of the I-head engine was retained with an extended drive for the overhead cams. Therefore, the geometrical aspects of this engine were closer to the automotive engines than the earlier designs used for optical visualisation. Through this design, a compression ratio of ~11 was successfully tested. A mechanical stroboscope was designed to synchronise with the camshaft speed. This enabled data acquisition for 3° CA in every cycle at a present crank angle position. This concept was similar to the synchronisation between high-speed cameras and encoders used in today's advanced optical diagnostics systems. The optical engine enabled the first engine combustion diagnostics under load at full throttle conditions of a gasoline engine (Bowditch 1961). In the 1960s, an in-direct injection swirl chamber design engine was used to study the effect of air motion on combustion. The role of counter-rotating vortices in promoting unburned charge combustion through mixing was witnessed for the first time. Advanced combustion research and CFD studies have proved crucial for studying these flow structures in diesel engine combustion. These revolutionised designs form the basis of today's advanced optical engines used for compression ignition and spark ignition combustion diagnostics. The optical engines have come a long way in almost a century since the first ever in-cylinder flames were visualised. Their compression ratios have increased significantly, and realistic piston bowl geometries have been developed for the optical diagnostics of the industry-grade engines.

12.3 Differences Between Optical and All-Metal Engines

Optical engines have contributed significantly to the evolution of diesel technologies in the last few decades. Efforts have been made to achieve realistic engine conditions in optical engines with similar compression ratios and geometrical features to the all-metal engine. However, a few key differences must be considered while interpreting the results or designing optical diagnostics campaigns. The optical windows are generally made of quartz or sapphire. In contrast, aluminium is extensively used in most light-duty engines in the passenger vehicle segment. The thermal conductivity of aluminium is ~100 times higher than quartz and ~5 times higher than sapphire. The thermal conductivity of the wall material affects the heat flux through the wall (Woods et al. 1992). Hence, heat losses from the walls are relatively lower in optical engines than the all-metal engines. The optical liner is not cooled by water, unlike metal engines. In optical engines, lubricating oil is not used at the piston–liner interface to avoid optical distortion during imaging. Piston cooling is also lesser due to the absence of oil jet cooling. The piston surface temperature for all-metal engines is ~150 °C, significantly lesser than the optical piston surface temperature. The temperature of the optical piston can be measured through thermal imaging. However, the soot deposition during firing conditions makes the task even more challenging (Fukui et al. 2016). A recent thermography study on the piston surface temperature of the optical piston revealed that the surface temperature could go up to 192 °C during motoring conditions. The piston temperature peaked at 543 °C during the firing of the optical engine (Mancaruso et al. 2018). The temperature gradient through the thermal boundary layer near the wall decreases due to higher surface temperature, resulting in lower heat flux through the wall (Jennings and Morel 1991). The surface temperature plays a crucial role in engine heat transfer. Therefore, the boundary conditions during the modelling of all-metal and optical engines should be decided wisely based on these experimental observations. A significant difference between the combustion characteristics of the all-metal and optical diesel engines has been reported (Kashdan and Thirouard 2009; Aronsson et al. 2008). The optical engine results in an earlier start of combustion due to higher bulk gas temperature, resulting from lower heat losses through the optical components. A shorter ignition delay decreased premixed combustion and extended mixing-controlled combustion in the optical engine compared to the all-metal counterpart. Therefore, the combustion duration becomes higher in optical engines, generally. The optical piston surface temperature plays a dominant role as the combustion becomes less sensitive to injection timing and inlet air temperature. Kashdan et al. (2004) compared the combustion phasing of an all-metal engine with an optical engine having transperent liner and metallic piston. They reported an excellent match in combustion phasing and HC emissions for both engines. In contrast, the phasing advanced significantly when the optical piston replaced the metal piston in the optical engine. The results further reconfirmed the effect of the thermal conductivity of the piston material on the engine combustion characteristics. The optical engine also led to higher combustion efficiency and NO_x emissions.

Apart from the thermal perspective, the optical engine has more subtle differences in dynamic loading compared to an all-metal engine due to the additional weight of the extended piston assembly. Arroson et al. (2012) showed that the squish height changes significantly (1.5 mm for 80 bar pressure at TDC) for heavy-duty optical engines. However, the effect was relatively lesser for light-duty engines. The equivalent spring constant of the system decides the deflection of the machine parts under loading. The loading on the engine component varies dynamically with speed and acceleration due to the inertia forces. The optical engine piston's spring constant was twice that of the all-metal engine. The dynamic stress-induced strain leads to a change in the compression ratio of the engine. This change will be larger in the optical engine than in the all-metal engine due to the higher spring constant value. Moreover, variations in squish height affect the squish flow, which affects the in-cylinder charge mixing. A significant change can also lead to a change in spray targeting location w.r.t. the piston lip. Therefore, relative geometrical differences between the all-metal and optical engines during firing need to be considered while designing experiments or setting up simulations. Perini et al. (2013) conducted numerical investigations to understand the effect of stiffness on the motoring pressure trace matching. They reported that blowby plays a major role in determining the in-cylinder pressure, which also needs to be considered.

The combustion characteristics of the all-metal and optical engines need to be as close as possible to fundamentally explain the trends in an all-metal engine. Several methods match the combustion phasing in optical and all-metal engines. The skip-firing method has a first-order effect on the combustion phasing (Kashdan and Thirouard 2009). In skip-firing mode, the optical engine is fired for a limited number of cycles and then motored for an extended number of cycles. This allows the piston to cool down, and overheating of the optical components can be avoided. The use of identical piston geometry is also recommended whenever possible. The piston geometry affects the spray impingement and evaporation phenomenon, thereby the combustion and emission characteristics. Kashdan et al. (2004) demonstrated that piston deformation could decrease the compression ratio, affecting thermodynamic conditions at TDC. The piston deformation results in a change in squish height, which affects emissions. Dynamic loading increases with an increase in engine speed. Therefore, the compression ratio may vary significantly at higher speeds and loads. To compensate for these geometrical changes, the compression ratio of the optical engine can be increased to match the pressure trace of the representative condition in the all-metal engine. Fundamental investigations are generally planned at lower speeds for best representative results and lesser compensation in the compression ratio. Significant structural modifications to provide better optical access further reduces the compression ratio. For example, Mueller and Musculus (2001) reported that increasing squish height resulted in a 3% higher mass in the optical engine for the same intake condition as the all-metal engine. To match the TDC conditions, the intake air was boosted, and its temperature was increased. As a result, the optical engine trapped a 46% higher mass at TDC than the all-metal engine. However, as long as the investigation conditions were far from the smoke limit, the higher trapped mass was not a problem. Therefore, adjusting the intake temperature and pressure to match the thermodynamic conditions at TDC remains a practical option. The engine coolant temperature can also be adjusted to ensure appropriate matching of the combustion phasing. The typical engine coolant temperature is ~85 °C, which can be reduced to increase the heat transfer from the optical engine. The intake air temperature can also be optimised if required (Aronsson et al. 2008). However, if the smoke limit is approached, the fuel mass flow rate must be the same for both engines.

12.4 Combustion Visualisation

12.4.1 Natural Luminosity-Based Investigations

Combustion is controlled by the chemical reaction between species resulting in the release of energy in the form of heat and light. However, combustion can also take place without any visible light emission. Visible combustion is common, where light emission falls under the visible wavelength domain. Flames of the cook-stoves, wood fire and coal fires are some common examples. The glowing soot particles contribute dominantly towards the luminosity of flames in typical diesel combustion. The radiations from the soot particles can be considered as black-body radiations, which fall under the visible wavelength range for the typical combustion temperature range. The flames can be seen with the naked eye and imaged using general-purpose imaging devices. However, high-speed cameras are required to understand the fundamental aspects of combustion in diesel engines. Figure 12.2 depicts the typical diffusion flames of a diesel engine. The images were captured using a high-speed camera at 10,000 fps.

Qualitative analysis of soot radiations can be done by analysing the emission spectrum of soot particles using specific wavelength filters. Considering soot particles as a black body, the emission spectrum can be described using Planck's law. The

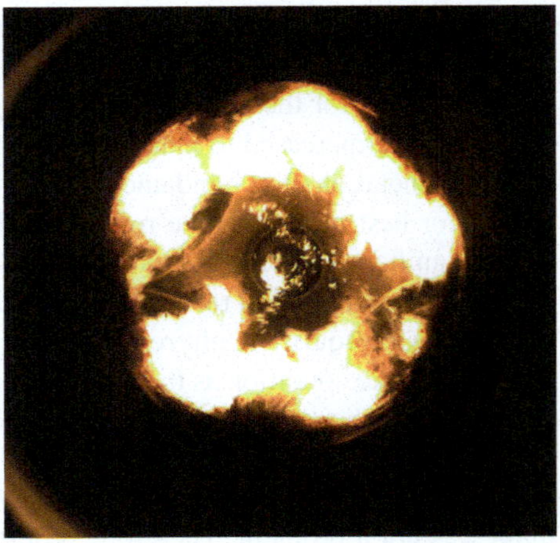

Fig. 12.2 Natural luminosity (NL) of combustion of diesel diffusion flames inside a light-duty optical diesel engine (from the unpublished dataset of Jena et al. (2022))

soot particles are black at room temperature and shift towards blue with increased temperature. Soot's natural luminosity has been extensively used to understand diesel combustion. However, several other species emit chemiluminescence during the combustion reactions. OH-chemiluminescence is one of the spontaneous emissions during high-temperature heat release. The OH radical concentration can be used to analyse the equivalence ratio, strain rate and heat release rate during combustion. The excited OH radical luminosity (306–310 nm) is captured using UV-sensitive high-speed imaging sensors (Higgins et al. 2001). Unlike the broadband spectrum of soot luminosity, the banded spectrum of OH-chemiluminescence has one of its peaks at ~310 nm. In sooty flames, common to diesel engines, the luminosity of glowing soot particles interferes with chemiluminescence (Karnani and Dunn-Rankin 2013; Dec and Espey 1998).

However, the black-body radiations from the soot are negligible in the UV range, where the emissions from OH radicals occur. As the temperature increases, the soot's emission shifts towards the UV range and is more likely to interfere with the chemiluminescence signal. A weaker chemiluminescence signal can be distinguished from the soot luminosity using filters. A camera equipped with a stereoscope having an interference filter at one end and a natural density filter at the other is suggested to resolve this interference issue. CH chemiluminescence has also been used to study auto-ignition in diesel-engine-like conditions (Pastor et al. 2009) (Dec and Espey 1998). Investigations of the fuel mixture preparation are beyond the scope of this chapter. However, it is worth mentioning that NL-based infra-red imaging can capture the vaporising hydrocarbons from fuels. The IR emission from hydrocarbons is dominantly from the CH stretch band, which peaks at ~3.4 microns. Typical high-speed cameras cannot capture any IR radiations beyond 1 micron. Therefore, special-purpose IR cameras and filters can be used for this purpose. A more detailed discussion on this topic can be referred to from the work of Musculus et al. (2017).

Combustion image velocimetry (CIV) is an advanced combustion diagnostic technique developed recently to study the flow dynamics of flames during combustion.

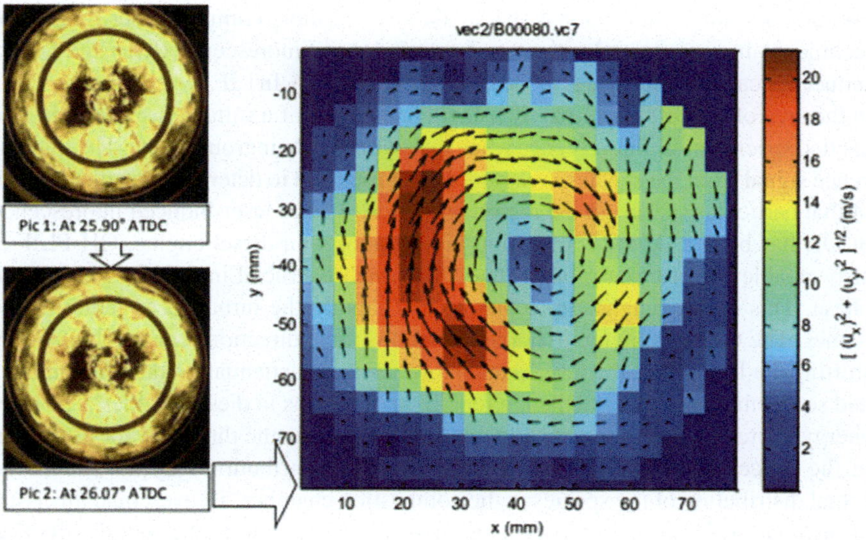

Fig. 12.3 In-cylinder flow field obtained through the CIV technique depicting the deviation of the flow field from the rigid-body rotation behaviour (Dembinski 2014)

The fundamental concept of this technique is similar to particle image velocimetry. Similar to PIV, two consecutive images are captured during an engine cycle. Unlike PIV, the special in-homogeneity in the natural luminosity is used to track the motion of turbulent structures. Therefore, illumination is not required. Seeds used in PIV burn out at high temperatures, making flow analysis difficult at elevated load and temperature conditions. CIV suffers from a low field of view and low vector resolution. CIV enables to resolve the flow under high-temperature diesel combustion conditions for an extended load range.

Several studies have used CIV to understand the flow dynamics during combustion. Figure 12.3 depicts the flow field during combustion in a small-bore diesel engine. One of the disadvantages of the natural luminosity method, which includes chemiluminescence and soot luminosity, is that it is attributed to line-of-sight measurements. The line-of-sight measurements provide an integrated result in the combustion chamber. Therefore, the details of the flame structure cannot be captured accurately. However, the natural luminosity methods have been extensively used to provide preliminary information about the combustion before advanced diagnostic techniques are used.

12.4.2 Laser-Based Diagnostics

The issues of line-of-sight measurement can be overcome by using laser-based combustion diagnostics. Laser-based diagnostics can be done over a point, a line, a plane or a three-dimensional space. Two of the most common laser diagnostic techniques in engine combustion are Laser-Induced Fluorescence (LIF) and Laser-Induced Incandescence (Le et al. 2016; Musculus 2006). In LIF, the chemical species in the area of interest is excited by using a laser pulse of a suitable wavelength. The excited species emits fluorescence when they return to their ground state. The fluorescence signal from the species is captured and processed to determine the distribution of that particular species in the region of interest. Planar laser-induced fluorescence of OH has been used for combustion visualisation in a diesel engine. OH-PLIF is also capable of detecting ground state OH, which is crucial in the late combustion phase. This planar diagnostic technique can capture the turbulent structure of the flames. The OH distribution in horizontal and vertical directions can be obtained by shifting the location of the laser sheet. However, beam attenuation due to aromatics and soot remains a major challenge for LIF applications in diesel engines. The laser energy decreases due to attenuation and may fall below the threshold for excitation of the concerned species. Therefore, the obtained distribution may not reflect the actual distribution of the species in the combustion chamber. To overcome this issue of attenuation, low sooting surrogates fuels are used to study the fuel distribution (Su et al. 2016; Kashdan et al. 2004; Donkerbroek et al. 2011).

Laser-induced incandescence (LII) is specifically used to study the in situ soot distribution in sooty flames. A pulsed laser increases the temperature of the soot particles to incandescence. The soot particles attain a significantly higher temperature than the surrounding gas. Thus, the signal is blue-shifted and collected as an LII signal. Sublimation of small molecules sets the upper limit of temperature (~4000 K) of the soot particles. The signals from LIF and LII often interfere with each other, which may affect the accuracy of the results. The laser heating soot particles also excite some species, which may emit a fluorescence signal. An Nd: YAG laser having a wavelength of 1064 or 532 nm can be used for LII studies. However, it has been observed that the 532 nm laser beam excites the polycyclic aromatic hydrocarbon (PAHs) in highly diluted conditions. Therefore, the 1064 laser beam is preferred to avoid noise in the LII signals.

12.5 Diesel Combustion Visualisation

12.5.1 Stages in Diesel Combustion

12.5.1.1 Ignition Delay and Premixed Combustion

A typical diesel engine works with a geometric compression ratio of 17:1. The compression work increases the temperature and pressure in the combustion chamber. The fuel is injected near TDC at high fuel injection pressure (FIP). The high

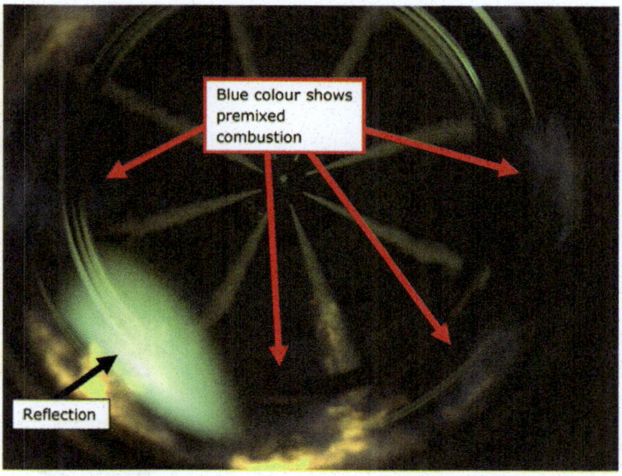

Fig. 12.4 Premixed combustion of wall-impinging sprays originating from the premixing charge near the wall jet vortex (Dembinski 2014)

FIP results in an ultra-high velocity diesel jet penetrating through the combustion chamber. The interaction of the spray with the high-density hot gases and in-cylinder turbulence results in atomisation, vaporisation and mixing necessary for the fuel's auto-ignition. In light-duty diesel engines, wall impingement of the spray is inevitable. The kinetic energy of the impinging fuel spray results in the formation of wall jets along the piston bowl surfaces. The ignition delay allows the initial fuel mass to form a premixed charge. At the end of the ignition delay, the premixed portion of the injected fuel mass auto-ignites, resulting in an intense heat release. In small-bore engines, the initial flame luminosity is observed near the piston bowl periphery, as shown in Fig. 12.4. Cool flames develop near the wall jet vortex due to enhanced charge mixing in this region (Le and Kook 2015). The blue flames indicate the premixed nature of the flames. The hot gases from the initial combustion entrain the incoming fuel spray droplet; thus, the second combustion stage begins.

12.5.1.2 Mixing-Controlled Combustion

The rate of injection governs the mixing-control combustion. The fuel from the spray core is vaporised and diffused into the reaction zone at the periphery of the spray. The turbulent diffusion rate determines the rate of combustion. Thus, the flames in diesel engines are dominantly diffusion flames. As the figure shows, the diffusion flames generate soot, producing intense soot radiations. The reaction zone in the periphery of the spray is also the location of the highest temperature in the combustion chamber. Most of the NO_x emitted from the diesel engine is produced in this zone. Moving upstream along the spray, the combustible mixture decreases. The flame is stabilised downstream from the nozzle during the mixing-controlled combustion phase. The distance of stabilised flames from the nozzle is called the lift-off length. Increasing the flame lift-off length decreases the mixture equivalence ratio at the flame lift-off location. A longer flame lift-off length is favourable for

Fig. 12.5 Mixing-controlled combustion visualisation in a diesel engine (Dembinski 2014)

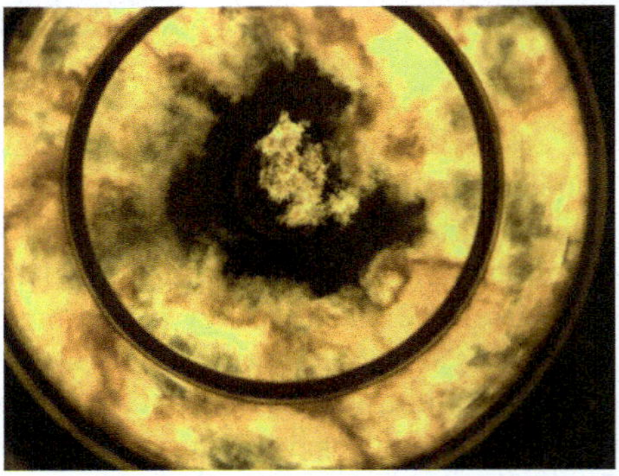

Fig. 12.6 Post-oxidation stage in diesel engine after the end of injection (Dembinski 2014)

lower soot production. Inlet temperature conditions have been shown to affect the flame lift-off length. However, the local temperature considered for the calculation as the temperature near the spray is higher than the mean cylinder temperature (Chartier et al. 2013). The flame lift-off length is also affected by the swirl direction. The swirl transports the hot gases downstream of the spray. This results in a slightly shorter flame lift-off length in the down-swirl direction of the spray. Hot gas entrainment in the spray increases with a reduction in inter-jet spacing, leading to shorter flame lift-off length (Chartier et al. 2013) (Figs. 12.5 and 12.6).

12.5.1.3 Post-oxidation Stage

The late combustion phase or post-oxidation is similar to diffusion combustion in some way. During this phase, the hot cloud of gases rotates along with the in-cylinder swirl, and oxidation of the unburned hydrocarbons occurs. The heat release during the post-oxidation stage ranges between 25 and 45%. Higher in-cylinder turbulence is desired for faster soot oxidation before the exhaust stroke begins. During the expansion phase, the gas temperature decreases along with the in-cylinder turbulence.

Thus, the chemical kinetics slows down during the later stage of combustion stroke and finally seizes.

12.5.2 Parameters Affecting Combustion

The diesel engine combustion system is one of the complex contributions of engineering to the field of combustion. The fascinating physics behind the flames has been understood greatly with the improvements in high-speed imaging and laser-based diagnostics techniques. This section focuses on understanding the underlying physics using ancient to state-of-the-art literature observations. A diesel engine injects fuel into the combustion chamber at high pressure. High-pressure direct injectors and a common rail system have been the standard for modern-day diesel engines. Due to low procurement and maintenance costs, many agricultural and Genset engines still rely on mechanical injection systems. However, the fundamental aspects of combustion in both systems remain the same. Before proceeding with the spray into the combustion chamber, one needs to understand the characteristics of common rail diesel injectors.

The common rail diesel injector is controlled by an engine control unit (ECU) to inject the required fuel quantity at the pre-set injection timings, which vary with load, speed, and ambient conditions. The injector pulse width (duration of the solenoid energising time) varies according to the injector flow sheet to meet the fuel demand. Figure 12.7 shows the injector current and mass flow rate with time. It can be observed that the fuel mass flow rate increases after a small delay (~0.4 ms) from the injector current. This delay is called the nozzle opening delay (NOD). NOD is typically the same for all FIPs and durations. This means that the physical start of injection is substantially delayed from the injector command. For instance, a 0.4 ms delay corresponds to a 4° CA delay at the start of injection with an engine speed of 1500 rpm. This delay will increase in terms of CAD with increasing speed. Similarly, the injector closing is also delayed by ~1 ms after the end of the injection and is called the nozzle closing delay (NCD). It can be observed that the NCD is higher than the NOD. This suggests that the injection continues even after the end of the injection command. This observation is crucial for interpreting the results from all-metal engines and comparing results among different engines. For spray modelling in diesel engines, the delay in the start of injection need to be considered. Several mathematical models for predicting the injector rate shape are available in open literature (Xu et al. 2018; Perini et al. 2020). It should be noted that the NOD and NOC may vary slightly depending on the injector and injection control modules. The injection rate affects the spray characteristics and, subsequently, combustion in diesel engines. Therefore, injector geometry has a critical role in diesel engine combustion. The injectors are characterised by the number of nozzle holes, their size and the umbrella angle of the spray. The number of nozzles and nozzle size varies with the size of the engine and the power requirement. Large-bore engines typically use a higher number of nozzles for better utilisation of the in-cylinder air. Reducing the nozzle size improves the

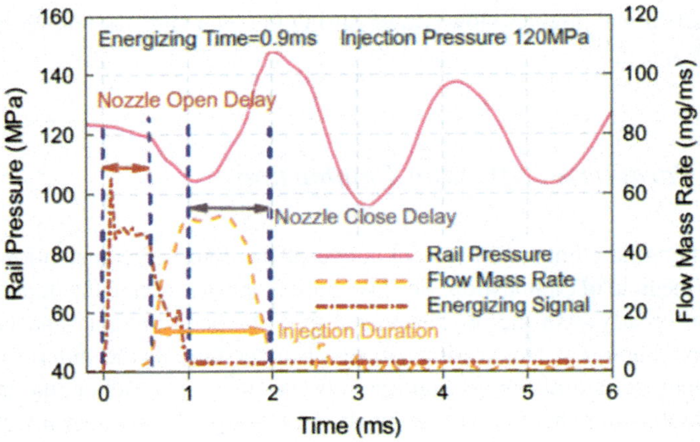

Fig. 12.7 Injector current and mass flow rate traces for a typical high-pressure diesel injector (Xu et al. 2018)

spray atomisation, leading to smaller droplet size distribution. However, a reduction in nozzle size will reduce the fuel mass flow rate.

A reduced injection rate leads to a longer injection duration and delays the combustion end. Allocca et al. (2012) studied the effect of nozzle hole size and injector flow characteristics on combustion in an optical diesel engine. The author reported that the local spray–air mixing improved with a reduction in the nozzle size. However, a reduction in nozzle hole size reduces the fuel mass flow rate, leading to a lower heat release rate. The comparison of spray penetration length showed that the smaller nozzles exhibit a higher contraction in the initial stage. Eventually, larger nozzles show higher spray penetration with time and superior air utilisation. Shundho et al. (1991) claimed that a combination of high pressure and lower size nozzle results in higher homogeneity in the spray plume. Biofuels have lower calorific value than conventional diesel, requiring higher injected fuel mass to match the engine power output. The fuel injection system can be modified by increasing the number of nozzle holes or their diameter. More nozzle holes are believed to improve the charge mixing characteristics in common practice. However, a recent study showed that the near-nozzle dynamics show non-linear characteristics with the nozzle numbers (Huang et al. 2019). Figure 12.8 shows that the highest turbulence intensity occurred for the five-hole case, which decreased with increasing nozzle holes. The turbulence intensity was minimum for eight nozzle hole cases. However, the spray became turbulent in the nine-nozzle hole cases.

Kang et al. (2022) investigated the effect of nozzle hole size and the number of nozzles for five different injectors. Two injectors with the same nozzle hole diameter (117 μm) having 8 and 10 nozzle holes were compared. Three 8-hole, 10-hole, and 12-hole injectors with a diameter of 110 μm were also investigated. From Fig. 12.9, it can be seen that, for the nozzle having a diameter of 117 μm, an increase in the number of nozzles resulted in a reduction in flame luminosity. With an increasing

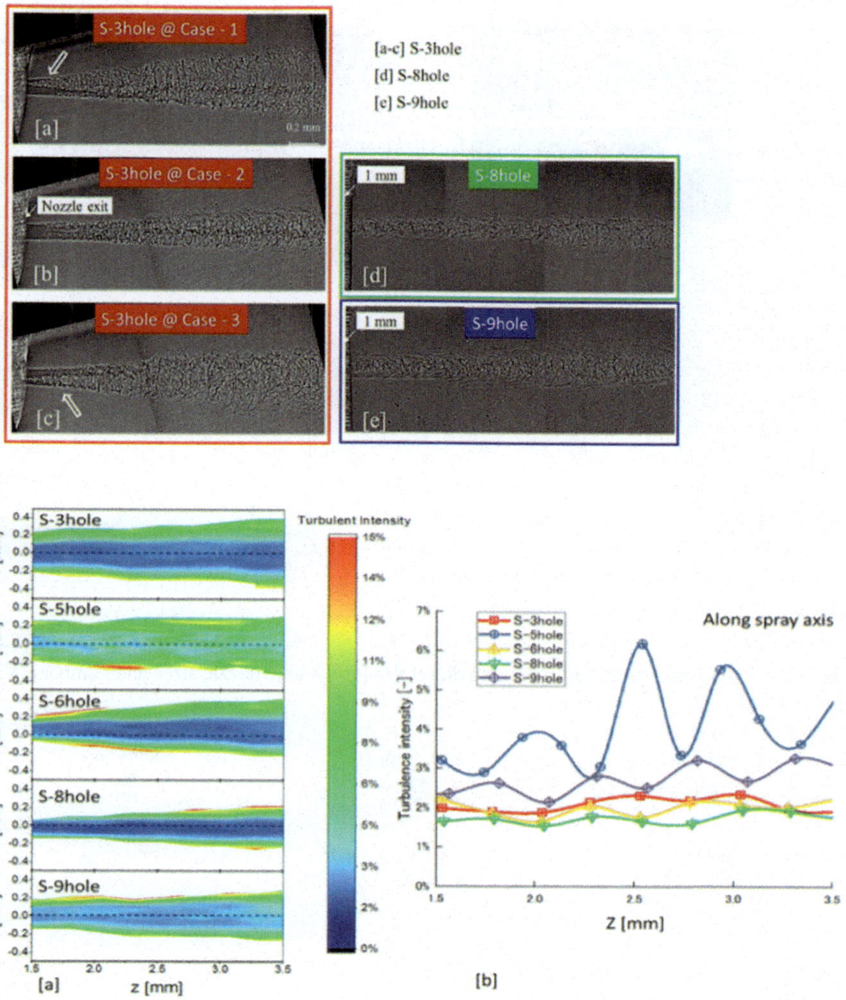

Fig. 12.8 Effect of the number of nozzle holes on turbulence intensity along the spray (Huang et al. 2019)

number of nozzle holes, the injected mass per nozzle decreased, adversely affecting diffusion combustion. However, an opposite trend was seen for the 110 μm nozzle cases. The soot luminosity increased with an increasing number of nozzle holes. The author argued that the air entrainment to an individual jet was affected by the nearby jet, deteriorating the mixing (Fig. 12.10).

Therefore, the benefit of improved spray atomisation due to smaller nozzle size was dominated by inferior charge mixing, resulting in higher soot formation. The flame lift-off length and flame penetration decreased with an increasing number of nozzle holes for both nozzle sizes. A recent study (Avulapati et al. 2021) reported

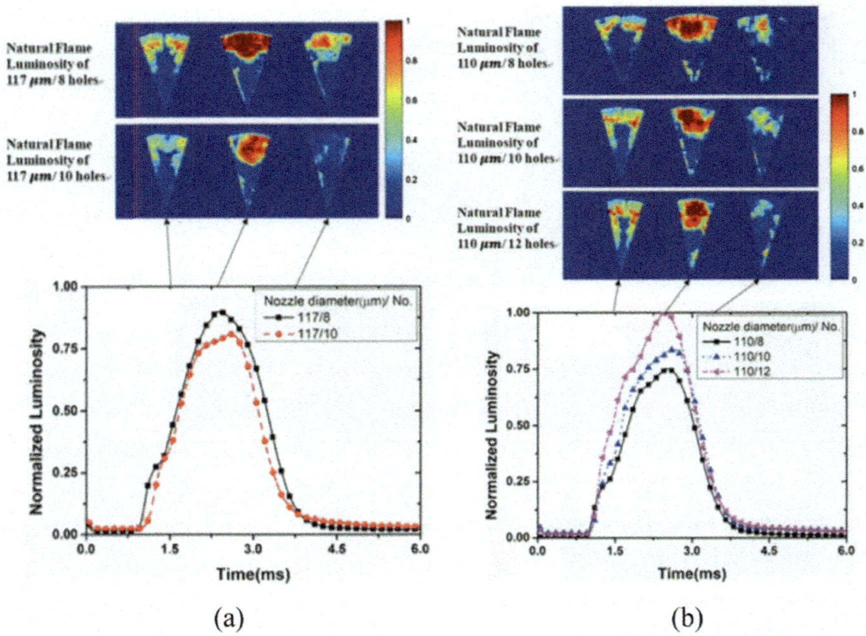

Fig. 12.9 Flame luminosity for impinging diesel jets for variable nozzle sizes and numbers (Kang et al. 2022)

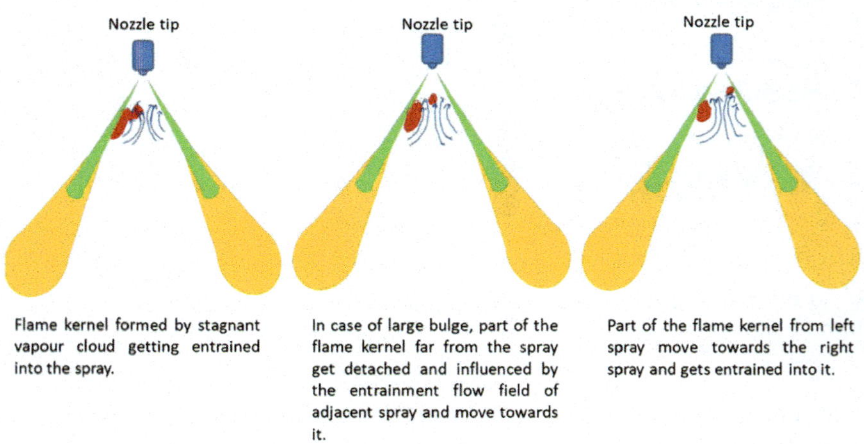

Fig. 12.10 Conceptual model of entrainment of flame kernel formed by the bulges (Avulapati et al. 2021)

that spray bulging occurs near the nozzle during the initial phase of spray evolution. The bulge gets entrained by the air, vaporises quickly and ignites before the ignition of the main spray. The hot gas from the bulge further entrains the spray, affecting the local equivalence ratio. A part of the bulge may get trapped by the adjacent spray. The reported bulging effect is likely to contribute to jet-to-jet variations and decreased lift-off length. Therefore, the nozzle size and number of nozzles need to be optimised simultaneously. Moreover, the spray swirl interaction and combustion geometry also play a vital role in determining soot formation.

Yang et al. (2021) studied the effect of inter-jet spacing on soot formation by investigating the turbulence intensity in an optical engine. The authors reported that the turbulence intensity decreased with a reduction in inter-jet spacing near the spray overlap location. Zhang et al. (2019a) studied the impinging spray behaviour in a small-bore optical diesel engine. A modified single-hole injector was used in this study to isolate the effect of a single spray plume, as shown in Fig. 12.11. Three sampling probes were flushed inside the combustion chamber at the spray's up-swirl, down-swirl and impingement locations. Immature soot particles were detected near the spray impingement locations. The authors found that the soot oxidation was higher at the up-swirl location than at the down-swirl location, which resulted in formation of more compact soot particles.

Jena et al. (2022) showed that a re-circulating vortex was formed near the up-swirl side of the impinging spray due to counter-flow conditions between the wall jet and the swirl flow. Enhanced mixing at this location was attributed to a lower soot concentration. FIP is one of the major tuning parameters for combustion and emission optimisation in diesel engines. Therefore, understanding the in-cylinder phenomenon with respect to pressure variations is crucial for diesel engine development. An increase in FIP results in a reduction in spray droplet size distribution. The fuel injection duration decreases with increased FIP for the same injected fuel mass. Therefore, the fuel injection velocity increases, leading to higher spray jet momentum. All these factors improve fuel–air mixing, resulting in shorter ignition delays and higher combustion rates. Figure 12.12 shows the OH (blue) and soot (LII signals, red) distribution for three different FIPs in a small-bore optical engine. It can be observed that the soot signal peaked at an earlier crank angle for the 130 MPa FIP case compared to lower FIP cases. The rate of soot formation increased with increasing FIP.

However, the OH radical build-up also increases simultaneously. The higher OH radical concentration results in faster soot oxidation. By comparing the three cases in Fig. 12.12, it can be concluded that the soot formation was higher for higher FIP; however, the increased soot oxidation rate resulted in reduced soot concentration in the later combustion stage (Rao et al. 2019). The fuel spray redistributes the angular momentum in the combustion chamber, which makes the flow unstable. The unstable flow structures contribute to turbulence at a later stage. This redistribution depends on the spray mass and momentum (Le and Kook 2015). Therefore, higher FIP can contribute to higher turbulence intensity. Higher pressure results in a shorter injection duration. The intensity of wall jet vortices increases with increased FIP. Longer mixing due to shorter injection duration and increased charge mixing results

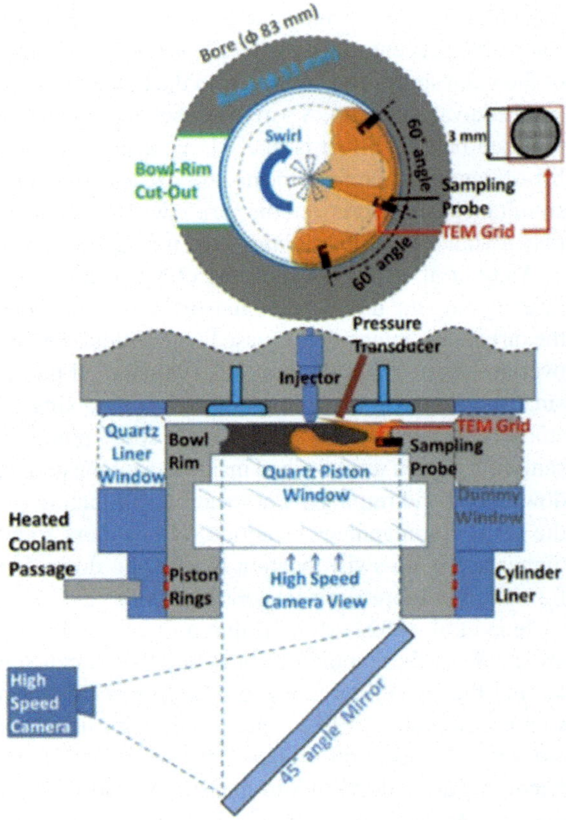

Fig. 12.11 Schematic of the optical engine for in-cylinder soot sampling and optical diagnostics (Zhang et al. 2019a)

in a higher degree of premixed combustion at higher FIP (Le and Kook 2015). The effect of FIP on the in-cylinder distribution of the pilot fuel mass was investigated by Sahoo et al. (2013). At lower FIP, the spray penetration of the injected fuel mass decreased, forming rich pockets in the combustion chamber. With an increasing FIP, the pilot mass became substantially premixed. The pilot injection was used to control the combustion rate in diesel engines. However, at lower FIP, the pilot mass led to soot formation. After-treatment devices can capture the soot. However, the particulate filter regeneration leads to fuel penalty due to post-injection. An unoptimised post-injection strategy leads to excessive wall wetting and diluting lubrication oil. The thermodynamic conditions of the engine affect the liquid penetration in case of post-injection. Bozic et al. (2010) showed that conventional and late post-injection results in significant wall impingement. However, wall wetting can be eliminated by employing early post-injection. Post-injection can be used to improve in-cylinder soot oxidation after the end of the main injection. However, the post-injection may increase or decrease the total engine-out soot, depending on the engine operating conditions. A short, closely coupled post-injection can increase the soot oxidation. For short post-injection, the soot from the post-injection is quickly oxidised before

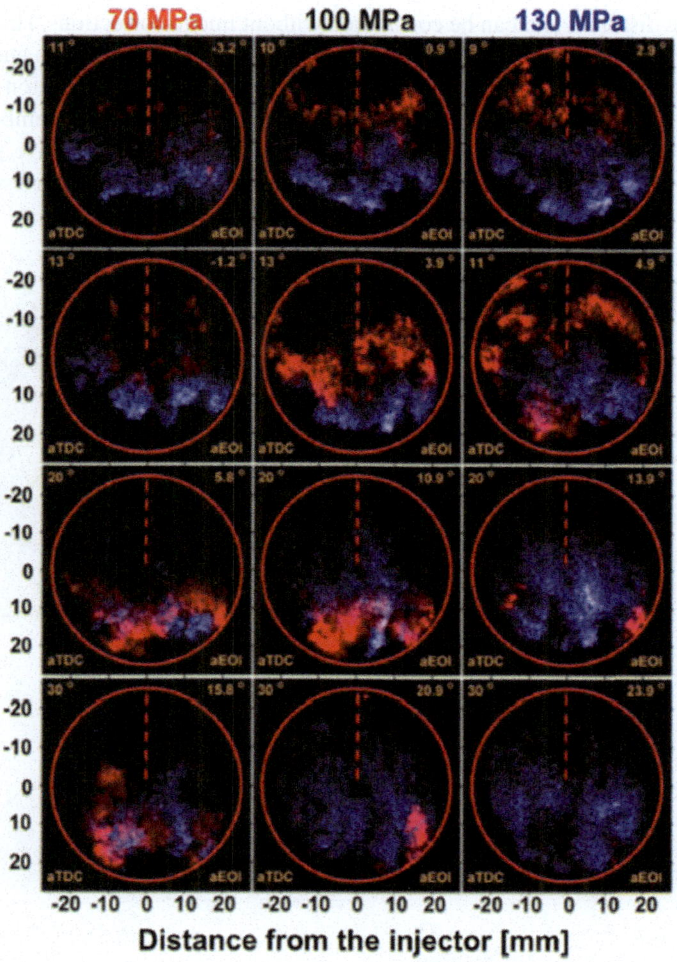

Fig. 12.12 Simultaneous imaging of OH and soot intensity distribution in an optical engine combustion chamber for different FIPs (Rao et al. 2019)

reaching the pitson wall. The entrainment wave assists in faster oxidation after the end of post-injection. With an increase in post-injection mass, the total soot will exceed the soot resulting from a single main injection case (Lind et al. 2018).

Combustion chamber shapes are equally important for efficient and clean diesel combustion. Several combustion chamber geometries have been investigated over the years. However, optical investigations with realistic piston geometry remain a challenging task. The irregular thickness of the piston cavity results in image distortion. The effect of piston curvature on the acquired images has been depicted in the figure. For Fig. 12.13a, the checked pattern was adhered to the cylinder head, while for Fig. 12.13b, the checked pattern followed the piston curvature. From the figure, it can be inferred that the images of the flame brushing over the piston surface will not be distorted and can be considered without image corrections. However, the planes far from the piston suffer optical distortion. The optical access and imaging complications have restricted the optical investigations with variable piston geometry.

CFD tools have been extensively used to optimise combustion chamber shapes. Recently, using combustion visualisation, stepped lip piston and wave protrusion piston shapes have been investigated. The piston geometry for the investigation is shown in Fig. 12.14.

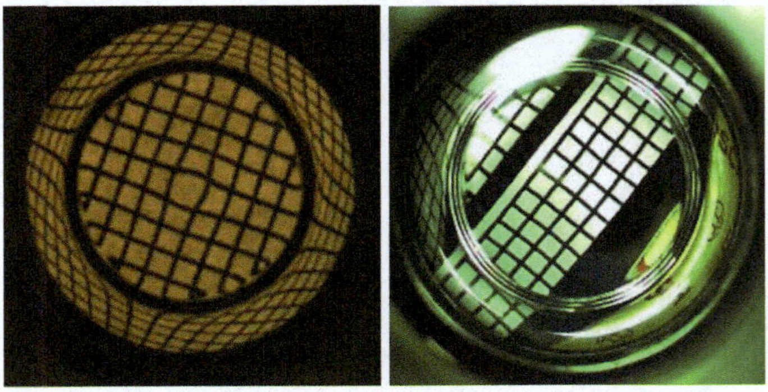

(a) (b)

Fig. 12.13 Image distortions due to piston curvature (Dembinski 2014)

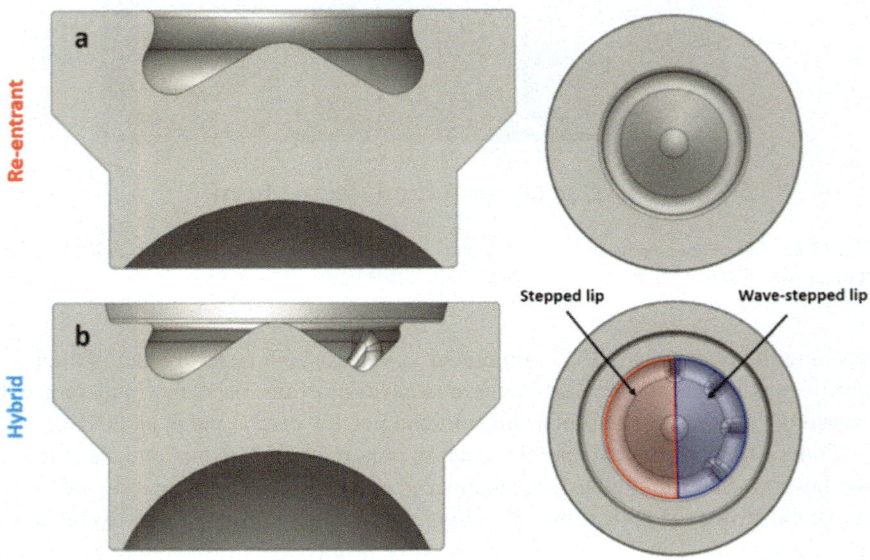

Fig. 12.14 Schematic of conventional (re-entrant) and innovative piston geometries (Pastor et al. 2021)

The re-entrant piston geometry has been the standard for most modern diesel engines. The stepped lip design has been developed to split the impinging spray, improving air utilisation and charge mixing. The latest development in the wave-stepped lip piston introduced a wave-like protrusion in the piston bowl near the impinging spray location. The crank angle resolved natural luminosity (NL), and OH radical intensity has been shown in Fig. 12.15 for different piston geometries (Pastor et al. 2021).

It can be observed that the hybrid piston resulted in a more symmetric distribution of flames in the early stages of combustion. At 10.11° CA, a dark zone can be observed in the natural luminosity image of the re-entrant piston geometry. The dark zone is attributed to cold soot, which develops near the wall region due to increased heat transfer to the wall. Quenching of the flames may occur near the wall, resulting in pockets of lower temperature. The soot luminosity radiations depend on their surface temperature. Therefore, as soot particles are cooled, their natural luminosity decreases. The dark zones were smaller for the hybrid piston case than the re-entrant piston. The wave side of the piston exhibited an even smaller dark zone than the other side of the wave piston. The higher OH radical intensity was observed in the squish region of the hybrid piston. The spray splitting due to the presence of the step directed a portion of the spray towards the squish, improving the air utilisation and faster combustion. Significantly higher OH radical intensity for the case of hybrid pistons suggested faster combustion and a reduction in fuel-rich regions. At 37° CA, the OH radical signals decayed in the case of the hybrid piston. However, the OH radical intensity persisted for longer duration in the case of the re-entrant piston. This suggests that the hybrid piston geometry resulted in a shorter combustion duration.

In the case of the conventional piston, the impinging wall jets overlap, forming local zones of a higher equivalence ratio. The wall jet vortices transport these pockets in the radial direction towards the centre of the piston bowl. The wave shape guides the impinging wall jets towards the bowl centre. Therefore, the head-on collision of the wall jet formed by the adjacent sprays can be avoided. This decreases the kinetic energy, which is utilised to increase the turbulence in the mixing zone (Eismark et al. 2019). Thus, the rich pockets in the radial mixing zones are suppressed, improving the soot oxidation in the post-oxidation phase. Moreover, Milano et al. (2022) demonstrated that the interactions of the swirl flow with the wave protrusion result in different turbulent structures, increasing the fuel–air mixing near the piston wall.

Fuel's chemical and physical properties also affect diesel combustion. However, a few studies cover alternative fuel combustion in a diesel engine. Fuel's cetane number varies with the fuel composition, which affects the in-cylinder charge reactivity. Zhang et al. (2021) investigated the effect of fuel's cetane number on an optical diesel engine's low-temperature and high-temperature heat release. These results showed that injection must be advanced for the low-cetane fuel for the same start of combustion. Test fuel with a cetane number of 30 showed an elongated cool flame development, starting at an earlier crank angle than a high-cetane test fuel. The HCHO-PLIF results revealed that low-cetane fuel's low-temperature stage was relatively slower. Higher charge premixing was also observed. Despite lower premixed

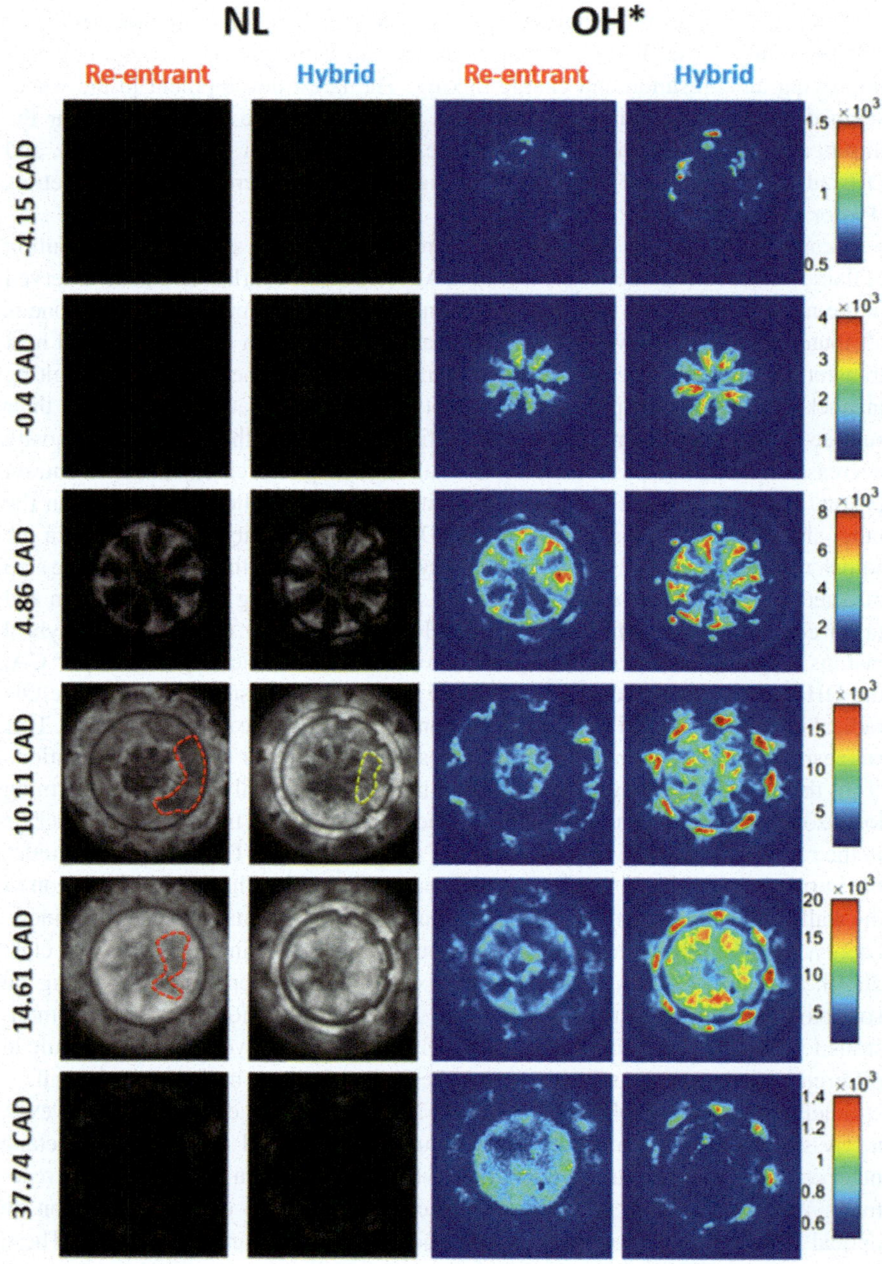

Fig. 12.15 Comparison of NL (left) and OH radical (right) distribution in the engine combustion chamber for different pistons (Pastor et al. 2021)

fuel, higher cetane fuel resulted in a higher heat release peak. Low-to-high temperature heat release shifts faster for cetane 40 and cetane 50 test fuels, resulting in more intense heat release. A lower degree of premixed results in a higher rate of soot production for cetane 40 test fuel. However, the higher OH radical intensity was also observed for test fuels with higher cetane numbers. Therefore, the overall soot reduction rate was higher for test fuel with cetane 40 than for test fuel with cetane 30. Interestingly, cetane 50 test fuel exhibited a lower soot oxidation rate than lower cetane test fuels despite a higher rate of OH radical build-up.

A significantly lower degree of premixing resulted in higher soot production and delayed oxidation. Therefore, the fuel cetane number test fuels need to be adjusted for optimum results. Tian et al. (2021) showed that the proportion of aromatics in the test fuel has an insignificant effect on the soot development and distribution. Higher aromatic fuels result in earlier soot inception, higher soot peak, and net soot production. Zhang et al. (2019b) investigated the effect of the carbon chain length of biodiesel fuel molecules on soot characteristics using Planar Laser Incandescence Intensity (PLII) technique in a small-bore optical diesel engine. The average PLII intensity for the two surrogates having different chain lengths is shown in Fig. 12.16. The long-chain biodiesel surrogate resulted in higher soot intensity. Lower oxygen percentage in longer chain biodiesel resulted in inferior soot oxidation. The higher ignition quality of the longer chain biodiesel resulted in higher soot intensity near the wall. In contrast, the soot intensity was restricted to the wall jet edge only. The degree of premixed combustion decreased with a longer chain length, similar to high-cetane fuel.

Low-carbon fuels such as methanol are crucial to achieving a lower carbon emission target. However, unlike biodiesel, methanol has low reactivity to auto-ignite in a typical CI engine environment. There are limited studies on the direct injection of methanol in CI engines. Higher latent heat of vaporisation leads to excessive charge cooling, which alleviates the low reactivity of methanol. The calorific value of methanol is also significantly lesser than conventional diesel; therefore, a higher fuel injection duration is required to match the same energy input in an unmodified diesel engine architecture. Longer injection coupled with low reactivity of methanol leads to a higher portion of the fuel energy release later in the power stroke. This leads to engine power derating. However, redesigning the fuel injection equipment with a modified injection strategy can resolve these issues. Several studies have argued that the cetane rating is not a reliable parameter to access the ignitability of methanol. The reactivity of methanol is more biased to the local temperature than its cetane rating. Therefore, hot surface ignition can be a promising technology for methanol-fuelled direct injection CI engines. Previously, glow plugs were explored to tackle the methanol reactivity issue. Mueller and Musculus (2001) conducted a comparative study of a diesel surrogate (CN45) with methanol to understand the methanol combustion process with glow plug assistance. The authors reported that methanol combustion was quieter than CN45 combustion. The optical NL imaging study revealed that the glow plug initiated the combustion for the methanol direct injection case. The flames propagated to the adjacent sprays in pairs from the spray

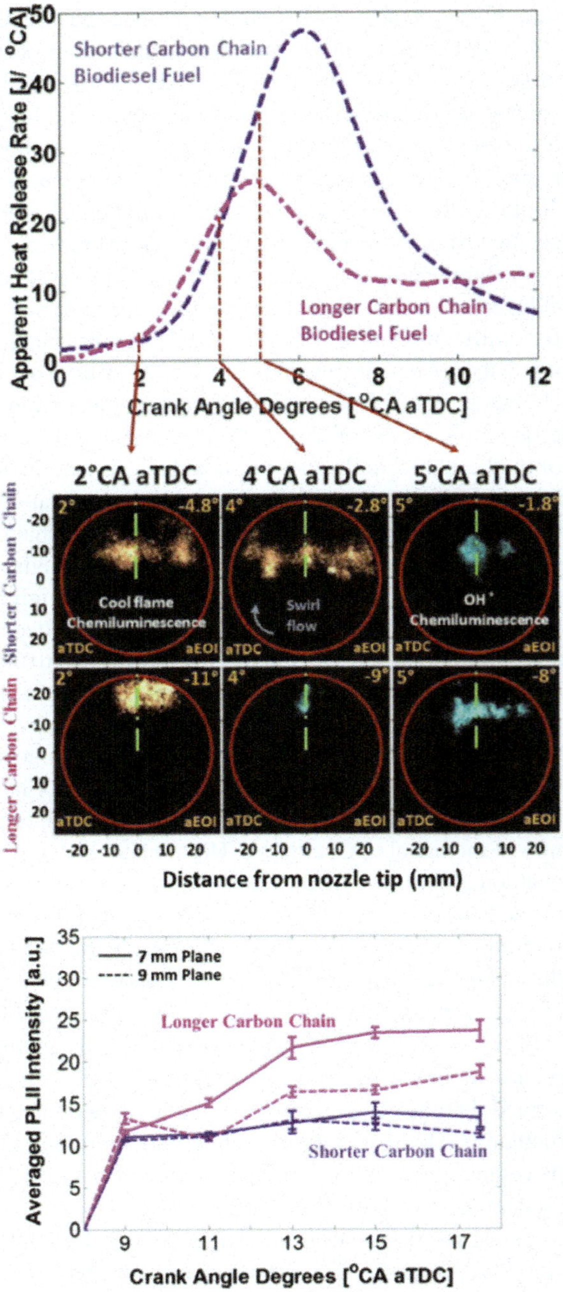

Fig. 12.16 Effect of biodiesel chain length on the soot intensity in a diesel engine (Zhang et al. 2019b)

interacting with the glow plug. As a result, the HRR showed multiple peaks corresponding to each pair of spray flames. The studies conducted before Mueller and Musculus (2001) had reported diesel-like heat release. The over-filtering of data could have omitted multiple peak features in these studies. The synchronisation between the optical images and the HRR profile highlighted the novel combustion mechanism in a glow plug-assisted methanol-fuelled direct injection CI engine. The authors further recommended carefully analysing the HRR profiles of alternative fuel combustion under similar conditions. The lower calorific value of methanol was attributed to the smoother combustion of the methanol-fuelled direct injection CI engine than CN45, where the natural soot luminosity was two orders of magnitude lesser than the CN45.

In a recent study by Matamis et al. (2021), the premixed combustion of directly injected methanol was compared with the spray-driven combustion in an optical engine. The authors increased the intake air temperature to achieve methanol combustion. The injection timing was advanced for the premixed combustion to 25° bTDC through 17° bTDC. Spray-driven combustion was investigated with an injection timing of 7° bTDC and 10° bTDC. LIF imaging was done to capture the fuel distribution in the combustion chamber. Combustion images overlapped the fuel distribution images, indicating the role of fuel distribution on combustion. The previous section shows that the conventional diesel combustion starts near the periphery of the piston bowl, and the flame stabilises downstream of the injector at the flame lift-off length. However, a striking difference was observed for the spray-driven methanol combustion. The combustion was initiated mid-way through the spray. The flames propagated towards the spray periphery and bowl centre from that point. The ignition behaviour was also different than expected for the advanced injection case. Figure 12.17 shows that the initial luminosity was near the bowl centre, where the fuel mixture is lean. The possible reason for such observation was linked to the higher heat of vaporisation of methanol. The methanol spray heads near the bowl periphery, resulting in higher charge cooling than the bowl centre. Therefore, the combustion started at a leaner site where the effect of charge cooling subsided at a faster rate. The authors also reported the ignition of remote spray plumes under fluctuating liquid lengths. The remaining plumes were ignited in random order. The physics behind this novel observation is yet to be established. The authors suggested high-speed imaging (frame speed of a few hundred kHz) to capture the intermediate phenomena.

Researchers have explored several ignition improvers to achieve stable combustion in methanol-fuelled direct injection CI engines. However, there is limited understanding regarding the combustion kinetics of such fuel blends. It would be interesting to note how the cetane improvers alter the combustion behaviour of methanol. Moreover, the glow plug-assisted combustion concept can be explored further by varying the number of nozzles and swirl to study flame convection. It is obvious at this point that the existing geometry of the diesel combustion chamber needs to be revisited based on the combustion characteristics of conventional and alternative fuels. Diesel–methane dual-fuel combustion has been of interest for heavy-duty and marine applications. Several optical studies have been carried out to understand dual-fuel combustion (Dronniou et al. 2014; Azimov et al. 2011; Schlatter et al. 2016;

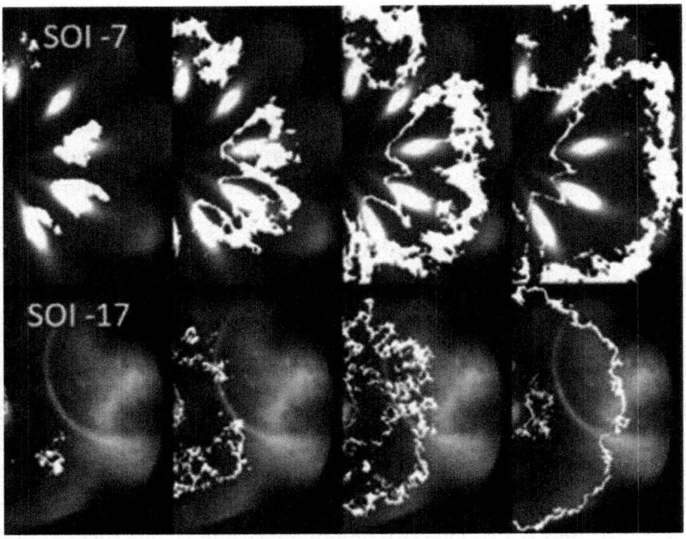

Fig. 12.17 Premixed (bottom row) and spray-driven (top row) combustion of methanol (Matamis et al. 2021)

Ishibashi and Tsuru 2017). A recent study (Mancaruso et al. 2020) investigated the effect of methane entrainment in diesel jets on low-temperature and high-temperature heat release patterns. Figure 12.18 shows the comparison of auto-ignition of diesel and diesel–methane dual-fuel combustion.

It can be observed from Fig. 12.18 that the auto-ignition occurred near the bowl periphery for both cases. The low-temperature heat release proceeds towards the centre of the combustion chamber from the initial ignition kernel along the spray axis. The authors argued that the low-temperature heat release is identical to conventional diesel and diesel–methanol combustion. Methane has very low reactivity compared to diesel; therefore, the auto-ignition is completely governed by pilot diesel. In such a case, the ignition probability depends on the mixing characteristics of the pilot. It has been reported that, unlike conventional diesel combustion, a lower pressure pilot injection can increase the probability of ignition due to higher reactivity stratification (Grochowina et al. 2018). However, the onset of the low-temperature ignition depends on the methane fraction entrained by the diesel jet. The chemical interactions of methane can particularly delay the low-temperature heat release in dual-fuel combustion (Srna et al. 2019). A similar chemical interaction with other alternative fuels will affect the ignition delay differently. Particularly, methanol will reduce the charge temperature due to the higher latent heat of vaporisation.

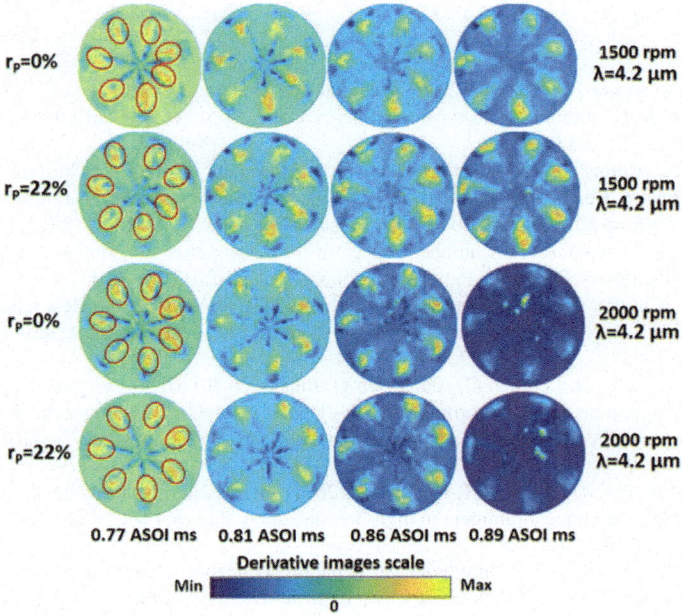

Fig. 12.18 The onset and propagation of low-temperature reactions in diesel (rp = 0%) and diesel–methane dual-fuel (rp = 22%) cases

12.6 Summary

Optical engines have contributed significantly to the fundamental understanding of diesel combustion and the evolution of diesel technology despite subtle differences between optical and all-metal engines. The differences have been continuously minimised via innovative designs. Bowditch piston revolutionised optical engine development and research. This chapter covers early optical engines developed for combustion visualisation in firing engines. State-of-the-art diagnostic techniques have also been introduced, along with their limitations. Several methods are discussed to approach the conditions of all-metal engines inside an optical engine. The stages of diesel combustion have been illustrated with flame images from optical engines. Various factors affecting the flame evolution in diesel engines are summarised based on the literature. There is limited research on the optical diagnostics of alternative fuels in the open literature. Some fundamental studies on the effect of cetane number on diesel combustion have been discussed. The combustion of alternative fuels such as methanol significantly differs from conventional diesel since its calorific value and thermophysical properties are different, requiring an increased fuel mass flow rate. The effect of FIP, number and size of nozzles on the spray has been discussed. These studies need to be extended to alternative fuels to develop tuned injection systems. This chapter develops an in-depth understanding of conventional diesel combustion for further research and development of alternative fuel-powered CI engines.

References

Agarwal AK, Gupta JG, Dhar A (2017) Potential and challenges for large-scale application of biodiesel in the automotive sector. Prog Energy Combust Sci 61:113–149. https://doi.org/10.1016/j.pecs.2017.03.002

Alalwan HA, Alminshid AH, Aljaafari HAS (2019) Promising evolution of biofuel generations subject review. Renew Energy Focus 28:127–139. https://doi.org/10.1016/j.ref.2018.12.006

Allocca L, Vaglieco BM, Montanaro A, Mancaruso E, Ciaravino C, Avolio G (2012) Investigation of diesel injector nozzle flow number impact on spray formation and combustion evolution by optical diagnostics. SAE Tech Pap. https://doi.org/10.4271/2012-01-0701

Aronsson U, Chartier C, Horn U, Andersson Ö, Johansson B, Egnell R (2008) Heat release comparison between optical and all-metal HSDI diesel engines. SAE Tech Pap. https://doi.org/10.4271/2008-01-1062

Aronsson U, Solaka H, Lequien G, Andersson O, Johansson B (2012) Analysis of errors in heat release calculations due to distortion of the in-cylinder volume trace from mechanical deformation in optical diesel engines. SAE Int J Engines 5:1561–1570. https://doi.org/10.4271/2012-01-1604

Avulapati MM, Pos R, Megaritis T, Ganippa L (2021) Insights into near nozzle spray evolution, ignition and air/flame entrainment in high-pressure spray flames. Fuel 293:120383. https://doi.org/10.1016/j.fuel.2021.120383

Awudu I, Zhang J (2012) Uncertainties and sustainability concepts in biofuel supply chain management: a review. Renew Sustain Energy Rev 16:1359–1368. https://doi.org/10.1016/j.rser.2011.10.016

Azimov U, Tomita E, Kawahara N, Harada Y (2011) Premixed mixture ignition in the end-gas region (PREMIER) combustion in a natural gas dual-fuel engine: operating range and exhaust emissions. Int J Engine Res 12:484–497. https://doi.org/10.1177/1468087411409664

Bowditch FW (1961) A new tool for combustion research a quartz piston engine. SAE Tech Pap. https://doi.org/10.4271/610002

Bozic G, Kook S, Ekoto IW, Petersen BR, Miles PC (2010) Optical investigation into wall wetting from late-cycle post injections used for diesel particulate filter regeneration. Am Soc Mech Eng Int Combust Engine Div ICE, ASMEDC 507–516. https://doi.org/10.1115/ICEF2010-35075

Chartier C, Aronsson U, Andersson Ö, Egnell R, Johansson B (2013) Influence of jet-jet interactions on the lift-off length in an optical heavy-duty di diesel engine. Fuel 112:311–318. https://doi.org/10.1016/j.fuel.2013.05.021

Dec JE, Espey C (1998) Chemiluminescence imaging of autoignition in a di diesel engine. SAE Tech Pap. https://doi.org/10.4271/982685

Dembinski H (2014) In-cylinder flow characterisation of heavy-duty diesel engines using combustion image velocimetry (Doctoral dissertation, KTH Royal Institute of Technology)

Donkerbroek AJ, van Vliet AP, Somers LMT, Dam NJ, ter Meulen JJ (2011) Relation between hydroxyl and formaldehyde in a direct-injection heavy-duty diesel engine. Combust Flame 158:564–572. https://doi.org/10.1016/j.combustflame.2010.09.024

Dronniou N, Kashdan J, Lecointe B, Sauve K, Soleri D (2014) Optical Investigation of dual-fuel CNG/diesel combustion strategies to reduce CO2 emissions. SAE Int J Engines 7:873–887. https://doi.org/10.4271/2014-01-1313

Duraisamy G, Rangasamy M, Govindan N (2020) A comparative study on methanol/diesel and methanol/PODE dual fuel RCCI combustion in an automotive diesel engine. Renew Energy 145:542–556. https://doi.org/10.1016/j.renene.2019.06.044

Eismark J, Andersson M, Christensen M, Karlsson A, Denbratt I (2019) Role of piston bowl shape to enhance late-cycle soot oxidation in low-swirl diesel combustion. SAE Int J Engines 12:03-12-03-0017. https://doi.org/10.4271/03-12-03-0017

Fukui K, Wakisaka Y, Nishikawa K, Hattori Y, Kosaka H, Kawaguchi A (2016) Development of instantaneous temperature measurement technique for combustion chamber surface and

verification of temperature swing concept. SAE Tech Pap. https://doi.org/10.4271/2016-01-0675
Gong C, Li Z, Yi L, Liu F (2020) Experimental investigation of equivalence ratio effects on combustion and emissions characteristics of an H2/methanol dual-injection engine under different spark timings. Fuel 262:116463. https://doi.org/10.1016/j.fuel.2019.116463
Grochowina M, Schiffner M, Tartsch S, Sattelmayer T (2018) Influence of injection parameters and operating conditions on ignition and combustion in dual-fuel engines. J Eng Gas Turbines Power 140. https://doi.org/10.1115/1.4040089
Guo Y, Ristovski Z, Graham E, Stevanovic S, Verma P, Jafari M et al (2020) The correlation between diesel soot chemical structure and reactivity. Carbon N Y 161:736–749. https://doi.org/10.1016/j.carbon.2020.01.061
Higgins B, McQuay MQ, Lacas F, Rolon JC, Darabiha N, Candel S (2001) Systematic measurements of OH chemiluminescence for fuel-lean, high-pressure, premixed, laminar flames. Fuel 80:67–74. https://doi.org/10.1016/S0016-2361(00)00069-7
Huang W, Moon S, Gao Y, Wang J, Ozawa D, Matsumoto A (2019) Hole number effect on spray dynamics of multi-hole diesel nozzles: an observation from three-to nine-hole nozzles. Exp Therm Fluid Sci 102:387–396. https://doi.org/10.1016/j.expthermflusci.2018.12.022
Ishibashi R, Tsuru D (2017) An optical investigation of combustion process of a direct high-pressure injection of natural gas. J Mar Sci Technol 22:447–458. https://doi.org/10.1007/s00773-016-0422-x
Jamrozik A (2017) The effect of the alcohol content in the fuel mixture on the performance and emissions of a direct injection diesel engine fueled with diesel-methanol and diesel-ethanol blends. Energy Convers Manag 148:461–476. https://doi.org/10.1016/j.enconman.2017.06.030
Jena A, Singh AP, Agarwal AK (2022) Optical and computational investigations of the effect of Spray-Swirl interactions on autoignition and soot formation in a compression ignition engine fuelled by diesel, dieseline and diesohol. Appl Energy 324:119677. https://doi.org/10.1016/j.apenergy.2022.119677
Jennings MJ, Morel T (1991) A computational study of wall temperature effects on engine heat transfer. SAE Tech Pap. https://doi.org/10.4271/910459
Kang S, Lee S, Hong D, Bae C (2022) Effects of nozzle orifice diameter and hole number on diesel combustion and engine performance. Int J Automot Technol 23:481–494. https://doi.org/10.1007/s12239-022-0044-8
Karnani S, Dunn-Rankin D (2013) Visualizing CH* chemiluminescence in sooting flames. Combust Flame 160:2275–2278. https://doi.org/10.1016/j.combustflame.2013.05.002
Kashdan JT, Docquier N, Bruneaux G (2004) Mixture preparation and combustion via LIEF and LIF of combustion radicals in a direct-injection HCCI diesel engine. SAE Tech Pap. https://doi.org/10.4271/2004-01-2945
Kashdan JT, Thirouard B (2009) A comparison of combustion and emissions behaviour in optical and metal single-cylinder diesel engines. SAE Tech Pap 2:2009-01-1963. https://doi.org/10.4271/2009-01-1963
Le MK, Kook S (2015) Injection pressure effects on the flame development in a light-duty optical diesel engine. SAE Int J Engines 8:609–624. https://doi.org/10.4271/2015-01-0791
Le MK, Zhang R, Rao L, Kook S, Hawkes ER (2016) The development of hydroxyl and soot in a methyl decanoate-fuelled automotive-size optical diesel engine. Fuel 166:320–332. https://doi.org/10.1016/j.fuel.2015.11.006
Liew WH, Hassim MH, Ng DKS (2014) Review of evolution, technology and sustainability assessments of biofuel production. J Clean Prod 71:11–29. https://doi.org/10.1016/j.jclepro.2014.01.006
Lin C (2003) The fuel properties of three-phase emulsions as an alternative fuel for diesel engines⋆. Fuel 82:1367–1375. https://doi.org/10.1016/S0016-2361(03)00021-8
Lind T, Roberts G, Eagle W, Rousselle C, Andersson I, Musculus MPB. Mechanisms of Post-Injection Soot-Reduction Revealed by Visible and Diffused Back-Illumination Soot Extinction Imaging. SAE Tech Pap 2018. https://doi.org/10.4271/2018-01-0232

Mancaruso E, Todino M, Vaglieco BM (2020) Study on dual fuel combustion in an optical research engine by infrared diagnostics varying methane quantity and engine speed. Appl Therm Eng 178:115623. https://doi.org/10.1016/j.applthermaleng.2020.115623

Mancaruso E, Sequino L, Vaglieco BM (2018) Temperature measurements of the piston optical window in a research compression ignition engine via thermography and templugs. SAE Tech Pap 2018. https://doi.org/10.4271/2018-01-0083

Matamis A, Lonn S, Luise L, Vaglieco BM, Tuner M, Andersson O et al (2021) Optical characterisation of methanol compression-ignition combustion in a heavy-duty engine. Proc Combust Inst 38:5509–5517. https://doi.org/10.1016/j.proci.2020.06.024

Millo F, Piano A, Roggio S, Pastor JV, Micó C, Lewiski F et al (2022) Mixture formation and combustion process analysis of an innovative diesel piston bowl design through the synergetic application of numerical and optical techniques. Fuel 309:122144. https://doi.org/10.1016/j.fuel.2021.122144

Mueller CJ, Musculus MP (2001) Glow plug assisted ignition and combustion of methanol in an optical di diesel engine. SAE Tech Pap. https://doi.org/10.4271/2001-01-2004

Musculus MPB (2006) Multiple simultaneous optical diagnostic imaging of early-injection low-temperature combustion in a heavy-duty diesel engine. SAE Tech Pap. https://doi.org/10.4271/2006-01-0079

Musculus MP, Eagle WE, Roberts G, Malbec LM, Sequino L, Mancaruso E (2017) Comparing infrared emission from hydrocarbon C-H stretch during direct injection with and without reaction in an optical heavy duty engine. 10th U.S. National Combustion Meeting College Park Maryland

Pastor JV, Payri R, Gimeno J, Nerva JG (2009) Experimental study on RME blends: liquid-phase fuel penetration, chemiluminescence, and soot luminosity in diesel-like conditions. Energy Fuels 23:5899–5915. https://doi.org/10.1021/ef9007328

Pastor JV, García A, Micó C, Lewiski F, Vassallo A, Pesce FC (2021) Effect of a novel piston geometry on the combustion process of a light-duty compression ignition engine: an optical analysis. Energy 221:119764. https://doi.org/10.1016/j.energy.2021.119764

Perini F, Busch S, Reitz RD (2020) A phenomenological rate of injection model for predicting fuel injection with application to mixture formation in light-duty diesel engines. Proc Inst Mech Eng Part D J Automob Eng 234:1826–1839. https://doi.org/10.1177/0954407019898062

Perini F, Dempsey A, Reitz R, Sahoo D, Petersen B, Miles P (2013) A computational investigation of the effects of swirl ratio and injection pressure on mixture preparation and wall heat transfer in a light-duty diesel engine. SAE Tech Pap 2. https://doi.org/10.4271/2013-01-1105

Puricelli S, Cardellini G, Casadei S, Faedo D, van den Oever AEM, Grosso M (2021) A review on biofuels for light-duty vehicles in Europe. Renew Sustain Energy Rev 137:110398. https://doi.org/10.1016/j.rser.2020.110398

Rao L, Zhang Y, Kook S, Kim KS, Kweon CB (2019) Understanding in-cylinder soot reduction in the use of high-pressure fuel injection in a small-bore diesel engine. Proc Combust Inst 37:4839–4846. https://doi.org/10.1016/j.proci.2018.09.013

Ren Y, Huang Z, Miao H, Di Y, Jiang D, Zeng K et al (2008) Combustion and emissions of a DI diesel engine fuelled with diesel-oxygenate blends. Fuel 87:2691–2697. https://doi.org/10.1016/j.fuel.2008.02.017

Russo C, Tregrossi A, Ciajolo A (2015) Dehydrogenation and growth of soot in premixed flames. Proc Combust Inst 35:1803–1809. https://doi.org/10.1016/j.proci.2014.05.024

Sahoo D, Miles PC, Trost J, Leipertz A (2013) The impact of fuel mass, injection pressure, ambient temperature, and swirl ratio on the mixture preparation of a pilot injection. SAE Int J Engines 6:1716–1730. https://doi.org/10.4271/2013-24-0061

Schlatter S, Schneider B, Wright YM, Boulouchos K (2016) N-heptane micro pilot assisted methane combustion in a rapid compression expansion machine. Fuel 179:339–352. https://doi.org/10.1016/j.fuel.2016.03.006

Shundoh S, Kakegawa T, Tsujimura K, Kobayashi S (1991) The effect of injection parameters and swirl on diesel combustion with high-pressure fuel injection. SAE Tech Pap. https://doi.org/10.4271/910489

Srna A, Bolla M, Wright YM, Herrmann K, Bombach R, Pandurangi SS et al (2019) effect of methane on pilot-fuel auto-ignition in dual-fuel engines. Proc Combust Inst 37:4741–4749. https://doi.org/10.1016/j.proci.2018.06.177

Su HC, Zhang R, Rao L, Kook S, Hawkes ER, Chan QN (2016) Overcoming beam attenuation issues in OH-PLIF diagnostics using an aromatics-free, low-sooting surrogate fuel in an optical diesel engine

Tian R, Zhang Y, Kook S, Kim KS, Kweon C-B (2021) Effect of jet fuel aromatics on in-flame soot distribution and particle morphology in a small-bore compression ignition engine. Fuel 305:121582. https://doi.org/10.1016/j.fuel.2021.121582

Wang X, Cheung CS, Di Y, Huang Z (2012) Diesel engine gaseous and particle emissions fueled with diesel-oxygenate blends. Fuel 94:317–323. https://doi.org/10.1016/j.fuel.2011.09.016

Wei J, Fan C, Qiu L, Qian Y, Wang C, Teng Q et al (2020) Impact of methanol alternative fuel on oxidation reactivity of soot emissions from a modern CI engine. Fuel 268:117352. https://doi.org/10.1016/j.fuel.2020.117352

Withrow L, Boyd TA (1931) Photographic flame studies in the gasoline engine. Ind Eng Chem 23:539–547. https://doi.org/10.1021/ie50257a018

Woods M, Bryzik W, Schwarz E (1992) Heat rejection from high output adiabatic diesel engine. SAE Tech Pap. https://doi.org/10.4271/920541

Xu L, Bai XS, Jia M, Qian Y, Qiao X, Lu X (2018) Experimental and modelling study of liquid fuel injection and combustion in diesel engines with a common rail injection system. Appl Energy 230:287–304. https://doi.org/10.1016/j.apenergy.2018.08.104

Yang J, Rao L, de Silva C, Kook S (2021) The influence of inter-jet spacing and jet-swirl interaction on flame image velocimetry (FIV) derived flow fields in a small-bore diesel engine. Int J Engine Res 146808742110384. https://doi.org/10.1177/14680874211038432

Zhang Y, Kim D, Rao L, Kook S, Kim KS, Kweon CB (2019a) In-flame soot particle structure on the up-and down-swirl side of a wall-interacting jet in a small-bore diesel engine. Proc Combust Inst 37:4847–4855. https://doi.org/10.1016/j.proci.2018.07.104

Zhang R, Pham PX, Kook S, Masri AR (2019b) Influence of biodiesel carbon chain length on in-cylinder soot processes in a small bore optical diesel engine. Fuel 235:1184–1194. https://doi.org/10.1016/j.fuel.2018.08.096

Zhang Y, Tian R, Meng S, Kook S, Kim KS, Kweon CB (2021) Effect of the jet fuel cetane number on combustion in a small-bore compression-ignition engine. Fuel 292. https://doi.org/10.1016/j.fuel.2021.120301

The manufacturer's authorised representative in the EU is Springer Nature Customer Service Centre GmbH, Europaplatz 3, 69115 Heidelberg, Germany. If you have any concerns regarding our products, please contact ProductSafety@springernature.com

Printed and bound by CPI Group (UK) Ltd, Croydon, CR0 4YY
01/10/2026
02225868-0008